Municipal Solid Waste Management

- Processing   - Energy Recovery   - Global Examples

Municipal Solid Waste Management

• Processing • Energy Recovery • Global Examples

P. Jayarama Reddy

BSP BS Publications

CRC Press
Taylor & Francis Group
A BALKEMA BOOK

Distributed in India, Pakistan, Nepal, Myanmar (Burma), Bhutan, Bangladesh and Sri Lanka by **BS Publications**

Distributed in the rest of the world by
CRC Press/Balkema, Taylor & Francis Group, an **Informa** business
Schipholweg 107C
2300 AK Leiden, The Netherlands
www.crcpress.com
www.taylorandfrancis.com

First published in India by
BS Publications, 2011

Hardback: ISBN 978-0-415-69036-2
E-book: ISBN 978-0-203-14569-2

Printed in India by
Sanat Printers, Kundli

Published by
BS Publications
A Unit of BSP Books Pvt., Ltd.
4-4-309, Giriraj Lane, Sultan Bazar, Hyderabad - 500 095 (A.P.) India.

Dedicated to

SAI

Foreword

The book by Dr. P Jayarama Reddy focuses on issues related to municipal solid waste treatment and management. It is a compendium of topics from cradle to grave. To the best of my knowledge this is the first of its kind effort. The author has addressed all issues related to municipal solid waste management. It covers in detail topics starting from characterization of the waste, modes of collection and transfer to technological advances in composting and landfilling. The methods related to producing energy from wastes are discussed at length and provide great insight to the readers.

The best practices followed in developed, developing and underdeveloped countries provides a truly global experience. Examples from Nigeria, Tanzania, Chile, Japan, and Thailand along with the Indian practice have been compiled with utmost care.

The topic of Energy from Municipal Waste is timely and would help practitioners, research scholars and teachers alike in promoting and propagating the concept of 4 R's (Reduce, recover, reuse and recycle). It was a pleasure for me to read this book and I am sure all the technocrats would be benefited by this book. I heartily congratulate Dr. P Jayarama Reddy on this excellent effort by him in bringing out the book on Energy from wastes and hope to see many more such endeavors from him.

Dr. Valli Manickam
Chairperson, Environment Area,
Administrative Staff College of India,
Hyderabad, India.

Preface

Recovery of energy in the form of heat or electricity and recyclables is an important benefit resulting from the processing of Municipal Solid Waste (MSW). Technologies have been developed utilizing principles of thermal, biochemical or chemical processing of solid waste to derive energy. These technologies are sensitive to the nature and quality of the waste collected. Hence the collected waste requires some kind of pre-assessment as well as treatment to turn into a suitable feedstock. These technologies are operative in many countries around the world to varying level of success.

Today, one of the major concerns of many municipal authorities and urban local bodies responsible for supervising public health and sanitation is the management of MSW. In developing countries, it is more complex and challenging due to many factors: inadequate infrastructure and financing, lack of definite responsibilities and roles of the authorities, insufficient rules, legal framework and poor enforcement. The uncollected waste in cities and towns and uncontrolled disposal of waste at the street corners, public places, city boundaries and the suburbs has threatened the public health and sanitation in several growing cites of the world. As a result, it is hard to find many 'clean and green cities' in most regions of the world. However, the situation has been improving in many countries due to public awakening, legal interventions and governments' initiatives.

The migration of rural people to urban centres has given rise to growth of small and large urban settlements and spreading of city suburban limits resulting in a stressed urban environment. This is particularly visible in the developing countries of the world primarily due to the increasing economic/industrial activity. Some of the regions are rapidly transforming from traditional rural economy to an urbanized one with increased production capacities of goods and services. Marked changes in consumption patterns of the people have been witnessed in recent decades.

Increased urban population has enhanced the demands for consumables resulting in larger generation of wastes both in volume as well as per capita with less recovery and reuse of solid wastes. This increasing volume of waste demands an effective waste management system in terms of collection capacity, treatment, disposal and aftercare. Thus the necessity for greater investments of human, technological and financial resources for waste management is pertinent in order to maintain a cleaner and sustainable environment. An integrated system of solid waste management in which the waste from its origin to disposal is considered is perhaps the answer.

The challenge of delivering effective and sustainable waste management is an issue which confronts all stakeholders including central and local Governments, the public and professional (private) waste managers and even citizens. Improving awareness of the various waste management options is crucial for the development of a more sustainable approach to waste management, linking public participation and the essential infrastructure expansion to recover '*value*' from the residual waste stream.

The process of 'Composting' waste has been the traditional way of treating solid waste. Composting is now practiced from a simple inexpensive type to a large and expensive type (centralized) depending on the waste quantity, composition and other factors.

'Energy from Waste' has a significant role to play in dealing with the residual municipal waste stream. An Energy from Waste plant, also known as Waste-to-Energy (WTE) plant, operates by taking the waste and converting its hidden energy into a type of usable energy – heat, electricity and transport fuels – just as coal, oil and gas are used as fuels in fossil-fired power stations. WTE can be used with all types of waste from domestic, commercial, industrial, construction and demolition, to sewage and agricultural and so on. The only criterion is that the waste fraction needs to be combustible and/or biodegradable. WTE is now an essential component of a sustained solid waste management programme.

Energy from Waste is the application of sound proven combustion engineering principles to a variety of technologies which reduce and sanitize the residual municipal waste fraction in order to recover '*energy.*' Several waste combustion systems capable of dealing with raw, processed or sorted fractions of MSW, fluidized-bed combustion systems, and processed waste energy recovery options such as refuse-derived fuel (RDF), gasification and pyrolysis are in operation in many countries, mostly developed countries. Biochemical processing technologies such as 'anaerobic digestion', also called 'biomethanation' of waste, wherein 'biogas' is produced is also a proven technology which is widely used worldwide. The Chemical processing – Esterification – to derive biodiesel from waste cooking oils is emerging as viable technology at commercial scale.

The major objective of this book is to introduce students of Science and Engineering, the waste managers, decision makers, planners and a wider audience to these technologies, the main components of the systems (plants), operational principles and requirements, strengths and weaknesses, working examples from around the world and so on. The treatment and uses of the post-combustion/gasification residues are also described.

In the developed countries as well as in a few developing countries, the MSW services including resources recovery and energy production, by and large, are systematized supported by proper legislations and regulations, pollution control policies and their strict enforcement. Therefore, these services are better placed in developed countries compared to developing countries where legislations and policies are inadequately present. This book covers the status of MSWM both in developed and developing countries, right from the generation of waste to the final step of disposal.

The MSW services in India including the Sources of funding, and Rules, legislation and legal provisions constitute one chapter. The desirability as well as the necessity of inviting Private Sector to participate in delivering MSW services along with the accompanying benefits and issues, the role of NGOs and Resident Welfare Associations (RWAs) is discussed in a separate chapter.

To set up a WTE plant with an appropriate technology, feasibility studies covering several issues, and proper Planning and Execution are essential. These aspects are fully explained to help a planner or an entrepreneur entering this sector. Initiatives taken by some States and the MSW services in three major cities of India – Chennai, Delhi and Greater Mumbai - as case studies are included in the book. Similarly, while briefly elaborating the state of affairs in all the continents - Asia, Africa, Latin America, Europe and North America - one or two countries/ Cities in each continent are chosen for detailed exposition.

The information for certain regions of the world are sufficiently available through survey reports, publications, research documents and websites of several organizations, for example, UNEP, USEPA, World Bank but are scanty in other regions. The conclusions drawn by several researchers are subject to this limitation.

The issues, challenges and opportunities are plenty in MSW which differ from country to country and within a country from city to city. An integrated solid waste management can address many of these challenges resulting in the lessening of air, water and soil pollution, and the associated public health problems. Further, resources of value and 'clean energy' can be recovered from the waste.

This book is a modest attempt by the author for a comprehensive presentation of the MSW issues with emphasis on '3Rs (Reduce, Reuse, and Recycle)' and recovery of 'Energy from Waste'. Annexure, Glossary and References are added at the end for the expediency of the reader.

- Author

Acknowledgement

The author conveys his sincere thanks to Dr. Valli Manickam, Chairperson, Environment Area, Administrative Staff College of India, Hyderabad, India for offering valuable suggestions and for writing foreword.

The author has drawn/quoted substantial material, photos, tables etc., from publications/Survey Reports/Documents released by several Universities/Institutes/Government departments worldwide, and Organisations such as UNEP, The World Bank, UNDP, and USEPA, and prominent NGOs in India and abroad. The author expresses his grateful thanks to the Authors of these references. These are also sincerely acknowledged at the appropriate places and at the end in the text.

The author is personally indebted to many of his close friends, former colleagues and students, and members of his family for their warmth, encouragement and support. He conveys his special love to his grand daughters, Hitha and Tanvi for their willing assistance in computer work and for picking up some photos and pictures; and to the little ones, Diya and Divija, for providing cheer at times of tiredness with their innocent queries.

It is a pleasure to acknowledge the general support given by BS Publications, mainly, Anil Shah and Nikhil Shah, and the special technical help by Naresh Davergave and his team to bring this book along with an awesome cover so nicely.

-Author

Contents

Foreword..(vii)

Preface ....(ix)

Acknowledgement..(xiii)

1. **BASICS**..1

    1.1   Introduction 1

    1.2   Types of Solid Waste..4

    1.3   Waste Management Concepts...10

    1.4   Health and Environmental Impacts.......................................11

    1.5   Global Warming 13

    1.6   Source Reduction...19

2. **WASTE GENERATION AND CHARACTERIZATION**.........24

    2.1   Waste Generation and Composition 25

    2.2   Waste Characterization...34

3. **WASTE REDUCTION AND RECYLING**................................38

    3.1   Recycling...39

    3.2   Status in Developing Countries 40

    3.3   Status in Developed Countries..44

    3.4   Case of Plastics...46

            3.4.1   Nature of Plastics...46

            3.4.2   Recycling of Plastic Waste 48

    3.5   Recovery and Recycling of E-waste......................................53

    3.6   Waste Trading..55

    3.7   Waste Picking as a Livelihood 55

4.   WASTE COLLECTION AND TRANSFER **58**

    4.1   Waste Collection in Developing Countries 58

    4.2   Waste Collection in Developed Countries..........................63

    4.3   Waste Processing and Disposal 66

5.   COMPOSTING **67**

    5.1   Process..67

    5.2   Benefits..69

    5.3   Composting Technologies 70

        5.3.1   Backyard or a Home Composting 73

        5.3.2   Vermi Composting 75

        5.3.3   Aerated (Turned) Windrow Composting............81

        5.3.4   Aerated Static Pile Composting 83

        5.3.5   In-Vessel Composting 85

    5.4   Biowaste Composting in Europe 88

    5.5   Composting Challenges 89

    5.6   Composting in Developing Countries................................89

    5.7   Composting in Developed Countries.................................95

6.   ENERGY FROM WASTE **98**

    6.1   Introduction 98

        6.1.1   Assessment of Energy Recovery Potential........100

        6.1.2   Environmental Impacts of the Technologies.....102

    6.2   Thermal Processing 103

        6.2.1   Combustion/Incineration 103

        6.2.2   Pyrolysis..129

        6.2.3   Gasification 131

        6.2.4   Plasma Arc Gasification....................................137

    6.3   Biochemical Processing..141

        6.3.1   Anaerobic Digestion (Biomethanation)............141

        6.3.2   Mechanical Biological Treatment 158

        6.3.3   Fermentation..161

6.4   Chemical Processing: Esterification 161

6.5   Recent Developments in WTE Technologies..................167

6.6   Planning and Execution of WTE Technologies...............169

6.7   Applications of Important Industrial Wastes...................173

7.   LANDFILLING...175

7.1   Introduction 175

7.2   Environmental Impact Study 177

7.3   Landfill Construction...178

7.4   Decomposition in the Landfill 182

7.5   Benefits..184

7.6   Recovery and uses of Landfill Gas 184

7.7   Associated Activities 186

7.8   Health and Environmental Impacts.................................186

7.9   An Example 187

7.10   Landfills in Developing Countries..................................191

7.11   Landfills in Developed Countries 197

8.   MSW MANAGEMENT IN INDIA...199

8.1   Introduction 199

8.2   Analysis of MSW 205

8.3   Storage and Collection of MSW.....................................210

8.4   Transfer Stations and Trasporation.................................216

8.5   MSW Treatment/Disposal 219

        8.5.1   Landfilling...220

        8.5.2   Composting 221

        8.5.3   Anerobic Digestion (Biomethanation) 223

        8.5.4   Incineration...227

        8.5.5   Gasification 228

        8.5.6   RDF Plants 228

8.6   Recovery of Recyclables Materials 230

8.7   Healthcare Waste Treatment...232

8.8   Hazardous Waste Management 233

8.9   E-waste Management..234

8.10   Rules, Legislation and Legal Provisions..........................236

8.11   Financial Resources 245

8.12   Future Scenario..251

9.   PRIVATE SECTOR PARTICIPATION IN INDIA 256

9.1   Introduction 256

9.2   Options in PSP Arrangement...258

9.3   Examples of PSP in MSW Services 261

9.4   Important Contractual Issues 264

9.5   Survey on Privatisation of MSW Services 268

9.6   Role of NGOs and CBOs..270

9.7   Initiatives by Some State Governments.............................271

9.8   Case Studies..277

**10.   MSW MANAGEMENT AND PLANNING –
        GLOBAL EXAMPLES...304**

10.1   Asia..304

10.2   Africa 312

10.3   Latin America 328

10.4   Europe...337

10.5   North America 350

ANNEXURES...357

A1   Waste Generation and Management Data by Country 357

A2   Waste-to-Energy Facilities in USA 363

A3   Waste-to-Energy Plants Operating in India.....................370

A4   Composting Plants in India..377

A5   Waste-to-Energy Status in China..379

A6   International Agreement and Commnitments to
      Environmentally Sound Management of Waste 386

A7   Types of Biogas Plants 389

A8   Zero-waste Approach...395

A9   Integrated Solid Waste Management...................................397

A10   Door to Door Refuse/Garbage Collection System in
      Surat Municipal Corporation City –
      A Project in Best Practice...402

A11   Centralised Co-digestion of Multiple Substrates: (CAD)
      Example of Denmark 407

Glossary **413**

References **423**

CHAPTER 1

Basics

1.1 Introduction

The start of civilization has seen human race generating waste such as bones and other parts of animals they slaughter for their food or the wood they cut to make their shelters, tools, carts etc. The advancement of civilization has witnessed the waste generation getting enhanced, and becoming more complex in nature. The beginning of industrial era has had enormous effect on the life styles of people which have started changing with the availability of many consumer products and services in the market. The manufacturing and usage of vast range of products as well as management of the resulting waste give rise to emission of greenhouse gases. This has led not only to the pollution of air and water but has affected the Planet Earth through global warming.

Rapid migration of rural populations to urban centres, in search of better opportunities of livelihood, has resulted in an overwhelming demographic growth in many cities worldwide. This situation is more pronounced especially in Asia and Africa. The projected growth rate in North America is less because it has already recorded a growth rate of > 70%. Also in Europe, the situation is similar. But in Africa and Asia, around 35% of the population presently is urban (Fig.1.1). Asian countries are experiencing an urban growth of approximately 4% per year. This growth rate is expected to continue for several more years, and by 2025, 52% of the Asian population is likely to be living in urban centres. As in Asia, Africa's population is mainly rural at present. However, Africa is also experiencing a high rate of urbanization at 4 to 5 % per annum, and by 2025, urbanization is likely to be similar to Asia. This high rate of urbanization can lead to serious environmental degradation in and around several cities.

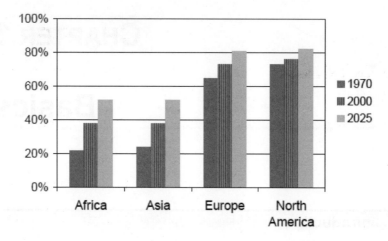

Fig. 1.1 Projected urban growth in different continents (source: UN 1996)

Cities and towns in India, a developing economy, have generated an estimated 6 million tonnes of solid waste in 1947. It has risen to about 48 million tons in 1997; and in 2001, to more than 91 million tons (taking the urban areas only), which comes to 0.12 to 0.6 kilograms per person per day. In contrast, in 2006, a developed country like US has generated more than 251 million tons of municipal solid waste which amounts to approximately 2.1 kilograms of waste/ person/ day. This is in addition to approximately 7.6 billion tons of industrial solid waste generated by industrial units annually. On an average, per capita waste generation in US (a developed country) is very *much higher* than in India (a developing country).

Significance of Waste management: Waste is any garbage or refuse or other discarded material including solid, liquid, semi-solid, or contained gaseous material arising from domestic, community, industrial, commercial, agricultural or human operations. The sludge from a wastewater treatment plant, water supply treatment plant, or air pollution control facility is also considered as waste.

Waste management is a global issue and requires maximum attention. It is highly obligatory to reduce the pollution of air and water, the dreadful effects on human health and to maintain a clean environment. Waste management sector can contribute to greenhouse gas mitigation in ways that are economically viable and meet many social priorities. The adverse effects of global warming are witnessed already around the globe to varying degree in different regions. A safe and sustainable environment is an absolute necessity for a healthy living. The civic society has, therefore, exclusive responsibility of considering waste treatment as a priority issue.

The management of waste involves waste collection, resource recovery and recycling, transportation, and processing or disposal. Of these, the most important one is processing/disposal of waste. The urbanized areas are concerned with the problem of developing cost-effective environmentally acceptable disposal methods of solid waste. The major advantages of a planned approach to waste treatment are (i) reducing pollution and the consequences such as global warming, (ii) keeping the human habitats ranging from small towns to big cities clean and green, (iii) recovering 'resources' which can be recycled into useful products for reuse, and more importantly (iv) processing of wastes into useful clean energy – heat and electric power.

Since the waste can be solid, liquid, gaseous or medical or hazardous substances, each category is treated with different and appropriate method(s). Waste management practices differ in developed and developing countries, in urban and rural areas, and for residential and industrial producers. Waste management has to be viewed as a central element in the sustainable development planning of a city or a town or a community.

Management of non-hazardous residential and institutional waste in metropolitan areas is usually the responsibility of local municipal authorities/urban local bodies, while management for hazardous commercial and industrial waste is usually the responsibility of the producer.

1.2   Types of Solid Waste

(Courtesy Photos: CPREEC website)

Domestic wastes are generated by household activities such as cooking, cleaning, repairs, interior decoration, and used products/ materials such as empty glass/ plastic/ metal containers, packaging stuff, clothing, old books, newspapers, old furnishings, etc. *Commercial wastes* are the wastes generated in offices, wholesale stores, shops, restaurants and hotels, vegetable, fish and meat markets, warehouses and other commercial establishments. *Institutional wastes* are generated from institutions such as schools, colleges, hospitals, research institutions. The waste includes mostly paper, cardboard, etc., and hazardous wastes. *Municipal wastes* are wastes generated due to municipal services such as street sweeping, and dead animals, market waste and abandoned vehicles or parts; also includes already mentioned domestic wastes, institutional wastes and commercial wastes.

Garbage includes animal and vegetable wastes due to various activities like storage, preparation and sale, cooking and serving; these are biodegradable.

Ashes: Residues from the burning of wood, charcoal and coke for cooking and heating in houses, institutions and small industries. Ashes consist of fine powders, cinders and clinker often mixed with small pieces of metal and glass.

Rubbish: Apart from garbage and ashes, other solid wastes produced in households, commercial establishments, and institutions.

Bulky wastes: Bulky wastes are large household appliances such as cookers, refrigerators and washing machines as well as furniture, crates, vehicle parts, tyres, wood, trees and branches. The bulky metallic wastes are sold as scrap metal but some portion is disposed as sanitary landfills.

Street wastes: Street wastes include paper, cardboard, plastic, dirt, dust, leaves and other vegetable matter collected from streets, walkways, alleys, parks and vacant plots. Municipal waste includes street waste also.

Dead animals: It includes animals that die naturally or killed by accident. It does not include carcass and animal parts from slaughterhouses as these are considered as industrial wastes.

Construction and demolition wastes: Some quantities of the major components of the construction materials such as cement, bricks, cement plaster, steel, rubble, stone, timber, plastic and iron pipes are left out as waste during construction as well as demolition. About 50% of the wastes are not currently recycled in India and 70% of the construction industry in India is not aware of recycling techniques.

Industrial wastes: These are non-hazardous solid material discarded from manufacturing processes and industrial operations, and are not considered as municipal wastes. However, solid wastes from small industrial plants and ash from power plants are frequently disposed of at municipal landfills.

Table 1.1 Source and quantum of some major industrial wastes in India.

Sl No.	Waste	Quantity (Million tonnes per annum)	Source
1	Steel and Blast Furnace	35.0	Conversion of steel
2	Brine mud	0.02	Caustic soda industry
3	Copper slag	0.0164	By product from smelting of copper
4	Fly ash	70.0	Coal based thermal power plants
5	Kiln dust	1.6	Cement plants
6	Lime sludge	3.0	Sugar, paper, fertilizer, tanneries, soda ash, calcium carbide industries
7	Mica scraper waste	0.005	Mica mining areas
8	Phosphogypsum	4.5	Phosphoric acid plant, Ammonium phosphate
9	Red mud / Bauxite	3.0	Mining and extraction of alumina from Bauxite
10	Coal washery dust	3.0	Coal mines
11	Iron tailing	11.25	Iron Ore
12	Lime stone wastes	50.0	Lime stone quarry

(Source: Manual on Municipal Solid Waste Management, CPHEEO, New Delhi)

Major producers of industrial wastes are the thermal power plants producing coal ash; integrated iron and steel mills producing blast furnace slag and steel melting slag; non-ferrous industries like aluminium, zinc and copper producing red mud and tailings; sugar industries generating press mud; pulp and paper industries producing lime, and fertilizer and allied industries producing gypsum. It is mandatory for the industries that generate wastes to manage by themselves. It is also mandatory to obtain prior permission from the respective state pollution control boards to start such industries under relevant rules. The industrial wastes, and the sources with quantities generated in India are given in Table 1.1.

Slaughter House Waste: India has the world's largest livestock population. According to the Ministry of Food Processing, Government of India, a total of 3616 slaughter houses exist. They slaughter over 2 million cattle and buffaloes, 50 million sheep and goat, 1.5 million pigs and 150 million poultry annually, for domestic as well as export purposes. Slaughtering of animals generates both liquid and solid wastes consisting of non-edible organs, stomach contents, dung, bones and sludge from waste water treatment. The large type of slaughter house generates 6-7 tonnes; the medium type generates 2-6 tonnes; and small type generates 0.5-1.0 t/day. Central Pollution Control Board in India has brought out "Draft guidelines for sanitation in slaughter houses" in 1998.

In broader sense of the term, the *Municipal solid waste* (MSW) covers decomposable wastes such as food and vegetable wastes (cooking waste), and non decomposable wastes such as *metals* (aluminum, steel, etc.), *glass* (clear, colored, etc.), *paper* (newsprint, cardboard, etc.), *natural polymers* (leather, grass, leaves, cotton, etc.), and *synthetic polymers* (synthetic rubbers, polyethylene terephthalate, polyvinyl chloride etc.),

Similarly, the *Industrial waste* is made up of a wide variety of non-hazardous materials that result from the production of goods and products. Commercial and institutional, or industrial waste is often a significant portion of municipal solid waste, even in small cities and suburbs.

Some of the wastes referred to as *Special wastes* include (i) Cement kiln dust, (ii) Mining waste, (iii) Oil and gas drilling muds and oil production brines, (iv) Phosphate rock mining, beneficiation, and processing waste, (v) Uranium waste, and (vi) Utility waste (i.e., fossil fuel combustion waste). These are generated in large volumes and are believed to cause less risk to human health and the environment than the wastes specified as hazardous waste.

Medical Waste (or **Hospital waste**): It refers to the waste materials generated at health care facilities, such as hospitals, clinics, physician's offices, dental practices, blood banks, and veterinary hospitals/clinics, as well as medical research facilities and laboratories. The Medical waste is defined as "any solid waste that is generated in the diagnosis, treatment, or immunization of human beings or animals, in related research, or in the production or testing of biological." For example, the following trash constitutes medical waste: blood-soaked bandages, culture dishes and other glassware, discarded surgical gloves, discarded surgical instruments, discarded needles used to give shots or draw blood, cultures, stocks, swabs used to inoculate cultures, removed body organs (e.g., tonsils, appendices, limbs), and discarded lancets. Several health hazards are associated with poor management of medical wastes like injury from sharps to staff and waste handlers associated with the health care establishments, Hospital Acquired Infection (HAI) of patients due to spread of infection, and Occupational risk associated with hazardous chemicals, drugs, unauthorized repackaging and sale of disposable items and unused/date expired drugs. This waste is highly infectious and can be a serious threat to human health if not managed in a scientific manner. It has been roughly estimated that of the 4 kg of waste generated in a hospital at least 1 kg would be infected.

Hazardous waste: The waste that is dangerous or potentially harmful to human health or the environment is called hazardous waste which can be in the form of liquids, solids, gases, or sludge. The discarded commercial products like cleaning fluids or pesticides, or the by-products of manufacturing processes can also be hazardous.

Degeneration times for wastes: The approximate times that different types of garbage take to degenerate are (source: Prakriti, Dibrugarh Univ.):

Organic waste, i.e., vegetable and fruit peels, leftover foods etc: A week or two,

Paper: 10–30 days,

Cotton cloth: 2–5 months,

Wood: 10–15 years,

Woolen items: 1 year,

Tin, aluminium, and other metal items(such as cans): 100–500 years,

Plastic bags: One million years?

Glass bottles: Undetermined

E-waste: Electronic waste or e-waste is referred to the end-of-life electronic and telecommunication equipment and consumer electronics; to be specific, computers, laptops, television sets, DVD players, mobile phones etc., which are to be disposed. UN estimates that between 20 and 50 million tons of e-waste is generated world-wide every year and approximately 12 million tons of this comes from Asian countries. (source: Electronic Waste Recovery Business).

Although much of the e-waste comes from developed countries, considerable quantities also originate from within India. As of March 2009, approximately 400,000 tons of e-waste was produced in India; 19,000 tons of this came from Mumbai, the largest e-waste generator in the country (source: Toxics Link).

E-waste is the fastest growing segment of the MSW stream. E-waste equals 1% of solid waste on average in developed countries which grew to 2% by 2010. In developing countries, like India, E-waste forms 0.01% to 1% of the total solid waste. Globally, computer sales continue to grow at > 10% rates annually. Sales of DVD players are doubling year over year. Yet the lifecycle of these products are shortening, shrinking to 10 years for a television set to 2 or 3 years for a computer. As a result, a high percentage of electronics are ending up in the waste stream releasing dangerous toxins into the environment. These are a division of WEEE (Waste Electrical and Electronic Equipment). The categories under WEEE are: large household appliances, small household appliances, IT and telecommunication equipment, consumer equipment, lighting equipment, electrical and electronic tools, medical devices, monitoring and control instruments and so on. Most of the equipment is made of components, some of which contain toxic substances. If proper processing and disposal methods are not followed, these substances affect human health as well as the environment. For example, cathode ray tubes contain large amounts of carcinogens such as lead, barium, phosphor and other heavy metals. If they are broken or disposed in an uncontrolled manner without taking safety precautions, it can result in harmful effects for the workers, and pollute the soil, air and ground water by releasing toxins.

Special care is warranted during recycling and landfilling of e-waste as they are prone to hazards.

1.3 Waste Management Concepts

There are a number of *concepts about waste management* which vary in practice between countries or regions as already mentioned. Some of the most general, widely-used concepts include:

(i) *Waste hierarchy:* Waste hierarchy proposes that waste should be managed by different methods according to its characteristics. The preference of the options represents the hierarchal structure. Thus, prevention, reuse and recycling are given the highest preference, while open burning is unacceptable. The hierarchy is designed to improve the environmental aspects of ISWM. Practices, which produce serious impacts on the environment, are the least accepted ones.The waste hierarchy is an accepted key element of ISWM (see Annexure 9 for ISWM). The waste management plans are to derive the most useful benefits from products and to generate the minimum amount of waste, and are listed according to their desirability in terms of *waste minimization*. The Waste hierarchy is schematically represented in Fig.1.2.

(ii) *Extended producer responsibility:* PR is a strategy designed to promote the integration of all costs associated with products throughout their life cycle (including the end-of-life disposal costs) into the market price of the product. Extended producer responsibility is meant to impose accountability over the entire lifecycle of products. This means that firms which manufacture and trade in products are required to be responsible for the products not only during manufacture but after their useful life also.

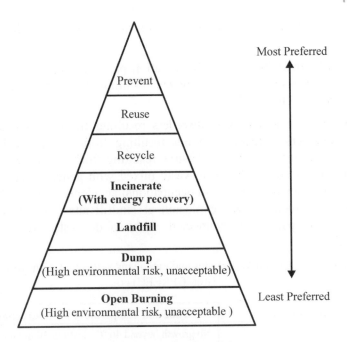

Fig. 1.2 Waste hierarchies.

(iii) Polluter Pays Principle relates to the polluting party paying for the impact caused to the environment. With respect to waste management, a waste generator is required to pay for the disposal of the waste.

1.4 Health and Environmental Impacts

A large number of components in MSW create health and environmental problems. Health impacts include exposure to toxic chemicals through air, water and soil media; exposure to infection and biological contaminants; stress related to odor, noise, pests and visual amenity; risk of fires, explosions, and subsidence; and spills, accidents and transport emissions.

The occupational hazards associated with waste handling according to UNEP Report (1996) are:

Infections: Skin and blood infections resulting from direct contact with waste, and from infected wounds; Eye and respiratory infections resulting from exposure to infected dust, especially during landfill operations; Different diseases that result from the bites of animals feeding on the waste; and Intestinal infections that are transmitted by flies feeding on the waste;

Chronic diseases: Workers at Incineration plants are at risk of chronic respiratory diseases, including cancers resulting from exposure to dust and hazardous compounds;

Accidents: Bone and muscle disorders resulting from the handling of heavy containers; infecting wounds resulting from contact with sharp objects; poisoning and chemical burns resulting from contact with small amounts of hazardous chemical waste mixed with general waste; burns and other injuries resulting from occupational accidents at waste disposal sites or from methane gas explosion at landfill sites. Some common parasites and pathogens connected with solid waste are given in Table 1.2.

Table 1.2 Common parasites and pathogens associated with waste
(Ref: CPREEC)

Organisms	Time and Temperature for destruction
S. Typhosa	No growth beyond 46 °C, death in 30 minutes at 55-60° and 20 minutes at 60 °C, destroyed in a short time in compost environment
Salmonella sp.	In 1 hour at 55 °C and in 15-20 minutes at 60 °C.
Shigella sp.	In 1 hour at 55 °C.
E. Coli	In 1 hour at 55 °C. and in 15-20 minutes at 60 °C.
E. histolytica cysts	In few minutes at 45 °C. and in few seconds at 55 °C.
Taenia saginata	In a few minutes at 55 °C.
Trichinella spiralis larvae	Quickly killed at 55 °C, instantly at 60 °C.
Br. Abortus or *Br. Suis*	In 3 minutes at 62-63 °C and in 1 hour at 55 °C.
Micrococcus pyogenes var. aureus	In 10 minutes at 54 °C.
Streptococus pyogenes	In 10 minutes at 54 °C.
Mycobactercum tuberculosis var. hominis	In 15-20 minutes at 66 °C. or after momentary heating at 67 °C.
Corynebacterium diptheriae	In 45 minutes at 55 °C.
Necator americanus	In 50 minutes at 45 °C.
A. lumbricoides eggs	In 1 hour at 50 °C.

The environmental impacts can be pollution and global warming, photochemical oxidant creation, abiotic resource depletion, acidification, eutrophication, and eco toxicity to water. The communities, industries, and individuals have, therefore, found several ways to reduce and better manage Municipal Solid Waste through a combination of practices not only to extract reusable components but to generate energy in the form of heat or electricity. These practices include source reduction, recycling, and processing/disposal through different technologies such as composting, combustion/incineration, gasification, anaerobic digestion, landfill and so on.

There are several factors that influence successful management of the solid waste, and the vital ones are: Awareness creation among people about the benefits of proper waste disposal, emphasis on waste reduction, long range self sustainability as well as technical feasibility, institutional arrangements, for example, ensuring market for the products, involving community as well as other stakeholders in the waste management programme.

Several Waste processing/treatment and disposal technologies are available for environmentally sound management of MSW. These are broadly grouped as:

Established waste treatment technologies, such as Recycling, Composting, Landfill, Incineration, and Windrow composting; and *Alternative waste treatment* technologies such as Gasification, Pyrolysis, In-vessel composting, Anaerobic digestion, Mechanical biological treatment, Mechanical heat treatment, Sewage treatment, Tunnel composting and Waste autoclave. These technologies enable us to derive clean energy (heat and/or electricity) and resources recovery from the waste before its proper disposal.

Before discussing these technologies, let us look at how a successful waste management system would help to combat global warming. Global warming is a subject of local, regional, national and international concern, because of its severe impacts on the humans and the environment in several ways. These impacts are currently witnessed, and are projected to affect the future generations as well.

1.5 Global Warming

Majority of climate experts feels that Global warming has been the most significant environmental issue that mankind has ever faced. It is an issue with implications for the future generations too. Over the last few decades, thousands of scientists worldwide have engaged in intensive research to understand the reasons and the reality of global warming, and its near- and long-term impacts on the people and the planet Earth. The scientific studies have led to the conclusion that the global warming is 'real' and is 'happening'. The 2007 Report of the Intergovernmental Panel on Climate Change (IPCC) has established that the warming is 'unequivocal', and the increase in global average temperatures is a result of pollution caused by human activities such as fossil fuels usage, agricultural operations, land-use change and deforestation.

What is global warming? The presence of greenhouse gases in the atmosphere from 'natural' and 'human-made' sources is essential because they trap heat and keep the planet Earth warm enough for the life to survive. This effect known as 'natural' greenhouse effect sustains life that includes humans, animals, insects, birds, all ecosystems etc., on the planet. It has been observed, however, that there have been significant fluctuations in the concentrations of greenhouse gases in the atmosphere over the millennium. Their concentrations have been on the rise since the start of Industrial revolution in the 1700s; and during the most recent two decades, their increase particularly that of carbon dioxide, methane and NO_x have been significant. This has been happening due to the growth of cement, paper, steel and other industrial units, power generation plants, cars, trucks and other vehicles, as well as agricultural activities. In addition, indiscriminate destruction of forests and trees has aggravated the problem. Consequently, the greenhouse effect is 'enhanced' and average surface temperature of Earth and oceans have recorded an increase. This is referred to as 'global warming'. Industrialised countries, particularly USA, account for most of the increase of these heat trapping gases. Fast developing countries like China and India, consequent to their increased development activities over the last few years to eradicate poverty and to improve the GDP have joined the group of top emitters of greenhouse gases. Due to the enhanced concentrations of these gases or 'enhanced greenhouse effect', the global average temperatures have risen by 1.3±0.32°F during 1906-2005, most of it occurring since 1975. The rise in temperatures is continuing at an accelerating pace, with 2008 as the 10th warmest year on record. The global warming is starkly evident not only in the rise of global average temperature, but also in global sea level rise, melting of glaciers and thinning of snow cover in the Arctic, Antarctic and other regions, and permafrost thawing, to mention a few.

The *impacts* of the global warming on the global climate and the environment have been extensive and varied (2007 IPCC Reports, Jayarama Reddy 2011). We are witnessing extreme hot days and cold nights, heat waves and wild fires, more frequent and severe storms, cyclones, and droughts, change in precipitation patterns, spread of disease to regions previously unknown, migration of birds and plants to cooler regions, extinction of certain plant and animal species etc., in different regions of the world. There have been many more early signals observed as a result of global warming in all the continents during the last two to three decades. The frequency and severity of these impacts have been differing over the regions of the globe.

The simulations of the specially developed *climate models* indicate a continual warming of the planet if the current rates of greenhouse gas emissions continue, with the temperatures rising by another 2.7 to 11°F by 2100. This huge rise could trigger a wide range of changes in the global climate in this century and beyond. The projected climate changes may occur at an enhanced rate compared to what we have been experiencing, and affect adversely the ecosystems, agriculture and food supplies, water resources, coastal regions, human health and settlements, and in general, the entire environment. The island-states and countries like Bangledesh are threatened by sea level rise. The observations over the recent decades point out that many aspects of climate change are happening *faster* and *with more severity* than what climate models have projected.

Both developed and developing countries have recognised that the increase of greenhouse gases in the atmosphere and the resulting climate change weaken the economies, disrupt the development of the countries, especially poorer countries which are more vulnerable to climate change, and adversely affect people and the environment. As early as in 1980s, very many countries met for the first time and approved an agreement called Montreal Protocol, to control CFCs which impact the protective ozone layer. The Earth Summit held in Rio de Janeiro in 1992 with the participation of over 180 countries was a mega event where a full range of environmental issues were addressed; and an international treaty, 'United Nations Framework Convention on Climate Change (UNFCCC)' was formulated to set a goal of 'stabilising' greenhouse gas concentrations in the atmosphere at safe levels. Under the Framework Convention, an agreement called 'Kyoto Protocol' that set targets for the industrialized countries (called Annex-1 countries) to curb their greenhouse gas emissions to an average of 5.2% below their 1990 emission levels came into force in 2005, and began to bind for ratified countries in 2008. More than 180 countries ratified the Protocol, which might be considered as the the first multinational step to limit greenhouse gas emissions. The largest emitter country, US and Australia, though signatories, did not ratify. The Kyoto Protocol (KP) will expire by 2012, and some of the countries that accepted the targets are unlikely to fulfill for two reasons: (i) KP may not provide adequate conformity incentives, and (ii) The most developed and the largest polluting country, US, is not bound by KP since it has not ratified.

Three market-oriented mechanisms, Clean Development Mechanism, Joint implementation, and Emissions Trading, were formulated to help the Annex-1 countries to reduce the costs of meeting their obligations under Protocol. The Clean development Mechanism (CDM) is a project-based mechanism where industrialized countries can purchase carbon benefits from projects implemented in developing countries to meet their emission reduction obligations. In the developing countries, these project investments help to promote projects attuned to sustainable development such as clean energy projects (examples: solar, wind, biomass, waste-to-energy, clean coal technology etc). The CDM enjoys solid support in developing countries; a large number of projects are undertaken in these countries in the areas of renewable energy generation and solid waste management. In US, despite non-ratification of the Protocol by the federal government, many States and several major industries have voluntarily designed policies and programmes to reduce greenhouse gas emissions. The European Union has unilaterally committed to higher targets of emission reductions than those specified in the Protocol and has come up with new green processes and technologies. Despite these efforts, the global greenhouse gas emissions have been steadily increasing.

Today, US and China each contribute 20% of world's greenhouse gas pollution, European Union 14%, Russia and India 5% each. The fast developing economies, China, India, Brazil and Russia currently figure among the major emitters of greenhouse gases as well as in the world's top ten consumers of energy due to the wide spectrum of their economic activities. If the trend continues, especially in China and India, their emission levels may exceed those of many developed countries. Yet they are not inclined to agree to time-bound targets for emission reductions citing their relatively low 'emissions per head' compared to those of developed countries. The per capita CO_2 emission for different countries, in metric tons, are: USA: 20.01, EU: 9.40, Japan: 9.87, Russia: 11.71, China: 3.60, India: 1.02, and World Average: 4.25 (taken from India's National Action Plan 2008). With large populations and with no sign of reverse in the population growth, the 'per head emissions' are likely to remain low for many more decades for China and India. They are, however, committed to be a part of international initiative in tackling climate change, as reflected in their energy, environmental and climate policies and actions. The eradication of poverty and human suffering take priority over the issues of global climate in the developmental plans of these countries.

Adaptation to the climate change by humans and systems is a way to reduce the cost and severity of impacts currently as well as in the future. So, the adaptive measures to climate change were initiated at local, regional, national and global levels in vital sectors like water resources, energy, agriculture, coastal communities, buildings, and human health. Mitigation actions to reduce greenhouse gas emissions help to stabilize or reverse their concentrations in the atmosphere. The mitigation plans are extensive and mostly sector-dependent; and differ in developed and developing countries because of dissimilar socio-economic conditions and other factors such as technology and infrastructure availability and public awareness. Both the adaptation and mitigation actions are essential; they are a combined set of actions in an overall strategy to reduce greenhouse gas emissions as well as to prepare the humanity to confront the impacts of climate change.

The global population is projected to reach 7.8 to 10.9 billion by 2050. This growth demands more energy, food, housing, goods and services, transport, so on. Burning of more fossil fuels to meet the energy needs, clearing of forests to provide for settlements, and growth in urbanization will not only affect economic and social development but also the environmental sustainability. To contain this trend, services such as family planning, and related health care and education must be extensively provided, especially in the developing countries. If the global population is not controlled, stabilisation of global warming may not be successfully manageable. Deployment of available low-carbon and energy-efficient technologies for manufacturing goods, increase of clean, renewable and advanced nuclear energy sources in the energy mix to reduce dependence on fossil fuels, and energy efficiency practices need to be undertaken on a greater scale.

The costs of implementing the new clean-coal technologies are currently high. While many developed countries have the means to implement clean energy sources and improve energy efficiency, the developing economies can hardly undertake, especially as a short term approach. These countries look for serious funding for adoption of these new technologies. CDM can be one source of funding; but it has to be remodeled by removing bottle-necks with the experience gained so far and implemented in the developing countries on a bigger scale.

The energy-intensive life-styles and behavioural trends of people in the rich countries and of affluent or high-income people in other countries enhance the carbon footprints. The people therefore have to get accustomed to low-energy and low resource consumption practices. Changes that stress on resource and energy conservation may hardly be achievable over shorter time scale, but can certainly lead to slow down greenhouse gas emissions, and to low-carbon economy. Global collaboration and cooperation among the countries which significantly emit greenhouse gases can ensure a sustainable environment and global security. The present generation has the obligation to help preserve the global environment and to promote health, education, and economic opportunity for everyone on the planet and for the generations that follow. Therefore, much needs to be done globally by the governments, organizations, and people to meet the challenges of global warming.

Waste management is perhaps a low-cost mitigation option to reduce emissions and promote sustainable development. For instance, Landfills are a major source of greenhouse gases (particularly methane, which warms the atmosphere more quickly than carbon dioxide), and also contaminate groundwater. Incinerators and other burning and thermal treatment technologies such as biomass burners, gasification, pyrolysis, plasma arc, cement kilns and power plants using waste as fuel, are a direct and indirect source of greenhouse gases to the atmosphere and convert resources that should be reduced or recovered into toxic ashes that need to be disposed of safely. Hence, all technologies available for treating waste to avoid emissions, such as Recovery of methane from landfills, Waste incineration with energy (electric power) recovery, Composting of organic waste, Biomethanation to produce biogas, Controlled waste water treatment for reuse of water, and Recycling and waste minimization need to be effectively deployed. Development of *Bio covers and bio filters to optimize CH_4 oxidation* would further help to reduce methane concentrations in the atmosphere.

These are not difficult to deploy; Governments must extend financial and institutional incentives for improved waste and waste water management, which may stimulate technology development and diffusion as co-benefits. The environment policies must include regulations regarding waste collection and disposal, enforcement strategies, and effective implementation at national level. The MSW Rules 2000, other Acts and legal provisions announced by Government of India in the last one decade are good enough to meet most of these aspects. More stringent rules and regulations are in place in the developed/ industrialized countries compared to developing countries. Most of the poorer countries are yet to develop scientific systems for waste management that include resource recovery and energy generation.

Waste related mitigation options have tremendous co-benefits in terms of improved *public health and safety, pollution prevention, soil protection and clean energy supply.*

1.6 Source Reduction

Chapter 21 of Agenda 21, a document adopted by the United Nations as a blueprint on action for environmental protection up to the twenty first century, unequivocally states that environmentally sound waste management must go beyond the mere safe disposal and recovery of waste that is generated (UNCED, 1992). Instead, it must seek to address the root cause of the problem by attempting to change unsustainable patterns of production and consumption. The problem of tackling waste starts with waste reduction at the manufacturing stage itself by going for innovative technologies and newer materials.

We use many materials and products in our day-to-day activities. These have specific lifetimes after which they become useless and we throw them as trash. Part of the trash can be recycled and reused, and the rest be disposed. The 'disposed' portion can be minimized or prevented at the stage of product manufacturing or its usage. The change in the design or manufacturing process or use of products and materials to reduce the waste prevention is known as "source reduction". Source reduction is the practice of designing, manufacturing, purchasing, or using materials or products and packaging in such ways that reduce the amount as well as the toxicity of the trash created.

Waste generation could be reduced if the local and national stakeholders (environmental and civic bodies) follow the concept of 'product stewardship'. This concept would persuade manufacturers to gear up towards environmental concepts of resource utilization with focus on costs and benefit of product development, consumption, disposal and resource recycle. Product stewardship follows the 'cleaner production (CP)' approach whereby waste generation at the upstream is targeted for reduction rather than abating downstream. Since all parties responsible during the life cycle of a product are involved, they have a role to play in managing the waste generated. The MSWM framework cannot achieve the adoption of this concept alone but has to coordinate it closely with the waste generators, manufacturers and the product middlemen until the consumer's end.

What is Product stewardship? It can be defined as a product life cycle where all parties responsible for the design, production, sale and use of a product assume responsibility for its environmental impacts throughout its life cycle. The concept of product stewardship incorporates the following principles:

1. All parties who have a role in designing, producing, selling or using a product or product components should assume responsibility for (a) reducing or eliminating toxic and/or hazardous constituents in products and/or its components, (b) reducing the toxicity and amount of waste that results from manufacture, use and disposal of the products; and (c) developing products that use materials, energy and water efficiently at every stage of a product's life cycle including manufacture, distribution, sale, use and recovery.

2. The greater the ability of a party to influence life-cycle impacts of the product, the greater the degree of responsibility the party should have to minimize them.

3. Those responsible for the design, production, sale or use of a product should have flexibility to determine how to reduce toxic and/or hazardous constituents in it and how to keep materials from becoming waste.

4. The costs of recovering resources and managing products at the end of their useful life should be internalized into the costs of producing and selling them.

5. Government should provide leadership in the area of product stewardship in all its activities, including, but not limited to, promoting it when it purchases products, making capital investments in green buildings and infrastructure, procuring services, and managing them at the end of their useful life (EPA 2003 at (http://www.epa.gov/epr/index.htm).

Green dot system: Many developed countries and some selected industries (heavy machinery, electronic and beverages to cite a few) in the developing countries have a buy-back system for recyclables which can effectively reduce the volume of waste generated at the consumer's end. Germany has implemented the 'green dot system' which makes it mandatory for the recycling industries to process the collected recyclables. Germany issued an ordinance on packaging in 1991 in an attempt to minimize the quantity of solid waste. The manufacturers are required to take back the packaging of their goods and reuse or recycle it.

The green dot system facilitated the industries to comply with waste management regulations, and its goals were set for collecting the waste and separating it. These two goals aimed to recycle 72 percent of glass, tinplate and aluminum packaging waste, and 64 percent of paper, plastic and composite packaging in Germany. The US, however, reported recycle rate of only 22 percent of glass and tinplate packaging in 1990. The regulations to achieve this ambitious goal would create incentives for the industries to minimize waste during manufacture and packaging. The green dot system concentrates on three types of packaging: transport packaging such as pallets and crates; secondary packaging in containers like boxes for commodities; and primary packaging - actual casing of the product. Companies/manufacturers were unable to meet the recycling quotas on their own and the Dual System was created for them with a membership. Members in the system put the 'green dot' trademark on their packaging that guaranteed recycling for their packaging if collected. Drop off and curbside collection for all packaging with the green dot trademark is also available. These recycling receptacles make it more convenient for households to recycle, helping the companies with greater chances to meet the required recycling quotas. As of September 1993, 12000 companies had signed for the green dot programme including 1900 firms based outside Germany.

The green dot system had proved that it could reduce the quantity of waste. In 1992-1993, the consumption of packaging has decreased by about 4%. Containers have been reused and the quantity of secondary packaging has dropped by 80%. The green dot system was responsible for the collection of 4.6 million tons of recyclables in 1993. But, there are some concerns with the system such as the oversupply of recyclable waste, and the necessity for creation of more markets for products made of recyclables. The success of the programme depends on whether the collected waste is fully reused/ recycled or not.

Source reduction is also applicable to domestic, commercial, and institutional sources of waste generation. Source reduction also refers to the 'reuse' of products or materials. Reuse can help to reduce waste collection, waste disposal and handling costs, because it avoids the costs of recycling or municipal composting or landfilling or combustion or such other processing methods. Source reduction also helps to conserve resources and reduces pollution. It helps to control emission of greenhouse gases that contribute to global warming, and toxicity of the material that is created.

The source reduction offers several benefits:

(a) ***Saves natural resources:*** Waste is not just created when items are thrown away after use. Throughout the life cycle of a product, from extraction of raw materials to transportation to processing and manufacturing and use, waste is generated. By reusing the items or by making those with less material, the waste will substantially decrease. Ultimately, fewer materials will need to be recycled or sent to landfills or waste combustion facilities.

(b) ***Reduces toxicity of waste:*** Selecting nonhazardous or less hazardous items is another important component of source reduction. Using less hazardous alternatives for certain items (e.g., cleaning products and pesticides), sharing products that contain hazardous chemicals instead of throwing out leftovers, reading label directions carefully, and using the smallest amount necessary are ways to reduce waste toxicity.

(c) ***Reduces costs:*** Apart from reducing dependence on methods of waste disposal, preventing waste also can mean economic savings for communities, businesses, institutions, and individual consumers.

(d) ***Benefits Industry:*** Industry also has an economic incentive to practice source reduction. When businesses manufacture their products with less packaging, they are buying fewer raw materials. A decrease in manufacturing costs can result in a larger profit margin, with savings that can be passed on to the consumer.

(e) ***Benefits Consumers:*** Consumers can share the economic benefits of source reduction. For instance, if products are bought in bulk or with less packaging or that are frequently reusable, then there will be cost savings. It means what is good for the environment can as well be good economically.

For example, look at the Source Reduction and Reuse facts in USA in 2000: More than 55 million tons of MSW were source reduced. Containers and packaging represented approximately 28 percent of the materials source reduced, in addition to nondurable goods (e.g., newspapers, clothing) at 17 percent, durable goods (e.g., appliances, furniture, tires) at 10 percent, and other MSW (e.g., yard trimmings, food scraps) at 45 percent. There are more than 6,000 reuse centers around the country, ranging from specialized programmes for building materials or unneeded materials in schools to local programmes such as Goodwill and the Salvation Army, according to the Reuse Development Organization.

Between two and five percent of the waste stream is potentially reusable according to local studies in Berkeley, California and Leverett, Massachusetts. Between 1960 and 2008, the amount of waste each person creates in USA has almost doubled from 2.7 to 4.5 pounds per day. The most effective way to stop this trend is by preventing waste in the first place. Since 1977, the weight of 2-liter plastic soft drink bottles has been reduced from 68 grams each to 51 grams. That means that 250 million pounds of plastic per year has been avoided in the waste stream.

CHAPTER 2

CHAPTER 2

Waste Generation and Characterization

Municipal solid waste disposal is an intricate issue in most of the developing countries unlike in developed countries. Poverty, population growth, and fast urbanization rates common to many developing

Fig. 2.1 Waste thrown on the streets in urban areas

countries are the major factors for concern. In addition, the Municipalities or the concerned authorities are not fully equipped to undertake the task of managing the wastes efficiently due to limited revenues/budget (UNEP 2002, Doan 1998, Cointreau 1982) through collection, transport, storage, treatment and disposal. As a result, a substantial part of MSW generated remains unattended and pile up not only at the collection points but at street corners and public vacant places (Fig. 2.1) creating problem for public health and environment.

2.1 Waste Generation and Composition

Solid waste generation rates and composition vary from country to country depending on the economic situation, industrial structure, waste management regulations and life style. However, there are several other factors in cities including 'degree of urbanization' that equally influence the generation rate. MSW generation in the East Asian region has been increasing at a rate of 3 to 7% per year as a result of population growth, changing consumption patterns, and the expansion of trade and industry in urban centers. The average rate of MSW generation in Asian cities ranges from 0.5 to 1.3 kg/capita/day, which is found to have a direct correlation with the per capita income of the city. In industrialized cities where the per capita income is high like the cities in Japan, the average rate of generation can be as high as 1.64 kg/capita/day, whereas in low-income cities like in India or Bangladesh or China or Sri Lanka, the rate is 0.64 kg on a per capita average. For middle-income cities like in Thailand or Malaysia, average rate is about 0.74 kg/capita/day. The waste generation rate in rural areas is found to be much lower compared to the urban areas in many of the countries in Asia. In African cities, the per capita waste generation rate is also in the same range of 0.45 to 1.3 kg/capita/ day. The waste generation rate in developed countries varies from 0.8 to 2.0 kg/capita/day.

In developing countries, the volume of waste generated varies from day to day and season to season. The rate is often less during the weekends compared to weekdays. This is primarily because a large number of people come to work in the big or medium cities from the suburb and nearby residential areas that lie beyond the city limits. For this reason the rate of generation remains high during the weekday and decreases during the weekends. Similarly, the quantity of waste generation increases during the festive days. For example, after a very popular festival in India, the amount of paper waste generated rises tremendously due to a huge amount of fireworks and crackers being used during the festival. Therefore, before a management plan to handle city's MSW can be adopted, the nature of city and future trends of change need to be studied in detail to project the future quantity of solid waste likely to be generated. Fig. 2.2 shows a prediction of quantity of solid waste expected to be generated during the next 25 years, for high-income, medium-income and low-income countries. It is clear from the figure that much attention is required for middle and low-income countries, where a high rate of increase is expected during the coming years.

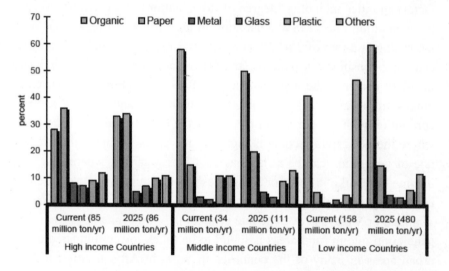

Fig. 2.2 Current and projected waste generation in high-, middle- and low-income countries (source: Martin 1996)

Tables 2.1 and 2.2 provide the waste generation rates in developing and developed countries. The differences in waste generation between developed and developing countries interestingly give rise to not only additional problems but also potential opportunities in waste management.

Table 2.1 Waste Generation Rates in Developing Countries

No.	Country	Urban MSW generation (kg/capita/day)
Low income		**0.64**
1.	Nepal	0.50
2.	Bangladesh	0.49
3.	Myanmar	0.45
4.	Vietnam	0.55
5.	Mongolia	0.60
6.	India	0.46
7.	Lao PDR	0.69
8.	China	0.79
9.	Srilanka	0.89
Middle income		**0.73**
1.	Indonesia	0.76
2.	Philippines	0.52
3.	Thailand	1.10
4.	Malaysia	0.81
High income		**1.64**
1.	Korea, Republic of	1.59
2.	Hong Kong	5.07
3.	Singapore	1.10
4.	Japan	1.47

Source: World Bank 1997a

Table 2.2 Per Capita Waste Generation in Developed countries

Country	MSW generation rate (kg/capita/day)
USA	2.00
Japan	1.12
Germany	0.99
Mexico	0.85
France	1.29
Turkey	1.09
Italy	0.96
Canada	1.80

Table Contd...

Country	MSW generation rate (kg/capita/day)
Spain	0.99
Poland	0.93
Australia	1.89
The Netherlands	1.37
Belgium	1.10
Hungary	1.07
Austria	1.18
Greece	0.85
Portugal	0.90
Sweden	1.01
Finland	1.70
Switzerland	1.10
Denmark	1.26
Norway	1.40

Source: OECD (1995), World Bank (1997a)

The composition of solid waste also differs between developing and developed countries. In developed countries, the composition varies with the size and affluence of the city. To anticipate changes in the size and composition of the MSW stream and to make decisions concerning its management, future projections of the MSW stream are made based on the impact of (a) demographics and (b) recycling, composting, and source reduction programmes. Although consumer behaviour and product composition also affect MSW generation and composition, the impact of these factors is difficult to predict and can have a vague overall effect.

In North America, some general trends are, however, evident. The percentage of paper and plastics in the waste stream is expected to continue increasing. Glass and steel containers are likely to continue to be replaced by lighter materials, such as aluminum and plastic. In addition, changes in local industry and commerce will affect the size and composition of the commercial and industrial waste streams. Over time, changes in MSW generation due to demographic factors have depended on population changes as well as on the per-capita generation. Per-capita generation depends on at least three major factors: household size, socioeconomic status, and the degree of urbanization. The effect of socioeconomic status on MSW generation is uncertain. As North Americans have become more affluent on average, they have purchased more of all goods contributing to increase in waste generation. Although the majority of the population lives in urban areas, the degree of urbanization has some effect on MSW generation. Rural areas often have lower per-capita generation rates for at least some components of the waste stream; for example, fewer newspapers because they are usually printed weekly rather than daily and are more often burned as fuel.

The European countries are considered more advanced in maintaining the waste composition and characterization data. Given variations in economic development, climate, food habits, and culture, it is hardly surprising that composition varies considerably from country to country in Europe. Composition of mixed waste also varies due to differences in source separation methods. However, some generalizations about composition are possible: (a) the largest waste streams in Europe, by far, are organics and paper. Depending on climate and country, these two waste streams combined account for around 50% to 80% of residential waste. It differs by country which of these two waste streams is the largest, but in most cases organics account for more waste by weight, ranging from 25% to 65% of residential MSW, while paper values are more constant, ranging between 20% and 40% by weight in most cases; (b) depending on packaging mix, the relative quantities of glass, plastics, and metals may shift from country to country. These three materials together may account for as little as 10% and as much as 25% of residential waste by weight. When textiles, usually in the range of 2 to 5% by weight, are added, these materials plus paper and organics largely account for the waste stream; (c) household hazardous wastes at 1 to 4% of residential waste by weight, present in European waste streams are far more important in terms of environmental implications than their quantity would indicate; and (d) the waste stream in Eastern European countries tends to be higher in putrescibles and lower in glass, plastics, and metals than the Western European stream. This is consistent with the generally less advanced economic development and status of Eastern European countries. Table 2.3 provides the physical composition of MSW in developed countries.

Table 2.3 Physical composition of MSW in Developed Countries

Country	Organic	Paper	Plastic	Glass	Metal	Other
Canada	34	28	11	7	8	13
Mexico	52	14	4	6	3	20
USA	23	38	9	7	8	16
Japan	26	46	9	7	8	12
Australia	50	22	7	9	5	8
Denmark	37	30	7	6	9	17
Finland	32	26	0	6	3	35
France	25	30	10	12	6	17
Greece	49	20	9	5	5	1
Luxembourg	44	20	8	7	3	1
Netherlands	43	27	9	4	5	8
Norway	18	31	6	4	5	36
Portugal	35	23	12	5	3	22
Spain	44	21	11	7	4	13
Switzerland	27	28	15	3	3	24
Turkey	64	6	3	2	1	24
Average	38	26	8	6	5	18

(*Source:* OECD 1995)

Note: Compositon of waste varies with the size of the city, season and income group

In Latin American countries, the data shows that quantities and quality of wastes are related to the economic conditions of the countries; the richer ones generating more wastes per inhabitant, and their wastes tend to contain more paper, glass, and metal containers than in the poorer countries. Waste quantities generated range between 0.3 and 1.0 kg/inhabitant/day (this includes commercial, market, and street-cleaning wastes). Densities vary between 150 and 200 kg/m^3 (when measured loosely), and from 400-500 kg/m^3 after compaction in the truck. The wastes are very humid (~ 45-50%) and have a high organics content (40-50%). Organics content tends to be higher in poorer countries. The higher consumption of manufactured products with the growth of per capita GDP is evident in the observed differences in the content of the waste generated between small towns (where the organics content is high) and large cities within the same country.

Many Asian cities report data on MSW composition; but it is often difficult to use the data, as the place and season for the sampling is not specified. Cities like Bangkok, Jakarta, Manila, Seoul, and Surabaya, however, collect and report very useable data. Paper and plastic contents are generally higher in cities like Tokyo and Singapore. In cities like Beijing and Shanghai, the ash/soil content is very high due to the burning of coal for space heating during the cold seasons; the proportion of papers, textiles, and other light materials is very low, due to recovery and recycling. In South and West Asia, the countries generally have high percentages of organic and inert matter in their disposed waste, whereas the northern and central areas have more synthetic and combustible materials, metals, glass, and toxic goods such as batteries. The traditional categories for solid waste analysis, namely, organic, inert, paper, glass, metal, plastics, textile, wood/garden wastes, food wastes/bones, ash /soil, are usually not really helpful for understanding the appropriate treatment systems. The synergistic effects of indiscriminate mixing in MSW are not yet clear. There is a need to develop appropriate analytic categories and testing procedures for these countries. National Engineering Environmental Research Institute in India has been working on these issues. The Central Pollution Control Board, India has recently commissioned a national study on solid waste generation and disposal in India which will be discussed in later pages. Large and bulky waste items such as abandoned motor cars, furniture, and packaging are found in the higher-income economies, for example in Israel, Saudi Arabia, and UAE; but they are not put out for municipal pick-up in the Indian subcontinent. In the oil-rich countries, used cars are often abandoned on desert roads outside of cities. The amount of human feces in the MSW is significant in squatter areas where "wrap and throw" sanitation is practiced or bucket latrines are emptied into waste piles and containers. The latter is very common in Kabul, for instance, where sewerage is nominal.

The information regarding waste generation or characterization in African countries, in general, is meagre. However, in cities of Accra, Ibadan, Dakar, Abidjan, and Lusaka, 'the as-delivered MSW (wet basis)' shows generation rates of 0.5-0.8 kg/capita/ per day (compared to 1-2 kg per person per day in the OECD); putrescible organic content ranging from 35-80% (generally toward the higher end of this range); plastic, glass, and metals at less than 10%; and paper at a low teens percentage. These average figures indicate a waste stream of limited potential commercial value for the recovery of metals, glass, plastic, and paper. Since most of the countries have upgraded or plan to upgrade the MSW systems with international aid, the availability of systematic information may improve in the coming years. Table 2.4 illustrates the composition of MSW in the developing countries.

Table 2.4 Composition of Solid Waste in Developing Countries,

	Compostable	Paper	Plastic	Glass	Metal	Other
Low income	**41**	**4.6**	**3.8**	**2.1**	**1**	**47.5**
Nepal	80	7	2.5	3	0.5	7
Bangladesh	84.37	5.68	1.74	3.19	3.19	1.86
Myanmar	80	4	2	0	0	14
India	41.8	5.7	3.9	2.1	1.9	44.6
Lao PDR	54.3	3.3	7.8	8.5	3.8	22.3
China	35.8	3.7	3.8	2	0.3	54.4
Sri Lanka	76.4	10.6	5.7	1.3	1.3	4.7
Middle income	**57.5**	**14.9**	**10.9**	**2.4**	**3.1**	**11.2**
Indonesia	70.2	10.6	8.7	1.7	1.8	7
Philippines	41.9	19.5	13.8	2.5	4.8	17.8
Thailand	48.6	14.6	13.9	5.1	3.6	14.8
Malaysia	43.2	23.7	11.2	3.2	4.2	14.5
High income	**27.8**	**36**	**9.4**	**6.7**	**7.7**	**12.4**
Hong Kong	37.2	21.6	15.7	3.9	3.9	17.7
Singapore	44.4	28.3	11.8	4.1	4.1	6.6
Japan	22	45	9	7	6	11

Source: World Bank (1999)

Industrial Waste generation: In some countries, significant quantities of organic industrial solid waste are generated. Industrial waste generation and composition vary depending on the type of industry and processes/technologies in the concerned country. Countries apply various classifications for industrial waste. For example, construction and demolition waste can be included in industrial waste, in MSW, or defined as a separate category. The default categorization used in the '2006 IPCC Guidelines for National GHG Inventories' assumes construction and demolition waste are part of the industrial waste. In many countries industrial solid waste is managed as a specific stream and the waste amounts are not covered by general waste statistics. OECD (OECD 2002) collects statistical data on industrial waste generation and treatment.

In most developing countries industrial wastes are included in the municipal solid waste stream; therefore, it is difficult to obtain data of the industrial waste separately.

Industrial waste generation data (total industrial waste generation and data for manufacturing industries and construction waste) are given in

Table 2.5 for some countries. The total amount includes also other waste types than those from manufacturing industries and construction.

Table 2.5 Industrial waste by country (1000 tons/yr)

Region/Country	Total	Manufacturing industries	Construction
Asia			
China	1 004 280		
Japan		120 050	76 240
Singapore	1 423.5		
Republic of korea		39 810	28 750
Israel	1 000		
Europe			
Australia		14 284	27 500
Belgium		14 144	9 046
Bulgaria		3 145	7
Croatia		1 600	142
Czech republic		9 618	5 083
Denmark		2 950	3 220
Estonia	1 261.5		
Finland		15 281	1 420
France		98 000	
Germany		47 960	231 000
Greece		6 680	1 800
Hungary		2 605	707
Iceland		10	
Ireland		5 361	3 651
Italy		35 392	27 291
Latvia	1 103	422	7
Malta		25	206
Netherlands		17 595	23 800
Norway		415	4
Poland		58 975	143
Portugal		8 356	85
Romania		797	

Table Contd...

Region/Country	Total	Manufacturing industries	Construction
Slovakia		6 715	233
Slovenia		1 493	
Spain		20 308	
Swedan		18 690	
Switzerland		1 470	6 390
Turkey		1 166	
UK		50 000	72 000
Oceania			
Australia		37 040	10
New Zealand		1 750	NR

(Source: 2006 IPCC)

{Refs: Environmental Statistics Yearbook for China 2003; Eurostat 2005; Latvian Environment Agency 2004; OECD 2002; National Environment Agency, Singapore 2001; Estonian Environment Information Center 2005; Statistics Finland 2005; Milleubalans 2005}

Although significant amounts of industrial waste are generated, the rates of recycling/reuse are often high, and the fraction of degradable organic material disposed at solid waste disposal sites is often less than that of MSW. Incineration of industrial waste may take place in significant amounts; however this will vary from country to country.

Composting or other biological treatment is restricted to waste from industries producing food and other putrescible waste. It has to be noted that the data in the Table 2.5 are based on weight of the wet waste. Also, the data are *default data* for the year 2000, although for some countries the year for which the data are applicable was not given in the reference, or data for the year 2000 were not available (IPCC 2006).

2.2 Waste Characterization

Solid waste characterization is very important because waste characteristics depend mainly on the type of source from which it is generated. We have seen earlier, there are several sources of generation of solid waste in a municipal area, namely domestic, industrial, agricultural, institutional, commercial, healthcare etc. The characteristics of solid waste from each of these sources vary widely. The quantity of these components is dependent on various factors that include the number of residences, industries, commercial and institutional units, hospitals in the city. There is no general rule by which the proportion of solid waste from different sources can be accurately estimated. Since each of the technologies (mentioned earlier) used in MSWM addresses distinct segments of the waste stream, the characterization data assist municipalities in (a) determining the best management methods for different materials; (b) planning recycling and composting programmes by identifying the amounts of recyclables and organic materials generated by different sources; (c) sizing WTE facilities based on the amount of wastes remaining in the waste stream after recycling and composting; and (d) estimating waste transportation and separation costs using local estimates of total municipal waste volume and weight (UNEP).

In the design of a WTE facility, it is important to consider the potential variations in both physical and chemical composition of MSW. Also, in the design of the furnace/boiler portion of WTE facilities, the MSW characteristics that are critical are the calorific value, moisture content, proportion of noncombustibles, and other components (such as heavy metals, chlorine, and sulfur) whose presence during combustion will result in the need for flue gas cleanup. The capacity of a WTE furnace/boiler is roughly inversely proportional to the calorific or heating value of the waste.

In a country like US where almost all technologies are employed in MSWM, characterization data is very vital. Proper local waste characterizations based on actual waste stream studies conducted at landfills, WTE facilities, materials recovery facilities (MRFs), or transfer stations can provide information about the amount of specific products and materials generated by each sector (i.e., residential, commercial, or industrial), the amount of waste recycled, seasonal variations in the waste stream, and differences between urban, suburban, and rural areas. Waste characterization studies should be updated periodically to account for changes in population density, industrial concentration, and community affluence. Comprehensive characterization studies are expensive in America. The commercial waste, on average, accounts for 40% of the municipal waste stream in North America, but percentages vary by community. In Los Angeles, wastes from commercial sources were nearly two-thirds of the city's MSW in 1989.

Most countries in Latin America and the Caribbean have reports on waste quantification and characterization based on research done in academic institutions. The waste management authorities characterize waste in order to estimate required landfill space and necessary infrastructure. University of Chile has recently published a three-year study on waste characterization which shows that the sampling methods presently being used in the region need to be reviewed and that available data may not be reliable. Mexico has an officially approved procedure for the analysis of wastes (UNEP).

Solid waste programmes in Europe routinely perform these studies in their development phase. The national statistics offices maintain annual standard composition figures, which are used as the basis for planning of the system. Nevertheless, some European composition studies only take a waste sample from one season, instead of all four seasons. The results of the waste composition studies are generally used in national and regional projections and in planning for collection systems and disposal capacity.

In industrialized cities of East Asia, MSW is quantified and characterized by municipal authorities at regular intervals. Engineering consultants and professionals from other bodies such as scientific institutes and the universities also carry out the characterization of MSW. In South and West Asian countries, the sampling and analysis of waste streams is not undertaken on a regular basis. Even for such studies as have been conducted, the methodology is often outdated. The reports do not record sampling procedures (e.g., from what points in the waste stream the samples were taken); usually, sampling is at one point only and is not repeated at different seasons of the year. Sampling at different points is important when significant amounts of recyclables are picked out of waste streams, or where animals eat organics.

Waste characterizations data in cities across Africa are not fully available because they are not commonly compiled. This may change as cities upgrade their MSWM systems with the aid provided by international agencies.

Climatic effect: Most Asian countries have tropical climates with high level of precipitation and humidity. High amount of precipitation adds a large quantity of moisture to the waste and increases the weight. The organic portion of the solid waste tends to decompose quickly due to the hot and humid condition and poses problems in handling and disposal of waste. Presence of moisture makes the waste unsuitable for incineration.

Africa's climatic zones consist of humid equatorial zone in the western, eastern and central part, dry zone in the northern and southern parts away from the equator and humid temperate in between the equator and the northern and southern parts. Rainfall is more in the central part and reduces in northern and southern parts. The moisture content in the solid waste in the central Africa is higher compared to the arid zones. In many of the countries situated in the central, western and eastern parts like the Republic of Congo, Tanzania, Madagascar hot and humid climate poses problems with the organic portion of the solid waste.

CHAPTER 3

Waste Reduction and Recyling

Municipal solid waste generation has both pros and cons, but the fact of the matter is that it is impossible to live without generating some trash. By using recycling methods, one can minimize the amount of municipal solid waste generated. Using reusable paper bags or cloth shopping bags, and taking steps to buy recycled products whenever possible are some of the desirable practices. These steps will decrease the amount of municipal solid waste that is generated, as we see in this section, making it a little easier to manage and dispose of.

Waste reduction and waste recycling are important components in waste management strategy. MSWM operations are consuming large fraction of municipal operating budgets; in developing countries, as much as 60% are for collection and transfer of the wastes for disposal. Recently in these countries, there is discussion of sustainable development through an integrated approach to waste management, including minimization of the production of wastes and maximizing waste recycling and reuse. Most of the cities have been practicing source separation and recycling formally and informally.

Recycling is often defined to encompass also waste-to-energy activities and biological treatment. For practical reasons a more narrow definition is used here: Recycling is defined as recovery of material resources (typically paper, glass, metals and plastics, sometimes wood and food waste) from the waste stream (IPCC 2006).

3.1 Recycling

Recycling of waste provides economic as well as environmental benefits, and also reduces reliance on virgin materials. Such programmes can reduce pollution, save energy, mitigate global climate change, and reduce pressures on biodiversity. Reusing items delays or sometimes avoid at item's entry in the waste collection and disposal system. Source reduction coupled with reuse can help to reduce waste handling and disposal costs, by avoiding the cost of recycling, municipal composting, land filling and combustion.

Recycling involves the reprocessing of waste into a usable raw material or product thus enabling materials to have an extended life in addition to reducing resource consumption and avoiding disposal costs. Transportation and collection of recyclable materials involve costs, resulting in an increased market price of such materials compared to virgin materials. The collection of materials that could be recycled or sorting and processing recyclables into raw materials (such as fibers) or manufacturing raw materials into new products are different steps. Collecting and processing secondary material's, manufacturing products with more recyclable content, and buying recycled products constitutes a cycle that ensures the overall success and significance of recycling.

(i) *Collection and Processing:* Collecting recyclables varies from community to community; however, there are four primary methods of collection: curbside, drop-off centers, buy-back centers, and deposit/refund programmes. Regardless of the method used to collect the recyclables, they are sent to a materials recovery facility to be sorted and prepared into marketable commodities for manufacturing. Recyclables are bought and sold just like any other commodity; the prices for the materials alter and fluctuate with the market.

(ii) *Manufacturing:* After cleaning and separation, the recyclables are ready to undergo the next stage of the recycling circle. More and more of today's products are being manufactured with total or partial recycled content. Common household items that contain recycled materials include newspapers and paper towels; aluminum, plastic, and glass soft drink containers; steel cans;

plastic water bottles, milk sachets, and plastic laundry detergent and cleaning liquid bottles. Recycled materials also are used in innovative applications such as recovered glass in roadway asphalt or recovered plastic in carpeting, park benches, and pedestrian bridges.

(iii) ***Purchasing Recycled Products:*** Purchasing recycled products completes the recycling cycle. By 'buying recycled' products, governments, businesses, institutions and individuals are contributing significantly in 'reusing' the materials. As consumers insist more environmentally sound products, manufacturers will continue to meet that demand by producing high-quality recycled products.

(iv) ***Benefits of Recycling waste materials***: Recycling process helps to reduce the need for disposal (landfilling and incineration), to allow cost benefit to the municipalities, to prevent pollution caused by the manufacturing of products from virgin materials, to save energy, to decrease emissions of greenhouse gases that contribute to global warming, to decrease exploitation of natural resources such as timber, minerals and water, and to help sustain the environment for future generation.

3.2 Status in Developing Countries

Materials recovery is very prominent in most urban areas/cities in the developing countries. This practice of materials recovery is done for the following reasons: (i) economic value of materials, (ii) the prevalence of absolute poverty, (iii) the availability of unskilled manpower willing to accept low wages, (iv) the frugal values of even relatively well-to-do households, and (v) the large markets for used goods as well as for products made from recycled plastics and metals.

The wastes which are considered uneconomical to recycle or of no use in affluent societies have a value in developing world, for example, coconut shells and cow dung are used as fuel. Every useful kind of waste that include clothes and rags, bottles, plastics of all kinds (especially milk pouches), metals, toys, and residue from coal fires from household or shop or institution is reused or traded. Second-hand markets thrive, and some are very large, such as those in Indian cities, Mumbai, Kolkata and New Delhi. Food wastes are sold to poultry and pig farmers, and food wastes from large hotels are auctioned in big cities like New Delhi. Construction wastes are reused and the residues are taken as fill for road repairs. In countries such as Vietnam, India and China, if one takes into account the use of compost from dumps sites as well as materials recovered, the major portion of municipal wastes of all kinds are ultimately utilized. Since manufacturers can readily use leftovers as feedstock or engage in waste exchange, residuals and old machines are sold to less advanced and smaller industries. In offices and institutions, paper, cardboard, glass bottles, plastics, and all salable materials are sorted out and sold, and the rest is sent to dust bins. At the household level, gifts of used clothes and goods to servants, relatives, charities and the needy are still significant. In the middle- and low-income cities of the East Asia/Pacific region, informal source separation and recycling of materials have always been practiced. Materials separated or picked out from mixed wastes include ferrous and nonferrous metals, papers/cardboard, glass, plastics, clothing, leathers, animal bones/feathers, books and household goods (which are repaired and sold in second-hand markets). In Bangkok, Jakarta, and Hong Kong there are some very large industries dealing entirely recyclables such as papers, ferrous metals, plastics, and glass. In the Pacific islands, repair and reuse are important and recycling industries are small-scale. In China and Vietnam, waste recovery and recycling has been organized at the city level and supported by national ministries. In China, especially, the major cities have large recovery companies which collect recyclables from offices, institutions, and factories. There are also neighborhood redemption centers where people can sell bottles, paper, and clothes.

In South Asia, resource recovery and recycling usually takes place in all components of the system predominantly by the informal sector, 'waste pickers' or 'a section of the solid waste management staff' for extra income. Such work is done for very low incomes in a very labor-intensive and unsafe way. These items then reach the recycling facilities through agents. In Cambodia, even though waste separation at the source is not practiced, the main items such as soft and hard plastics, glass, steel, paper, cardboard, aluminium and alloys etc., are collected and sent to Vietnam and Thailand for recycling for lack of such facilities in the country. Similarly, most of the recyclable wastes collected in Nepal and Bhutan are sent to India, due to insufficient recycling factories in the countries (Glawe et al 2006). The importance of recycling activities in reducing waste volume, recovering resources and its economic benefits is being amply acknowledged in these countries in recent years. Table 3.1 describes the prevailing 3R (Reduce, Reuse, Recycle) activities carried out by informal and formal sector in South Asian countries.

Table 3.1 3R activities in South Asian Countries:

Country	3 R practice
Afghanistan	Informal
Bangladesh	Informal
Bhutan	Informal
India	Formal + Informal
Maldives	Informal
Nepal	Informal
Pakistan	Formal + Informal
Sri Lanka	Formal + Informal

(Source: Visvanathan and Norbu 2006)

In West Asia, the Palestinian settlements in Gaza and the West Bank have been sites of intensive repair, reuse, and recycling since their inception. Waste separation at the household level and selling the recyclable materials to roaming buyers still continue. There are also small-scale recycling industries for plastic, paper, and glass in the region; scrap metal is recycled locally or transported to re-rolling mills of other countries. In Israel, the high quantity of recyclables reaching waste streams has compelled the government to sponsor the recovery and recycling of the materials due to the decline of the informal sector. Recycling has been emphasized since 1993 to reduce waste quantities at landfills. A number of recycling centers (drop-off centers) are running successfully in Israel, where people can leave textiles, paper, and cardboard; in Tel Aviv and Jerusalem there is curbside pickup of papers in boxes supplied by the authorities.

Different experiments in several countries are tried with source separation of household waste. For example, in Dubai, drop-off centers accept paper, cardboard, aluminum, and PVC bottles. Incentives are given to large-scale recycling units for paper and glass. In Saudi Arabia, an industry has been supported to convert waste paper into egg trays. In Sharjah (UAE), a company processes plastic into rubbish sacks. The governments in developing countries support recycling industries in principle, with financial support such as tax rebates varying from country to country (UNEP).

In Africa, the informal recovery and reuse of materials from the waste stream occurs at several levels. At the household level items are reused before entering the waste stream, thereby extending their useful life. Waste pickers also recover materials for personal and commercial purposes. The extent of commercial recycling of paper, metals, glass, and plastic depends on the presence of industrial or other end uses for these materials. While such industries may be found in some primary cities, they are largely absent in secondary cities and in rural areas. Even in those cases where they are found, they do not consistently stimulate recycling in their host cities. The official statistics on MSW generation and recycling, waste reduction or materials recovery in Africa are very few.

Materials recovery is prevalent in Latin America and the Caribbean. Recycling occurs in all large cities and in most medium-sized cities. Paper and cardboard, glass, metals (mostly aluminum) and plastics are the materials most often recovered and recycled (except plastics) by large-scale industries. Large scale recycling programs of non-hazardous industrial solid wastes have been established in Colombia, Mexico, and Venezuela. In some large cities in Argentina, Brazil, Mexico and Colombia, recycling bins have been set up outside supermarkets, where glass and paper products can be deposited.

The greatest amount of materials recovery and recycling is achieved through several networks such as roving buyers, small and medium traders, wholesaling brokers, and recyclers. Typical examples of the informal recycling industries are those which recycle broken glass into bottles, waste plastics to toys and shoes, and waste paper to paper board. The activities are mainly driven by the scarcity and expense of raw materials. From the point of view of waste reduction, the traditional practices of repair and reuse, and the sale, barter, or gift of used goods and surplus materials, are a benefit to the poorer countries.

Animals also play a significant role in the reduction of organic wastes in many places, especially smaller cities and towns in the developing world. Cattle, pigs, goats, dogs, cats, poultry, and crows feed regularly from garbage piles and open vats; animals such as goats, sheep, and pigs, kept in squatter areas are fed household vegetable wastes. In cities like Hong Kong, Bangkok, Manila, Cebu, and most cities and towns of China and Vietnam, pig and poultry farmers routinely collect food wastes from households and restaurants for animal feed. In some small Chinese cities, pigs are released on garbage dumps to reduce organic wastes. It has been estimated that up to 50% of domestic and restaurant organics are fed to animals. There is likely to be some kind of opposition to schemes for "wet-dry" separation and composting in areas where household organics are needed as animal feed. If the animals' population declines with the modernization of cities, the organic fraction of MSW will increase.

3.3 Status in Developed Countries

The operational aspects of waste reduction and recycling of materials are different in developed countries. The waste stream and how it is treated in typical cities of developing and developed countries is shown in Fig.3.1.

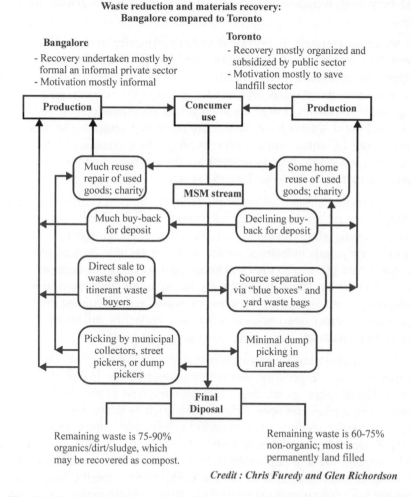

Waste reduction and materials recovery:
Bangalore compared to Toronto

Bangalore
- Recovery undertaken mostly by formal an informal private sector
- Motivation mostly informal

Toronto
- Recovery mostly organized and subsidized by public sector
- Motivation mostly to save landfill sector

Remaining waste is 75-90% organics/dirt/sludge, which may be recovered as compost.

Remaining waste is 60-75% non-organic; most is permanently land filled

Credit : Chris Furedy and Glen Richordson

Fig. 3.1 Waste reduction and materials recovery in typical cities of developing and developed countries (Source: UNEP).

In the urban centers of Australia, Hong Kong, Japan, Korea, and New Zealand which are economically advanced, a high degree of waste reduction, separation at source, and recycling is being achieved through public education, curbside collection, and volume-based collection. In many European Union (EU) countries, materials recovery and recycling have been acknowledged as an indispensable aspect of integrated solid waste management systems. This is reflected in policy documents and is evident in practice. Nevertheless, there are considerable variations in European practices in source reduction, materials recovery, and recycling. European countries often have completely unrelated systems for collecting different recyclable materials. Paper is often collected curbside from row- or single-family houses, and in half- or one cubic meter collection containers from apartments. Glass collection, on the other hand, is more likely to be performed using closed recycling collection containers, sometimes called igloos.

In European Union, the concept of 'producer responsibility' dominates the process of materials recovery; i.e, financial responsibility for recycling and disposal is placed on the product and/or package manufacturer, and incentives for recovery, reuse, and recycling through taxes, fees, and deposits are internalised. In Southern and Eastern Europe and also in France and the UK, the presence of recycling containers or collection programmes is more sporadic. Italy, which has made considerable progress in materials recovery, is an exception.

The rising disposal costs, the difficulty of siting landfills and incinerators, and the current public concern for the environment have made recycling a top priority in North America. Volunteer groups organize neighborhood collection drives for papers, bottles, and cans. A well-established network of haulers, brokers, and reclaim yards recover paper and metals from businesses. Waste reduction in North America includes source reduction, reuse, recycling, and composting, all of which divert significant quantum of materials from disposal facilities. Source reduction involves reducing the amount and toxicity of materials before they enter the waste stream, and can include product reuse, reduced material volume, reduced toxicity, increased product lifetime, and decreased consumption. These programmes have been implemented through education, research, financial incentives and disincentives (e.g., volume-based fees), regulation, and technological advances. North American recycling programmes include source separation, curbside collection, centralized drop-off or buy-back facilities, materials recovery facilities, and mixed-waste processing facilities. Typical materials recycled in North America include paper (e.g., cardboard, office paper, and newsprint), bottles and cans (e.g., steel, glass, aluminum and plastic), ferrous scrap, batteries, tires, used oil, appliances, and construction and demolition debris. Composting is often considered a form of recycling in North America.

Two main collection methods are used: central collection, where waste generators transport materials to a drop-off or buy-back center; and curbside collection, where recyclables are collected at the point of generation (usually households). Central collection centers have been in place for many years. The local governments in order to achieve higher recycling rates have started curbside collection also. Central collection centers accept materials from homes and small businesses. These are commonly known as drop-off centers and buy-back centers. Both the centers want waste generators to bring recyclables to a central facility, but only buy-back centers pay for the material. Both types of centers are less expensive to operate than curbside collection programmes (UNEP).

3.4 Case of Plastics

3.4.1 Nature of Plastics

Consider 'plastics', a recyclable material found in fair amounts in municipal solid waste. The quantity of plastic has exceeded the recovery capacity of even the high-recycling cities. When plastics were started using increasingly in packaging in China, any piece of plastic was valued by waste buyers and pickers. But, currently, the larger cities of China are experiencing the proliferation of plastic waste that is so problematic in Hong Kong, Indonesia, the Philippines, South Korea, and Thailand. Even in Yangon, where non-organic wastes are minimal, increasing numbers of small plastic bags are found in open drains (UNEP). The situation is similar in many developing countries in Asia.

Plastic has unique qualities: strong though light, durable, economical, and can be easily molded and used for different requirements. Plastic bags are popular with consumers because they are functional, lightweight, strong, inexpensive and hygienic. In addition, the environmental impact of plastic bags in landfills is low due to their inert (or un-reactive) nature. In fact plastic bags may have some benefits to landfills such as stabilizing qualities, leachate minimization, and minimization of greenhouse gas emissions (EPHC, 2002) as we see in the later pages. However, the very problem with plastic bag waste emanates from some of their advantages. First, because they are cheap there is excessive consumption and a tendency for misuse. Second, most of the plastic bags produced are too thin and fragile to be re-used. This characteristic of plastic bags lends them to inadvertent littering, which has become a serious problem in urban centres the world over. Simply, it is a multipurpose material due to its non-biodegradable nature; but is now considered a serious threat to global envirnmental health. Its environmental impact is extremely wide ranging: (i) it causes visual pollution that affects such sectors as tourism; (ii) plastic wastes block gutters and drains creating serious storm water problems. Bangladesh, for instance, imposed a ban on plastic bags in March 2002 following flooding caused by blockage of drains (EPHC 2002), (iii) plastic wastes that enter the sea and other water bodies kill aquatic wildlife when the animals ingest the plastics mistaking them for food, (iv) consumption of plastic bags by livestock can lead to death, and (v) plastics take 20 to 1000 years to break down. Dioxin is a highly carcinogenic and toxic by-product in the manufacturing process of plastics. If plastics, especially PVC, are burned dioxin and furan are released into the atmosphere. Conventional plastics have been associated with reproductive problems in both wildlife and humans. Studies have shown a decline in human count and quality, genital abnormalities and a rise in the incidence of breast cancer. If plastic bags are not properly disposed, they choke drains, block the porosity of the soil and create problems for groundwater recharge. Plastic disturbs the soil microbial activity. Plastic bags can also contaminate foodstuffs due to leaching of toxic dyes and transfer of pathogens.

In India, the plastic industry is growing phenomenally. Plastics have use in all sectors of the economy – infrastructure, construction, agriculture, consumer goods, telecommunications, and packaging. But, along with a growth in the usage, a country-wide network for collection of plastic waste through rag pickers, waste collectors and waste dealers and recycling enterprises has sprung all over the country over the last decade or so. More than 50% of the plastic waste generated in India is recycled and used in the manufacture of various plastic products. The rest remains strewn on the ground, littered around in open drains, or in unmanaged garbage dumps. Though only a small percentage lies strewn, it is this portion that is of concern as it causes extensive damage to the environment. India, however, has put plastic waste on a restricted list and the Department of Chemicals and Petrochemicals, Government of India has been considering ways of reducing plastic wastes; Bangladesh too has been discussing limits on plastic packaging.

Eco-friendly, biodegradable plastics would, however, be quite useful. Though partially biodegradable plastics have been developed and are under use, wholly biodegradable plastics based on renewable starch rather than petrochemicals are in the early stages of commercialization in India. More incentives have to be provided for the development of these useful plastics. If the safety is assured, it is trouble-free as well as economical to deliver milk in plastic bags rather than in bottles. Safe plastics have wide range of uses : as carry bags, pet bottles, containers, and trash bags in the houses; as disposable syringes, glucose bottles, blood and uro bags, intravenous tubes, catheters, surgical gloves in health and Medicare; as packaging items, mineral water bottles, plastic plates, cups, spoons in hotels and catering and travel.

The plastic industry in the developed world has realized the need to develop environment-friendly methods for recycling plastics wastes, and has set out targets and missions. Prominent among such missions are the Plastic Waste Management Institute in Japan, the European Centre for Plastics in Environment, the Plastic Waste Management Task Force in Malaysia. Manufacturers, civic authorities, environmentalists and the public have begun to acknowledge the need for plastics to conform to certain standards as well as a code of conduct for its use.

3.4.2 Recycling of Plastic Waste

The recycling process of different plastic waste generated in India, a developing country, and USA, a developed country, are briefly discussed.

(i) *Managing Plastic waste in India (source: CPCB):* It is estimated that post-consumer plastic waste constitutes approximately 4-5 percent by weight of Municipal Solid Waste (MSW) generated in India, compared to 6-9 percent in developed countries. Thermoplastics constitute about 80 percent and thermoset approximately 20 percent. A machine was developed for recycling of plastics in an environmentally sound manner. The aim of green recycling of plasticwaste was to develop a system which would have zero adverse impact on the environment. This has been achieved by assigning right motor of minimum capacity, selecting optimum L/D ratio, heat sealing and right temperature for the processes and trapping all the emission in pollution control equipment and treating the pollutants to produce byproducts. The Extrusion and Pelletization processes have been redesigned to reduce the pollution to a minimum and to enhance the process efficiency.

Another novel approach implemented was reusing plastic waste in road construction. The plastic waste (bags, cups, thermocole) made out of Polyethylene (PE), Polypropylene (PP), and Polystyrene

(PS) are separated, cleaned if needed and shredded to small pieces (passing through 4.35 mm sieve) The aggregate (granite) is heated to 170°C in the Mini hot Mix Plant and the shredded plastic waste is added, it gets softened and coated over the aggregate. Immediately the hot Bitumen (160°C) is added and mixed well. As the polymer and the bitumen are in the molten state (liquid state) they get mixed and the blend is formed at surface of the aggregate. The mixture is used for laying roads. This technique is extended to Central Mixing Plant too.

The Indian Council for Plastics in the Environment (ICPE) says that 1.2 million tones of plastics are recycled. In spite of recyclables like paper, glass, tin etc., are sorted at homes, 13 to 20 percent of them are again found from MSW collected by the concerned authorities.

(ii) *In America*: Plastics constitute about 14 to 22% of the volume of solid waste. Recycling is a best solution to handle this amount of waste. However, a major problem in the reuse of plastics is that during the reprocessing, the properties get degraded and makes it difficult to reuse for the same application. For example, the 58 gram, 2-liter polyethylene terephthalate (PET) beverage bottle consists of 48 gm of PET, and the rest 10 gms being a high density polyethylene (HDPE) cup base, paper label and adhesive, and molded polypropylene (PP) cap. The cup base, label, adhesive and cap are therefore contaminants in the recycling of the PET.

The problem of contaminants in plastic recycling are dealt by designing plastic products as "reuse-friendly" because recyclability can be a practical means for disposal. Many organizations are reevaluating the use of recycled plastics. For example, plastic beads are currently used to remove paint from aircraft employing a "sand blasting" type method. The use of recycled plastics is limited by the end-users of the plastics.

Other important reason for not discarding plastics is the conservation of energy. The energy value of polyethylene (PE) is 100% of an equivalent mass of #2 heating oil. Polystyrene (PS) is 75%, while polyvinyl chloride (PVC) and PET are about 50%. With the energy value of a pound of #2 heating oil at 20,000 B.T.U., discarding plastics to 'land fills' results in a waste of energy. Some countries, notably Japan, tap into the energy value of plastic and paper with waste-to-energy incinerators.

Another factor in the recycling process is the trend of increasing tipping fees at landfills. In northeastern states of USA, tipping fees have progressively increased, but in western states the fees have remained low due to the local government subsidies to landfills. As the cost of land filling of solid waste increases, the incentive to recycle also increases. When the cost of land filling exceeds the cost of recycling, recycling will be a sensible alternative to land filling. These factors have led to the following recommendations by the US Environmental Protection Agency: source reduction, recycling, thermal reduction (incineration), and land filling. Each of these has its problems. Source reduction calls for the redesigning of packaging and/or the use of less, lighter, or more environmentally safe materials. This approach could mean reduced food packaging with the possibility of higher food spoilage rates. There would be fewer plastics, but more food in solid waste to be disposed. Whatever disposal method is chosen, the choice is complex, and whatever the costs, the consumer will bear them.

Recycling of Different Plastics (in USA)

(a) *PET (polyethylene terephthalate):* In 1989, a billion pounds of virgin PET were used to make beverage bottles of which about 20% was recycled. Of the amount recycled, 50% was used for fiberfill and strapping. The reprocesses claim to make a high quality, 99% pure, granulated PET. It sells at 35 to 60% of virgin PET costs. The major reuses of PET include sheet, fiber, film, and extrusions. When chemically treated, the recycled product can be converted into raw materials for the production of unsaturated *polyester* resins. If sufficient energy is used, the recycled product can be depolymerized to ethylene glycol and terephthalic acid and then repolymerized to virgin PET.

(b) *HDPE (high density polyethylene):* Of the plastics that have a potential for recycling, the rigid HDPE container is the one most likely to be found in a landfill. Less than 5% of HDPE containers are treated or processed in a manner that makes recycling easy. Virgin HDPE is used in opaque household and industrial containers that are used to package motor oil, detergent, milk, bleach, and agricultural chemicals. There is a great potential for the use of recycled HDPE in base cups, drainage pipes, flower pots, plastic lumber, trash cans, automotive mud flaps, kitchen drain boards, beverage bottle crates, and pallets. Most recycled HDPE is a coloured opaque material that is available in a multitude of tints.

(c) *LDPE (low density polyethylene):* LDPE is recycled by giant resin suppliers and merchant processors either by burning it as a fuel for energy or reusing it in trash bags. Recycling trash bags is a big business. Their colour is not critical; therefore, regrinds go into black, brown, and to some lesser extent, green and yellow bags.

(d) *PVC (polyvinyl chloride):* There is much controversy concerning the recycling and reuse of PVC due to health and safety issues. When PVC is burned, the effects on the incinerator and quality of the air are of great concern. The burning of PVC releases toxic dioxins, furans, and hydrogen chloride. These fumes are carcinogenic, mutagenic, and teratagenic. This is one of the reasons why PVC must be identified and removed from any plastic waste to be recycled. Currently, PVC is used in food and alcoholic beverage containers with approval from competent authority. The future of PVC rests with the plastics industry which has to resolve the issue of the toxic effects of the incineration of PVC. Interestingly, PVC accounts for less than 1% of land fill waste. When PVC is properly recycled, the problems of toxic emissions are minimized. Various recyclers have been able to reclaim PVC without causing health problems. Recycled PVC is used for aquarium tubing, drainage pipe, pipe fittings, floor tile, and nonfood bottles. PVC is combined with other plastic waste to produce plastic lumber.

(e) *PS (polystyrene):* PS and its manufacturers has been the target of environmentalists for several years. The manufacturers and recyclers are working hard to make recycling of PS as common as that of paper and metals. One company, Rubbermaid, is testing reclaimed PS in service trays and other utility items. Amoco, another large Corporation, currently has a method that converts PS waste, including residual food, to oil that can be re-refined.

Prospects: Recycling is a viable alternative to all other means of dealing with consumer plastic waste. Western European companies, especially the German firms Hoechst and Bayer, have entered the recyclable plastic market with success. With a high tech approach, they are devising new methods to separate and handle mixed plastics waste.

Table 3.2 Major Plastic Resins and their use

Resin Code	Resin Name	Common Uses	Examples of Recycled Products
1	Polyethylene Terephthalate (PET or PETE)	Soft drink bottles, peanut butter jars, salad dressing bottles, mouth wash jars	Liquid soap bottles, strapping, fiberfill for winter coats, surfboards, paint brushes, fuzz on tennis balls, soft drink bottles, film
2	High density Polyethylene (HDPE)	Milk, water, and juice containers, grocery bags, toys, liquid detergent bottles	Soft drink based cups, flower pots, drain pipes, signs, stadium seats, trash cans, re-cycling bins, traffic barrier cones, golf bag liners, toys
3	Polyvinyl Chloride or Vinyl (PVC-V)	Clear food packaging, shampoo bottles	Floor mats, pipes, hoses, mud flaps
4	Low density Polyethylene (LDPE)	Bread bags, frozen food bags, grocery bags	Garbage can liners, grocery bags, multipurpose bags
5	Polypropylene (PP)	Ketchup bottles, yogurt containers, margarine, tubs, medicine bottles	Manhole steps, paint buckets, videocassette storage cases, ice scrapers, fast food trays, lawn mower wheels, automobile battery parts.
6	Polystyrene (PS)	Video cassette cases, compact disk jackets, coffee cups, cutlery, cafeteria trays, grocery store meat trays, fast-food sandwich container	License plate holders, golf course and septic tank drainage systems, desk top accessories, hanging files, food service trays, flower pots, trash cans

A potential use for recycled materials includes plastic lumber. The recycled plastic is mixed with wood fibers and processed into a substitute material for lumber, called Biopaste. The wood fibers would have to go to land fill if not reused. This is expected to eventually become a multi-million dollar enterprise. Research and development are continuing to improve this process and product. Recycling is a cost effective means of dealing with consumer plastic waste. But the cost of recycling needs to be brought down through innovation to reach the level of the recycling processes of paper and some metals. Automatic methods of recycling need to be developed. InTable 3.2, different types of plastics and their uses before and after they are recycled are listed.

3.5   Recovery and Recycling of E-waste

In the recent years recovery and recycling of E-waste has become very significant. The industrialized countries generate tremendous amounts of E-waste. Most of it is exported to developing countries for disposal.

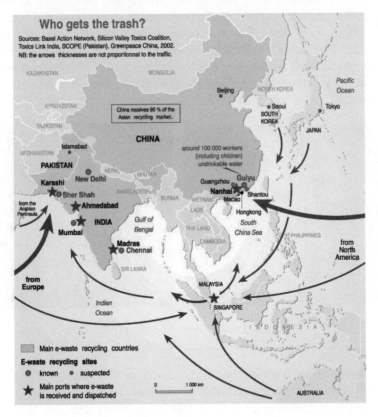

Fig. 3.2 Flow of e-waste to Asian countries.

According to Financial Express (2005), about 80 percent of the Electronic waste generated in the US is exported to India, China, Pakistan, Bangladesh and Sri Lanka (Fig.3.2). The reasons are:

availability of cheap labour, no stringent law on imports, and lack of healthcare awareness. In addition, it offers an easy income. China has banned the import of E-waste legally, but it is still getting through. E-waste recycling is done in India and Pakistan, and the same practice is prevalent in most developing countries.The recent ban on importing E-waste to China has diverted much of it to Bangladesh and other neighboring countries due to cheap labour and recycling businesses

Recycling of e-waste is a dangerous process. This waste consists of components containing hazardous ingredients such as lead, beryllium, mercury, cadmium, halogenated chlorides and bromides used as flame-retardants in plastics, and when these are combusted at temperatures, $600°$ to $800°C$, release dioxins and furons which are dangerous to human health. To recover copper and other metals, they burn the electrical components (including PV sheathing wires) releasing deadly toxins (Fig.3.3). The other electronic and electric waste are dismantled and sorted manually to fractions of printed wiring boards (PWB), cathode ray tubes (CRT), cables, plastics, metals, condensers and other materials like batteries, LCDs or wood.

Fig. 3.3 'E-waste recycling shop in Pakistan
Burning of WEEE to recover metals (Source: Toxics Link).

Burning printed wiring boards in India or Pakistan or Bangladesh showed shocking concentrations of dioxins in the vicinity of open burning places reaching 30 times the Swiss guidance level, according to a study. But, to date, industry, government and consumers have only taken small steps to deal with this frightening problem.

Especially in developing countries, electronic waste is the highly sought-after item for scavengers and local recyclers. The informal sectors in the urban areas of developing countries are now targeting more on the WEEE issue without realizing the level of toxicity involved in their actions. In Pakistan, 'Sher Shah' in Karachi is one of the principle markets for second hand and scrap materials where all sorts of electronic and electrical spare parts, computers and smuggled goods arrive by sea and land for sale or further distribution to other cities in Pakistan. Sher Shah serves as an open informal market, without state controls of any kind (source: Toxics Link).

3.6 Waste Trading

The waste trading enterprises are not generally regulated, and the extent of registration is more formal in Latin America and Asia than in Africa. The system is adaptive to market fluctuations; the casual labour shift to other work when there is reduced demand for the recovered materials. The activities of materials recovery, processing, and recycling employ a large number of people, with the result the governments and social welfare organizations are often more sensitive to the employment needs than to environmental considerations, and are prepared to trade off some environmental and public health risks.

Recyclables are extensively traded, even internationally, particularly in the Indian subcontinent. Almost all the recyclables of Nepal are exported to India, and Indians control this trade. Surplus materials from Kolkata are exported to Bangladesh. The most lucrative cross-border trade in used war materiel is that from Afghanistan to Pakistan. India imports large quantities of waste paper from western countries. There are proposals to expand importing, and such trade is very much wanted by some western countries. Environmentalists are concerned that this trade may allow the import of hazardous wastes in violation of the Basel Convention. Social advocates fear that people who make their living through picking or trading local wastes would suffer, since the demand for these materials might decline.

3.7 Waste Picking as a Livelihood

The picking of recyclables from mixed garbage at street bins, transfer stations, and dumps is very common in many *developing countries*. In cities, recovery and recycling is an income generation activity for many thousands of poor, uneducated and under-privileged people. In societies with significant levels of absolute poverty, poor people depend upon wastes for fuel, clothing, building materials, and even some food. In a study at the Bisasar Road landfill in Durban, South Africa, it was found that picking of waste supported 200 families, "earning" the equivalent of $15,500 per month, or $77 per family per month (Johannessen 1999). This practice is particularly risky where municipal wastes contain human excreta, biomedical and industrial wastes, and where pickers do not have protective clothing or access to washing facilities. It is a matter of great concern that children and pregnant women are plenty among pickers. In China, a few decades ago waste pickers used to be relatively low; now their number is increasing. The influx of rural migrants to the cities has contributed to this trend. Most of the pickers are from poor families with little or no formal schooling. They are often illegal immigrants from rural areas or even foreign countries. The incomes of these pickers range from as low as US$0.40 to around US$3.00 a day. Some cities in Indonesia have introduced licensing of waste pickers at dump sites in an attempt to control the practice. Licensing has met with mixed success, in some cases being welcomed at first but then encountering resistance (UNEP).

Most municipal authorities do not intend to enforce prohibition of picking, as it deprives the livelihood of poor people. Any major change to the waste disposal structure through technology decisions must take into account the urban poor, many of whom may be dependent on waste picking for their entire subsistence. Organizations working with pickers argue for better recognition of the usefulness of waste recovery to developing societies, and for humane treatment of pickers. In 1989, the President of Indonesia pleaded for the recognition of the useful role of waste pickers. Since then in the cities of Java, many schemes were initiated with international aid to assist pickers in various ways. These schemes are meant to overcome the social prejudice which restricts pickers to improve their status, acquire new skills, or simply move up the ladder to become buyers, dealers, or processors. In order to help the pickers from health and exploitation risks, NGOs and social activists have suggested to (i) provide free or subsidize protective clothing to reduce the health risks, (ii) provide access to basic health care and inoculations against tetanus, (iii) regulate picking, by the provision of designated picking areas at transfer stations and dumps, (iv) enable pickers to organize cooperatives to improve their earnings and working conditions, and (v) control harassment of street pickers and itinerant buyers.

Certain Cooperatives and NGOs in some developing countries implement some of these suggestions. NGOs are developing simple cleaning and processing methods for cleaning materials taken from mixed garbage to improve the prices that pickers can get for them. NGOs in these countries such as the Self Employed Women's Association in India have assisted waste pickers in forming cooperatives to obtain source-separated wastes. Cooperative organization has helped some pickers to become buyers of source-separated, clean materials, particularly in the Andean countries of Latin America. There are several NGOs who assist picker families in African and Asian cities. The provision of gloves and boots to pickers, however, in Kolkata and other places failed, as they preferred to sell them, and continued to work as before. In Mexico City and Ciudad Juare, conveyer belts are setup to facilitate sorting at dump sites. The conditions of waste pickers have been improved through their organization and training particularly in Colombia, and later Argentina, Brazil, Panama, Peru, and Venezuela have adopted the system. In Seoul, NGOs have assisted dump-side communities by providing housing, sanitary facilities, medical care, and education to children. NGOs or municipal administrations can help picker groups by offering facilities such as access to water, simple drying and baling machines and other needs at sorting sites.

Due to severe health risks that the pickers face, the societies should explore the possibilities to reduce the attractiveness of picking by creating employment opportunities for low-skilled people and by taking steps to lessen their poverty.

CHAPTER 4

Waste Collection and Transfer

Waste Collection is a key link in the chain of MSWM from the point of generation to ultimate disposal. In any initiative to strengthen or upgrade waste management service, sustainable, contextually appropriate collection should be a major focus of attention. In developed countries, the system, by and large, is in place and operates satisfactorily. In developing countries, the approach to collection differs from region to region. In some countries, collection involves a direct transaction between generator and collector. The level of service is low, and the generators often have to bring their wastes long distances and place them in containers that are sometimes difficult to use.

4.1 Waste Collection in Developing Countries

MSW collection and transfer in the East Asia region is, in general, the responsibility of the public authority. Both door-to-door collection and indirect collection are carried out with containers/communal bins placed near markets, in apartment complexes, and in other appropriate locations and moved to transfer stations and disposal sites by vehicles. The collection services are being privatized increasingly in recent years. Nearly 20% of collection service is now contracted out to private parties in this region and the practice is gaining momentum in Korea,

Indonesia, Malaysia, Singapore, and Thailand (Fig.4.1). In all medium and large cities of the region there are administrative structures for MSW collection and transfer services.

Fig. 4.1 Men who belong to the Yellow Brigade, move waste from households to a transfer point in the two-tier collection system of Surabaya, Indonesia (credit: Antonio Fernandez, source: UNEP)

In the poorer countries of the region, collection rates can be < 50%. However, in Bangkok, Jakarta, Kuala Lumpur, Seoul, and Shanghai, more than 80% of the waste generated is collected. There are, of course, disparities in collection service between rich and poor areas. Shanghai and Beijing cities in China have well structured systems for collection, transport, treatment, and reuse of MSW. However, a weak link in collection and transportation in Chinese cities is the lack of high-quality transfer stations (UNEP).

There is considerable variation in collection and transport systems in South and West Asia, not only from country to country, but within sections of one urban area. But the most important common issue is irregularity or lack of municipal service for squatter settlements or congested low-income areas. The frequency of scheduled collection is partly governed by climate and the system in use. In the Indian subcontinent, temperatures are high and the system is often "open" (i.e., the street containers and transfer points are not covered and waste is exposed). In such circumstances daily or twice-daily collection is required. In central and northern areas of the region where there is curbside or door-to-door collection, collection may be less frequent, although regular. In many cases in south Asia, municipal authorities arrange the collection from the bins located at many locations in a city/town. Fig.4.2 shows the collection efficiency of various capital cities in South Asia (UNEP 2003; DoE 2004; UNEP 2001a; AIT 2004; UNEP 2001b; WWF-Pakistan 2001; Visvanathan and Glawe 2006).

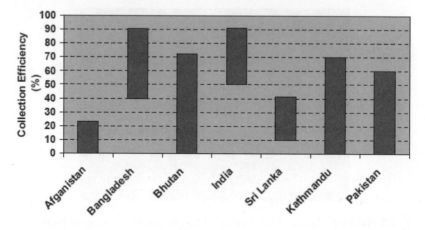

Fig. 4.2 Collection efficiencies (source: Visvanathan and Glawe, 2006)

The informal sector, groups of poor individuals, sort out valuable materials from the waste stream. The human-powered vehicles, animal-drawn carts, or rickshaws, are commonly used for transport from households to a transfer point. Payment is often done either for service and/or for materials. The collection is primarily manual and the collection schedules are not rigid. The transfer stations are generally insufficient in many cities. From the transfer point, the municipal or private party transports the waste from there to the ultimate disposal location. Transport relies on operational vehicles, and frequent breakdowns coupled with shortage of parts can immobilize collection vehicles for extended periods of time.

Collection and transfer in the low-income and northern middle-income economies in South and West Asia is still labor-intensive, although in certain large cities there are hydraulically lifted or movable containers to reduce handling. There are a few experiments with curbside and house-to-house collection in elite areas. In areas of Tel Aviv and Jerusalem, the authority supplies containers to households; these containers are lifted hydraulically and emptied into compactor trucks. In the larger cities of countries like Jordan and Syria, waste is generally disposed of in plastic bags and collected from houses, except in very congested areas.

Fig. 4.3 Waste in Ouagadougou, Burkina Faso
(credit:Romanceor, Wikipedia, Free Encyclopedia).

Many major cities in Africa have an established municipal waste collection system where collection is carried out by human- and animal-drawn carts (wheelbarrows, pushcarts), open-back trucks, compactor trucks, and trailers (Fig.4.3). Collection rates across the continent range from 20-80% with a median range of 40-50%. UNEP (1996) estimates that in West African cities, up to 70% of collection/transfer vehicles may be out of action at any one time affecting collection. In most cities collection is provided by the municipality, though pre-collection is done by community groups in some areas. Private operators also provide service on a fee basis to households and commercial establishments. In Cairo, a private group with experience in MSW collection is authorized to undertake this service, though it is uncommon in other African cities. Since the mid-1970s international aid has promoted initiatives to improve the coverage of MSW collection services, especially vehicular services, in Africa. In some West African cities, such as Dakar and Cotonou, local initiatives have focused on service to neglected areas surrounding urban centers. Street sweeping is also performed by municipal public works staff manually. In some cities the streets are swept at dawn prior to the opening of the market place and commercial center and prior to the first pass of the MSW collection service and in some cities at dusk. Transfer stations are not common in African cities. One such facility, operated by the City of Abidjan is shut down. The disposal point of the MSW is generally located on the perimeter of the city, within easy reach of vehicles and collection staff. The collection vehicles go directly from their point of last pickup to the disposal site.

Fig. 4.4 In Latin America, women are often involved in small-scale collection enterprises (credit: Alvaro Cantanhede, source: UNEP)

Large cities in Latin America and the Caribbean have fairly efficient waste collection coverage. Buenos Aires, Santiago, and Havana claim to collect essentially all of their wastes while São Paulo, Rio de Janeiro, Bogot, Medellú‹, Caracas, Montevideo, La Paz, and Port of Spain claim more than 90% coverage. In many cities the collection and transfer are carried out using conventional equipment and compactor trucks which are expensive to maintain. By entrusting the collection services to a private party who can administer funds more efficiently, this problem has been solved to a large extent. In Latin America, women are often involved in small-scale collection enterprises (Fig.4.4). The frequency and efficiency of waste collection in this region is a major problem. Frequency varies from daily to once a week (not including the many areas of cities which are not serviced at all), and the frequency of collection in an area is not determined by technical considerations such as putrefaction rates of the wastes, weather, vehicle availability, and routing necessities, but rather by its affluence. In cities such as Rio de Janeiro, Mexico City, Caracas, and Buenos Aires, more than 50% of the wastes go through a transfer station, and the need for the transfer stations has grown significantly in the recent years as the distance between the city and the disposal sites grows.

In areas where collection services which remove waste from individual households or streets exist, often there are no standardized containers used to store waste prior to pickup. Headley (1998) states that in Barbados, there are no containers designated by municipalities or collection companies to "set out" waste for collection; the individual residences design some sort of collection container which is quite often, plastic barrel or discarded oil drum. However the majority of households simply place shopping bags full of waste on the street and wait for collection. There may be physical dangers to waste workers in such situations; weather, animals, and other disturbances prior to collection may threaten (Zerbock 2003). In a study of these problems in Kenya, Mungai (1998) has agreed that the first step in "sanitary and efficient" waste management must be to ensure that all households use some form of corrosion-resistant container with lids in order to facilitate collection. Lidded containers would exclude most animal pests, reduce the amount of rainfall soaking into garbage and help to reduce trash blowing about on the street.

In most developing countries and cities, there are many residential areas where no collection of waste takes place. For example, collection may miss large extents of poor or squatter settlements; hilly areas; neighborhoods with unpaved or blocked streets; or whole areas where houses are too close together for collection vehicles to get through.

The recoverable materials may be separated during the collection process itself. The waste collection groups, waste pickers, or independent buyers may be involved in both collection of waste and separation, and recovery of materials. A common disadvantage of collection in developing countries is the acute lack of adequate service, particularly in poorer areas. This weakness combined with the relatively large volume of human and animal fecal matter means that waste collection process operates inadequately. Consequently the collection failure leads directly to human disease and suffering.

4.2 Waste Collection in Developed Countries

There are important technical and institutional differences in the collection of waste between industrialized (developed) countries and developing countries. Some countries which are 'transition' economies like Eastern European and the Balkan countries resemble industrialized countries more than developing ones in terms of their approach to collection, the role of the municipal authorities, informal collectors, and private-sector operators, and demographic and social factors relevant to collection. The transition countries are therefore generally included in the discussion of MSWM aspects concerning industrialized countries.

Waste collection in Europe differs considerably among regions and countries, based on population densities and extent of economic development. Yet, it is possible to notice a common approach to waste collection. Most Western European countries organize waste collection in twice-weekly, weekly, or biweekly routes using 120- or 140-liter rolling carts. These are collected with semi-automated compactor trucks usually having dual self-dumping lifts. In more southern European countries, these compactor trucks may be loaded from ordinary garbage cans and/or bags. On the other hand, Scandinavian countries collect household waste in tall 120-liter Kraft paper bags in a stationary metal frame. In Eastern Europe and in areas where multi-family apartment houses are prevalent, such as Southern Finland, residents may be offered one- to two-cubic yard containers for depositing their household waste. These are emptied, usually by rear-loading compactors, once or twice a week. In certain housing complexes in Eastern Europe, kitchen wastes are still collected in open containers and taken away for swine feed. Waste collection vehicles may go directly to disposal facilities (landfills or incinerators), or they may go to a transfer facility, where the waste is compacted into larger vehicles for longer haul distances. In most cases, transfer means that a compactor truck or other type of collection vehicle (such as an open truck, pickup truck, or wagon) arrives at the transfer facility and dumps its load of waste into a pit or onto a tipping floor. A front-end loader or bulldozer usually loads the waste onto a conveyor or a chute, from which it goes into a special compacting container. These are usually of large capacity and have high compaction ratios, and are used to compact the waste for more efficient long-haul transportation. Bulky waste and/or recyclables, especially corrugated cardboard, are separated on the tipping floor, both for their market value and to make the compaction more efficient. The baling of trash for long-distance hauling is not well developed in Europe; but long hauls are becoming more common as companies look to cheaper disposal opportunities in Eastern Europe. In general, transfer stations dealing in residential waste are run by some public body, but private operators also have been entering.

Collection and transfer of waste in North America is different and more systemized. MSW is typically stored by residents or offices in either metal or plastic trash cans, plastic or paper bags or special containers designed for mechanized collection to effect collection easy. Residential waste in America is collected in at least four ways: (a) at the curbside or passage; (b) from the backyard; (c) from a drop-off collection point; or (d) it is directly hauled by residents to the disposal site. The most common method is curbside or passage collection, where the resident places full waste containers at the curb or in the passageway and retrieves them after being emptied.

Backyard collection is more labour-intensive, more costly, and therefore less common in America. Collection usually occurs at least once per week and even more frequently in urban areas where storage space is limited. Drop-off collection centers are utilized in areas such as rural where individual collection is impractical and in communities where cost savings are more important than service provision. Drop-off sites typically have house dumpsters or even larger roll-off containers, which may be equipped with a compactor.

Special pick-up dates are usually established for bulky items such as old appliances, furniture, and tree stumps. Commercial and institutional waste is usually collected from a dumpster located at the establishment. These generators often hire a collection company to handle their waste, but some local governments take on this responsibility.

Several types of trucks such as rear, side, and front loaders, roll-off and tilt frames, transfer trailers, and vehicles designed for collecting recyclables are currently used for waste transportation. Rear and side loaders are the most common collection vehicles for residential collection and can be loaded automatically or by hand. Front loaders are typically used to pick up large dumpsters for the collection of commercial or institutional waste.

Roll-off container collection is more commonly used in rural areas. The containers are placed strategically throughout the region to allow the residents to drop off their waste. The containers are then collected and transported to disposal sites. Much larger transfer trailers, either open-top or enclosed, are used for bulk transport of compacted waste from transfer stations to more remote disposal facilities.

Because collection programmes are often the most expensive component of local waste management systems in America, the design and management of collection systems have undergone revision and redesign. One of the main planning decisions is whether the collection system should be publicly operated or through government contract to private firms, or by freely operating private firms. Regarding increased recycling activity, while separate collection of source-separated recyclables has extended the capacity of regular refuse collection trucks, it has also demanded the purchase or modification of additional vehicles.

Due to the shortage of acceptable sites, new landfills and waste-to-energy plants often serve several communities or an entire region. Regional MSWM facilities in America are thus making transfer stations a vital component of many waste management systems. Transfer stations are centralized facilities where waste is unloaded from smaller collection trucks and loaded into larger vehicles for hauling to waste processing facilities. Their design typically includes a tipping floor and either bulldozers for pushing waste into transfer trailers or a compactor for packing waste into trailers. In addition, more recyclables are now being sorted and processed at transfer stations.

The longer distance from the place of collection of MSW to regional waste processing facilities often makes transfer stations cost-effective. Transfer trailers carry larger volumes of MSW than regular collection trucks thereby lowering fuel costs, increases labour productivity, and reduces maintenance costs of collection vehicles. These advantages, however, must be balanced against the time spent transferring waste from collection trucks to transfer trailers and the capital costs of purchasing trailers and building transfer stations.

4.3 Waste Processing and Disposal

The important phase in the MSWM is 'Waste Processing/ treatment and Disposal.' There have been several methods - traditional and modern - developed and evolved over time, and the correct choice or choices is decided by MSW-related factors detailed so far. The interesting aspect, however, is that this phase offers benefits of economic and societal value to the communities. 'Waste' is not waste if properly handled and utilized. Opting the appropriate technology or combination of technologies, the cooperative involvement of the concerned stakeholders by scrupulously discharging their obligations, adequate finances, and proper planning and management ensure the success of this phase of MSWM. The technologies along with their appropriateness, merits and demerits are discussed in the next chapters on 'Composting', 'Energy from Waste' and 'Landfills'.

CHAPTER 5

Composting

5.1 Process

This is an old method of treating solid waste which overtime has evolved into an environmentally sound system of waste processing. Natural composting, or biological decomposition, began with the first plants on earth and has been happening ever since. As vegetation falls to the ground, it slowly decays, providing minerals and nutrients needed for plants, animals, and microorganisms. A large range of organisms are available to start and sustain composting.

Composting is the biological decomposition of organic waste (biodegradable materials) consisting of complex animal and vegetable materials into their constituent components. It is a natural process in which micro organisms consume what they like in warm, moist, aerobic and/or anaerobic environment. The most common form of composting - aerobic composting - takes place in the presence of oxygen. The end product is known as 'compost' which is an organic soil-like stable material that is dark brown or black, and is earth-smelling (Fig. 5.1). Compost is a rich nutrient-filled material (Fig. 5.2) which can be used as a soil amendment or as a medium to grow plants.

Under the action of microorganisms in the presence of adequate moisture and oxygen, a biodegradable material breaks down into carbon dioxide, moisture and compost. It may take a very long time for certain materials to biodegrade depending on the environment, but ultimately total breakdown happens.

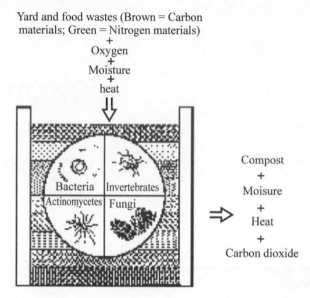

Fig. 5.1 Microbial decomposition organic materials into compost
(source: Vogel 2003)

Among the organic waste, some food products are barred because they can attract pests or affect the quality of compost. The materials to *include* are fruit and vegetable waste, egg shells, tea bags, coffee grounds, leaves, grass, wool and cotton rags, sawdust, non recyclable paper and yard clippings; and the materials to *exclude* are dairy foods, meat, oils and fats, grease, chicken, lard, fish, mayonnaise, cat manure and dog manure.

Composting is suitable for organic biodegradable fraction of MSW, garden waste or waste containing high proportion of lingo celluloses materials which do not degrade under anaerobic conditions, waste from slaughter houses and daily waste. This method is not suitable for wastes that may be too wet. In principle, the compost is created by combining organic wastes in proper ratios into piles, rows, or vessels, and adding bulking agents such as wood chips to accelerate the breakdown of organic materials, and allowing the finished material to undergo a curing process to fully stabilize and mature.

The microorganisms used in composting require an ideal environment with right quantities of air, water, right food and temperature in proper conditions which are discussed below.

Fig. 5.2 Handful of Compost (Photo Courtesy: Jepson Prairie Organics)

Asian countries in particular have a long tradition of making and using compost, and farmers in India have been customarily using compost made out of cow dung and other agro-waste. In Western Europe, compost is produced using a range of modern technologies.

5.2 Benefits

Composting Process offers the benefits of resource efficiency and producing a useful product of economic value from the organic waste, without sending to a landfill. The 'compost' offers a wide range of environmental, economic, and other benefits.

(a) *Compost enriches soils*: Compost is used as a soil amendment because it regenerates poor soils. This material when added to soils increases the nutrient content. That is, compost helps to improve soil texture and in augmenting micronutrient deficiencies. It helps soils to retain moisture and in maintaining soil health. Compost helps to control plant diseases and pest infestation, reduce the need for chemical fertilizers and water, and promote higher yields of agricultural crops. Compost, if not adequately matured, may cause chemical burns on plants or compete with them for use of soil nitrogen.

Fresh compost should not be used for starting sensitive seedlings such as tomatoes and peppers because they may succumb to damping-off disease. These seedlings should be started using a sterilized potting mixture.

(b) *Compost helps remediate contaminated soil*: The composting process absorbs odors and degrade volatile organic compounds (VOCs), including heating fuels, poly aromatic hydrocarbons (PAHs), and explosives. It also binds heavy metals and prevents them from being absorbed by plants or entering into water resources. The compost process degrades and, in some cases, completely eliminates wood preservatives, pesticides, and both chlorinated and non-chlorinated hydrocarbons in contaminated soils. It provides a less costly alternative to conventional methods of remediating (cleaning) soils with contaminants.

(c) *Compost helps prevent pollution*: Composting organic materials ultimately avoids the generation of methane and leachate formation in the landfills. Compost prevents pollutants in storm water runoff from reaching surface water resources. It also prevents erosion and silting on embankments parallel to creeks, lakes, and rivers, as well as erosion and turf loss on roadsides, hillsides, playing fields, and golf courses.

(d) *Compost facilitates reforestation:* wetlands restoration and habitat revitalization efforts by amending contaminated, compacted, and marginal soils

(e) *Compost is a low-cost alternative:* to standard landfill cover; Landfill operators use compost to cover landfills and carry out reclamation projects. Composting also extends municipal landfill life by diverting organic materials from landfills.

(f) *Other uses*: Public agencies use compost for landscaping highway median strips, parks, golf courses, athletic fields, recreational areas, and other public property, while home owners use mature compost to enrich gardens, improve the soil around trees and shrubs, as soil additive for house plants and as protective mulch for trees.

5.3 Composting Technologies

There are several Composting methods producing the same end-product. Choosing a proper composting process is very critical. There are *five primary factors* that must be 'controlled' during composting process:

(i) *Feedstock and nutrient balance*: Controlled decomposition requires a proper balance of 'green' organic materials such as grass clippings, food scraps, manure etc., which contain large amounts of nitrogen, and 'brown' organic materials such as dry leaves, wood chips, and branches which contain large amounts of carbon but little nitrogen. It requires a lot of experimental trials to obtain the right nutrient mix which is very critical to the success of composting process. The carbon-to-nitrogen ratios for different biodegradable wastes are given in the Table 5.1.

Table 5.1 C: N for different biodegradable materials

Material	C:N (by weight)
Materials with high nitrogen values	
Vegetable wastes	12-20:1
Coffee grounds	20:1
Grass clippings	12-25:1
Cow manure	20:1
Horse manure	25:1
Horse manure wit litter	30-60:1
Pultry manure (fresh)	10:1
Pultry manure (with litter)	13-18:1
Pig manure	5-7:1
Material with high carbon values	
Foliage (leaves)	30-80:1
Corn stalks	60:1
Straw	40-100:1
Bark	100-130:1
Paper	150-200:1
Wood chips and sawdust	100-500:1

(Source: Dickson et al 1991)

(ii) *Particle size:* Grinding, chipping, and shredding materials increases the surface area on which the microorganism can feed. Smaller particles produce a more homogeneous compost mixture and improve pile insulation to help maintain optimum temperatures. If the particles are too small, however, they might prevent air from flowing freely through the pile.

(iii) *Moisture content:* Microorganisms living in a compost pile require an adequate amount of moisture to survive. Water helps transport substances within the compost pile and allows the nutrients in organic material accessible to the microbes. The organic material generally contains moisture in varying amounts and the right moisture content should be between 40 and 60%. If it is too wet, anaerobic conditions result; if too dry, the decomposition process will slow down.

(iv) *Oxygen flow:* Composting is an aerobic process, occuring in the presence of optimal amount of oxygen. When oxygen is too little, it will go anaerobic resulting in unpleasant odours. Actions such as turning the pile, placing the pile on a series of pipes, or including bulking agents such as wood chips and shredded newspaper help aerate the pile. Aerating the pile allows decomposition to occur at a faster rate than anaerobic conditions. Care must be taken, however, not to provide too much oxygen, which can dry out the pile and impede the composting process.

(v) *Temperature:* As the microorganisms start decomposing waste, heat is generated. When temperatures rise above 140° F, the organisms start to die. So the pile has to be turned when temperatures reach this point to prevent overheating, which can result in drastic population fluctuations and odours. Eventually, the microorganisms will use up most of the readily decomposable waste, and the composting process will slow down. The temperatures drop, and the compost takes on a dark, granular texture. At this point, the compost can be placed in large stockpiles for curing, and the curing continues to improve until it is ready for use.

The optimal composting conditions are: Oxygen: > 10%; Moisture: 40 to 60%; C/N = 30:1; Temperature: 90-140°F.

Pre-processing: Pre-processing is necessary to create the conditions for bacterial action. Pre-processing involves three types of operations: (i) separation or removal of oversize, non-compostable, or dangerous materials; (ii) size reduction through chipping, grinding, or shredding, to create many small particles suitable to sustaining bacterial action (mentioned above), and (iii) blending and compounding to adjust the carbon-nitrogen ratio, moisture content, or structure of the materials to be composted. Mechanical pre-processing is often the most costly part of a composting system, as well as the most vulnerable to breakdown. Therefore pre-processing is minimized to the greatest extent possible by pre-selecting the waste streams to be composted through source separation and separate collection. If waste segregation at source is not properly carried out there is possibility of toxic materials entering MSW, and the compost produced may be unsafe for use. The time of composting depends on the process used, the compost ingredients, and how the system is managed. For example, under optimum conditions, thermophilic composting with frequent mixing or turning can produce useable compost within a month or two; a worm bin requires three to six months to turn food scraps to compost; and an unmanaged leaf pile may take more than a year to break down. In general, even after it appears finished, it is advisable to let compost "cure" for several months. During this additional time, degradation occurs at a slower rate, resulting in a more 'chemically' stable end product. Stability of compost can be tested by re-wetting the material and observing if it heats up again, which indicates that there are still un-composted materials in the pile. If the composting is completely done, the initial ingredients are no longer recognizable, and the end product is an earthy-smelling substance similar to a rich organic soil. These aspects will be seen in the following pages.

Composting technologies are many, from simple and inexpensive 'backyard or onsite' composting method to more expensive and high-tech methods such as 'in-vessel' composting. Composting varies as much in its complexity as in the range of organic materials recovered. The most common composting methods are listed in the order of *increasing costs* and *levels of technology required* and are described in detail on the following pages:

(a)   Backyard Composting,

(b)   Vermicomposting,

(c)   Aerated (Turned) Windrow Composting and Static Pile Composting, and

(d)   In-vessel Composting.

5.3.1 Backyard or Home Composting

Backyard or Home composting can be practised by residents and others who generate small quantities of organic waste on their own premises (Fig.5.3). Through home or onsite composting, homeowners and small businesses can significantly reduce the amount of waste to be disposed, and thereby save disposal costs.

This method is suitable for converting yard trimmings and food scraps into compost, and unsuitable to compost animal products or large quantities of food scraps. Households, commercial establishments, and educational institutions, hospitals etc can leave grass clippings on the lawn—known as "grasscycling"—where the cuttings will decompose naturally and return some nutrients back to the soil.

Fig. 5.3 Man Digging in Compost Bin.

Since this method is typically suited to small quantities of organic waste, climate and seasonal variations do not present major challenges. For example, if a rainy season approaches, the process can be adjusted accordingly without many problems. The conversion of organic material to compost can take up to two years, but manual turning can hasten the process considerably, by 3 to 6 months. The resulting natural fertilizer can be applied to lawns and gardens to help condition the soil and supply nutrients. Compost, however, should not be used as potting soil for houseplants because of the presence of weed and grass seeds.

Backyard/Home composting represents the smallest scale of composting and is a sound approach when a significant number of households have individual or collective yards or gardens. This type of composting is culturally familiar to most people. This composting process is highly preferable if the waste contains primarily vegetable matter, because it is easier to control rodents and insects when little animal matter is present. Backyard composting requires very little time or equipment. Knowledge and training, however, are the most critical aspects of backyard or onsite composting. Local communities or voluntary organisations can organise composting demonstrations and training programmes to encourage homeowners or businesses to compost on their own premises.

Backyard composting and mulching programme have been operated successfully in Northern Europe, North America, Australia, and New

Zealand and have proved to be much less expensive to a community than centralized compostable collection programmes. They have participation rates of nearly 30 per cent, with significant results in terms of reduction of considerable quantities of waste from the municipal waste that has to be disposed.

5.3.2 Vermi Composting

Vermi composting is a process in which worms are used in the composting process to produce compost. Vermiculture (also called sericulture) is worm farming for the production of worms. In recent years, worm farming has been practiced (Fig.5.4) on both a small and large scale with multiple goals: waste diversion, vermin composting, and sericulture.

Fig. 5.4 Dairy Manure Vermicomposting, Worm Power, Geneseo, NY
(Source: Sherman, R 2009)

In sericulture, two main types of earthworm are raised (Eisenia foetida and Lumbricus rubellis) which are commonly used to produce vermicompost, as well as for fish bait. Both are referred to by a variety of common names such as red worms, red wigglers, tiger worms, brandling worms, and manure worms. These two species are often raised together and are difficult to sort out, though they are not believed to interbreed. Several other species have also been successfully bred in recent years. Worms have been used as a protein and enzyme source for various products, including animal feed and biodegradable cleansers. Worms have also been used to manage agricultural wastes such as dairy manure. They convert waste into worm manure, a nutrient-rich, biologically beneficial soil product.

The requirements for Vermicomposting are (a) simple and inexpensive: worms, (b) worm bedding (e.g., shredded newspaper, cardboard), and (c) a bin to contain the worms and organic matter. Worms will regulate their own population according to their environmental conditions that include space, moisture, pH, temperature, bedding material, and amount of food among others. A typical household worm bin might start with one pound of worms (approximately 1,000 adults), which would soon multiply to 2,000 to 3,000 under suitable conditions. Conversely, if one or more of the above conditions are unacceptable, the worms may crawl out of the bin or die off. Preparing bedding, burying garbage, and separating worms from their castings are the aspects to be taken care of. The requirements are briefly described.

Worm Bin: A suitable bin can be constructed using untreated, non-fragrant wood, or a plastic container can be purchased. A wooden box is better to keep the worms outdoors, because it will keep the worms cooler in the summer and warmer in the winter. The bin size depends on the amount of food produced by the household. The general rule is that one square foot of surface area is required for each pound of garbage generated per week. If a plastic container is used, it should be thoroughly washed and rinsed before the worms and bedding are added. Red worms (the type generally used for vermicomposting) thrive in moist bedding in a bin with air holes on all sides. For aeration and drainage, nine holes of half-inch size in the bottom of the 2 feet × 2 feet bin, or 12 holes in the 2 feet × 3 feet bin are drilled. A plastic tray under the worm bin is placed to collect any moisture that may seep out. Drilling holes on the upper sides of the bin will also help the worms to get needed oxygen and prevent odours in the worm bin. A lid on the bin is kept as the worms prefer to work in the dark. The worm bin is located where the temperature remains between 55° and 77°F.

Bedding: The worms need bedding material to burrow and to bury the garbage. It should be a non toxic, fluffy material that holds moisture and allows air to circulate. Suitable materials are shredded paper (news papers, paper bags, card board or computer paper), composted animal manure (cow, horse, or rabbit), shredded and decaying leaves, peat moss (which increases moisture retention), or any combination of these. Glossy paper or magazines are not generally used. Two handfuls of soil are added to provide fiber for the worms. Crushed eggshells can be added to provide not only fiber but also calcium for the worms, in addition to lowering acidity in the bin. About 4 to 6 pounds of bedding is needed for a 2 feet × 2 feet bin, and 9 to 14 pounds of bedding for a 2 feet × 3 feet bin. Since the worms eat the bedding, more of it has to be added within a few months. The bedding must be kept moist by adding 3 pints of water for each pound of bedding to enable the worms to breathe. If the bedding dries out, some water is sprinkled on it with a plant mister. It is very vital to get the right type of worms such as red worms or "wigglers" (Eisenia foetida) that thrive in a worm bin.

Organic waste such as vegetables and fruits cut into small bits, eggshells, tea bags, coffee ground, paper coffee filters, and shredded garden waste are fed to the worms. Worms specially like cantaloupe, watermelon, and pumpkin. Citrus fruit waste may be added in limited amounts to prevent from becoming too acidic. Meat scraps or bones, fish, greasy or oily foods, fat, tobacco, or pet / human manure *should not be* mixed. All this food material has to be covered completely with the bedding to discourage fruit flies and molds. One pound of worms will eat about four pounds of food scraps a week. If more food than the worms can handle is added, anaerobic conditions will set in and cause odour. If adding food is stopped for a while, the odour dissipates. Red worms can tolerate temperatures from 50° to 84°F, though 55° to 77°F is ideal.

Procedure: The first step is choosing a proper location for the worm bin. The location should ensure the required temperature and humidity. It is preferable to place worm bin outside in the shade during the hot summer and indoors in winter to protect from the cold. The preparation of the bedding is the next step. Commonly available material like newspaper is soaked in water for a few minutes; then wringed it out like a sponge and fluffed before adding to the worm bin. The bedding need to be very damp, but not soaking wet (only two to three drops of water should come out when you squeeze the bedding material). The bedding is evenly spread until it fills about three-quarters of the bin. A couple of handfuls of soil (from outdoors or potting soil) are sprinkled into the bedding to bring in beneficial microorganisms and aid the worms' digestive process. The bedding is fluffed up roughly once a week to enable the worms gets plenty of air and freedom to move.

The worms are then gently placed on top of the bedding. If the bin lid is not placed for a while, the worms will burrow into the bedding, away from the light and will not try to crawl out of the bin due to the light overhead. After the worms have settled in their place, the collected food wastes are placed in a hole in the bedding, and cover it with at least an inch of bedding. The next feeding may be done after a week. The worms are allowed alone during this time to get used to their new surroundings. The food scraps are buried in a different area of the bin each time. Worms may be fed any time of the day. Other organisms may be seen in the worm bin, as they help break down the organic material. Most of the organisms will be too small to see, but white worms, springtails, pill bugs, molds, and mites can be easily spotted. In about six weeks, a soil-like material called '*worm castings*' is ready for use. The castings can be used to boost plant growth. In three or four months, it will be time to harvest the castings. The castings will be mixed with partially decomposed bedding and food scraps, in addition to worms; this is called *vermi compost.*

The vermi compost may be harvested by one of the two methods:

1. Food scraps may be placed on only one side of the worm bin for several weeks, and most of the worms will migrate to that side of the bin. Then the vermin compost can be removed from the other side of the bin where food scraps are not added. Then, fresh bedding is added. This process is repeated on the other side of the bin. After both sides are harvested, food scraps may be added to both sides of the bin again.

2. The contents of the worm bin are emptied onto a plastic sheet or tarpaulin in strong sunlight or artificial light. In 20-30 minutes, the worms will burrow down to escape the light. Then the top layer of vermicompost can be scrapped off, and more of it can be scrapped off every 20 minutes or so. After several scrapings, piles of worms are left which can be returned gently to the bin in fresh bedding. The harvested castings can be mixed into potting soil soon after harvest to give best result on indoor plants. For storing or for use for outdoor plants, curing in an aerobic environment is required to make it dry and to eliminate the possible introduction of new species.

How to choose a Vermicomposting system: A variety of methods may be used to process large volumes of organic residuals with earthworms, ranging from land and labor-intensive techniques to fully automated high-tech systems. Types of systems include windrows, beds, bins, and automated raised bioreactors. Choosing which vermicomposting system to use will depend upon: (a) Amount of feedstock to be processed, (b) Funding available, (c) Site and space restrictions, (d) Climate and weather, (e) State and local regulatory restrictions, (f) Facilities and equipment on hand, and (g) Availability of low-cost labor (Rhonda Sherman 2009).

Vermi composting can be ideal for residents of apartments/small houses or small offices or establishments. Small bins can be placed and operated in individual homes or schools or offices where food waste is likely to accumulate. By doing composting, additional benefit of reduction of solid waste can be achieved. It can be undertaken as a cottage industry as well.

It is likely that problems arise during the process; the causes and solutions for the probable troubles are listed in Table 5.2.

Table 5.2 Problems, causes and solutions in composting

Problems	Causes	Solutions
Bin smells bad	Over feeding, Non-compostables present, Food scraps exposed, Bin too wet, Not enough air.	Stop feeding for 2 weeks, Remove non-compostables, Bury food completely, Mix in dry bedding; leave lid off Fluff bedding; drill holes in bin
Bin attracts flies	Food scrap exposed, Rotten food, Too much food; especially citrus	Bury food completely, Avoid putting rotten food in bin, Don't overfeed worms
Worms are dying	Bin too wet, Bin too dry, Extreme temperatures, Not enough air, Not enough food	Mix in dry bedding; leave lid off Thoroughly dampen bedding, Move bin where temp. between 55 and 77°F, Fluff bedding; drill holes in bin, Add more bedding and food scraps
Worms crawling away	Bin conditions not right	See solutions above, Leave lid off and worms will burrow back into bedding
Mold forming	Conditions too acidic	Cut back on citrus fruits
Bedding drying out	Too much ventilation	Dampen bedding; keep lid on
Water collecting in bottom	Poor ventilation, Feeding too much watery scraps	Leave lid off for a couple of days; bin add dry bedding, Cut back on coffee grounds and food scraps with high water content

Health effects: Vermi culture does not necessarily kill all pathogens. Rather, some viruses and parasites can survive the process. Therefore, if the input materials contain pathogens, the finished product could still contain pathogens. This may be of particular concern in developing countries, where wastes used in vermi composting may not be source-separated.

Effect of Climate: Worms are sensitive to climatic variations. Extreme temperatures and direct sunlight are not healthy for the worms and may not keep them alive. In hot, arid areas, the bin should be placed under the shade. Many of the problems posed by hot or cold climates can be avoided by vermicomposting indoors and maintaining optimal temperatures.

Uses of vermin compost: Vermicompost is worm manure. Vermicompost improves soil structure, reduces erosion, and improves and stabilizes soil pH. In addition, vermicompost increases moisture infiltration in soils and improves its moisture holding capacity.

The worm compost can be mulched or mixed into the soil in the garden and around the trees and plants. It can also be used as a top dressing on outdoor plants or conditioner on the lawns. For indoor plants, vermicompost is mixed with potting soil. In horticulture, worm castings are the very best soil amendment available. The nutrient content of castings is dependent on the material fed to the worms, and the wormcastings provide these nutrients in a readily available form to plants. The biology of the worm's gut facilitates the growth of fungus and bacteria that are beneficial to plant growth. In addition, many chemical compounds that promote plant growth are found in castings. In essence, plant growth is significantly increased by vermicompost, whether it is used as a soil additive, a vermicompost tea, or as a component of horticultural soilless container media. Vermicompost causes seeds to germinate more quickly, seedlings to grow faster, leaves grow bigger, and more flowers, fruits or vegetables are produced (Fig. 5.5). These effects are greatest when a smaller amount of vermicompost is used—just 10-40 percent of the total volume of the plant growth medium in which it is incorporated. Vermicompost also decreases attacks by plant pathogens, parasitic nematodes and arthropod pests (Rhonda Sherman 2009).

Fig. 5.5 Turnips: 0%, 10%, 20% vermicompost by volume added to field plots, Biological & Agricultural Engineering, NC State University (source: Rhonda Sherman 2009)

5.3.3 Aerated (Turned) Windrow Composting

Fig. 5.6 Huge Compost Pile (Photo Courtesy: Campaign Recycle Maui Inc.,/Compost Maui Inc).

In this method, Organic waste is shaped into rows of long piles called "windrows" (Fig. 5.6) which form the basic environment for compost bacteria and other organisms to perform decomposition.

The aspects to be considered in planning windrows are:

(i) the size of the windrows must be of ample mass to allow for heat build-up,

(ii)   the windrow size is determined by the composition of the wastes and the climate ,

(iii) the shape of the windrows is related to the type of aeration that is used and the kind of equipment used to aerate,

(iv)   whether the windrows are open or covered depends on the climate and the moisture content of the waste, and

(v)   the spacing of the windrows is dependent on the size of the site and type of equipment used.

The ideal height, between 4 and 8 feet, is large enough for a pile to generate sufficient heat and maintain temperatures, while small enough to allow oxygen to flow to the windrow's core. The ideal pile width is between 14 and 16 feet. Active pile systems require manual or mechanical turning of the windrows. The turning aerates the piles, blends the materials, brings about additional size reduction, and prevents excessive buildup of temperature that may lead to spontaneous combustion. An active pile system has relatively high land use requirements, and low capital cost and low-to-moderate operating cost. It can be developed without purchase of specialized equipment because mechanical turning can be done with loaders or bulldozers, which are present in almost any municipality; it requires limited site infrastructure, and imposes very limited requirements for site modification. Windrow turning machines of different sizes have been developed in Asia, America, and Europe. The ones built in India are low cost and work effectively with the waste stream. Windrow turning machines enable production of more uniform compost. They decrease labour costs but increase the capital costs of active pile systems. They may increase land requirements, as the design of turning machines limits the size of the piles as well as influence pile spacings. Specialized windrow turning machines, however, are more effective in aerating windrows compared to bulldozers, and also a cost-effective alternative.

This technique of composting can accommodate large volumes of diverse wastes, including yard trimmings, grease, liquids, and animal byproducts (such as fish and poultry wastes); but frequent turning and careful monitoring are required. Since this method is suited for large quantities, it is appropriate for entire communities/ local authorities, and high volume food-processing businesses like restaurants, cafeterias, and packing plants.

Effect of Climate: In a warm, arid climate, windrows are covered or placed under a shade to prevent water from evaporating. In rainy seasons, the shapes of the pile can be adjusted so that water runs off the top of the pile rather than being absorbed into the pile. Windrow composting can also work in cold climates; the outside of the pile might freeze, but in its core, the temperature can reach 140 °F. Fig.5.7 shows 'locally designed windrow turning machines, like this one in New Delhi (credit: Chris Furedy, source: UNEP)'

Fig. 5.7

Environmental Concerns: Leachate (a liquid contaminant) is released during the composting process. It has to be collected and treated, otherwise the local ground-water and surface-water supplies can be polluted. Samples of the compost should be tested in a laboratory for bacterial and heavy metal content. The odours released during turning early in the composting process have to be controlled. A large buffer zone between the composting plant and neighboring residences is desirable, especially if the windrows are infrequently turned. This method yields significant amounts of compost requiring an arrangement to market the product.

5.3.4 Aerated Static Pile Composting

The operating principle of Static pile composting is same as in the 'aerated turned windrow composting'. The only difference is that the organic waste is mixed together in one large pile instead of rows. In these composting systems, the windrows are not turned. Instead, they are aerated continuously or periodically. Static piles require a site with aeration channels built into the pad on which the piles sit. Layers of loosely piled bulking agents (e.g., wood chips, shredded newspaper) are added so that air can pass from the bottom to the top of the pile. The piles can also be placed over a network of perforated pipes that deliver air into or draw air out of the pile during composting. The installed air blowers might be activated by a timer or a temperature sensor.

Aerated static piles are appropriate for a relatively *homogenous mix* of organic waste and work well for larger quantity generators of yard trimmings and compostable municipal solid waste (e.g., food scraps, paper products).

Fig. 5.8 Steam Rising from Compost Pile
(Photo Courtesy: Jepson Prairie Organics)

This method, however, *does not work* well for composting animal byproducts or grease from food processing industries. They are used more frequently in sludge composting than in composting of bio-waste or yard waste.

This method requires equipment such as blowers, pipes, sensors, and fans, which might involve significant costs and technical assistance. Construction of large piles which require less land than the windrow method is possible if a controlled supply of air is arranged. This method produces compost relatively fast, within 3 to 6 months.

Effect of climate: The climate affects the process of composting. In a warm arid climate, aerated static piles are covered or placed under a shelter to prevent water from evaporating. In the cold climate, the core of the pile retains its warm temperature. However, the aeration may be more difficult due to passive air flowing rather than active turning. Some aerated static piles are placed indoors with proper ventilation.

Environmental Concerns: Since there is no physical turning, this method requires careful monitoring to ensure that the outside of the pile heats up as much as the core. Applying a thick layer of finished compost over the pile which can help maintain high temperatures throughout the pile can lower bad odours that are emnating. Another method to reduce odour is to filter the air drawn out of the pile through a bio-filter made from finished compost.

5.3.5 In-Vessel Composting

In-vessel composting is an *industrial form* of composting biodegradable waste in enclosed reactors. In-vessel composting represents a higher-technology sound approach to composting.

Fig. 5.9 In-vessel composting

The reactors generally consist of metal tanks or concrete bunkers in which moisture, air flow and temperature can be controlled to create the optimal conditions for composting using the principles of a 'bioreactor' (Fig. 5.9). Generally mechanical mixing and/or forced aeration are used to turn or stir up the material for proper aeration. The buried tubes allow fresh air to be injected under pressure and the air circulation is metered. The temperature and moisture conditions are monitored using probes in the mass, to maintain optimum aerobic conditions. A bio filter is attached to the exhaust.

Almost all in-vessel systems require a residence time (time physically in the vessel) of 3 to 30 days, followed by a period of 21 to180 days of active composting in an active or static pile. Once the active composting is completed, the material is stored in piles or windrows for curing that may extend up to two years.

Municipalities generally use this technique for organic waste processing which includes final treatment of sewage bio-solids, to obtain safe stable compost. In-vessel composting can also refer to aerated static pile composting with the addition of removable covers that enclose the piles. This type of systems is extensively used by farmer groups in Thailand, supported by the National Science and Technology Development Agency there.

Unpleasant odours are caused by putrefaction (anaerobic decomposition) of nitrogenous animal and vegetable matter gassing off as ammonia. This is controlled with a higher carbon to nitrogen ratio, or increased aeration by ventilation, or using a coarser grade of carbon material to allow better air circulation. A biofilter is used to prevent and capture any gases naturally occurring (volatile organic compounds) during the hot aerobic composting. As the filtering material saturates over time, it can be used in the composting process and replaced with fresh material.

In-vessel composting systems vary in size and capacity. In-vessel systems have several advantages; they offer protection from weather conditions, better odour control, and shorter periods of active processing. However, they are expensive to build and operate. The developing countries, as of now, have to import the equipment and parts. *Variations in In-vessel systems*:

(a) Modular in-vessel systems represent the best practice in most cases where in-vessel composting is desired. These systems have a series of smaller vessels or divisions within the vessel. The modules can generally be obtained separately or added on to the system later. They are also set up to compost more than one waste stream at a time. Modular systems have *moderate* capital costs and *low-to-moderate* operating costs. Odour and leachate are very well controlled in these systems. Modular systems are a sound option in areas where siting is difficult and land is scarce.

(b) In drum systems, compostables are introduced into a rotating horizontal drum for a relatively short residence time. It is then followed by a long period of active-pile composting. These systems are called Dano-type systems after the original Danish design. The large metal drums are up to 30 meters long and may be divided into separate chambers. Some have trommel screens in the first chamber to remove designated materials during composting. Dano-type systems require *a large stretch of land* for the active pile composting.

(c) In tower systems, the compostable material is introduced into a vertical tower and composted under forced aeration. Some tower systems also mechanically turn or stir up the material during its residence. The residence time in tower systems is typically 2 to 5 weeks, and composting is essentially complete when the material is removed to curing piles. Tower-type systems offer more odour control during composting and require much less land, since the period of active composting takes place in the vessel. They represent *a particularly sound practice* for sludge composting or co-composting of sludge and yard wastes.

Types of feedstock: In-vessel composting can process large amounts of waste without taking up as much space as the windrow method. In addition, it can accommodate virtually any type of organic waste such as meat, animal manure, biosolids, food scraps. Some in-vessel composters can fit into restaurant kitchens while others can be large enough to accommodate large food processing plants.

Effect of Climate: In-vessel composting can be used year-round in virtually any climate because the environment is electronically controlled. This method can even be used in extremely cold weather if the equipment is insulated or operates indoors.

In this method, very little odour and minimal leachate are produced, and conversion of organic material to compost takes as little as a few weeks. However, the compost that comes out of the vessel requires a few more weeks or months for the microbial activity to stabilize and to cool the pile. Northern European countries, in particular Denmark, Germany, and The Netherlands have developed a particularly sound practice in their system approach to composting of separate kitchen and yard waste, which they call Òbiowaste.Ó; this practice entails the use of a modular in-vessel composting system followed by a period of composting, either in aerated static piles or active windrows. In either case, several months of curing are needed prior to processing for market. The finished biowaste compost can be used for agricultural, horticultural, and civil engineering applications.

Health and Environmental impacts: The impacts due to composting process include noise, odour, and ugliness (Garrod and Willis 1998). In addition, many of the microorganisms found in compost are known respiratory sensitizers that can cause a range of respiratory symptoms including allergic rhinitis, asthma, and chronic bronchitis (Swan et al. 2002). In this aerobic process, the main gas produced is carbon dioxide which contributes to global warming.

5.4   Biowaste Composting in Europe

Wet waste: Collecting wet waste separately from kitchen waste for composting was tried in the late 1970s and early80s in Europe. It was given up in favour of the biowaste systems, which deliver higher quality compost. To label wet/dry systems for collection is considered an unsound practice.

Mixed solid waste: Composting of mixed solid waste is a debatable topic. In industrialized and transition (fast growing economies) countries, the waste stream is generally too diverse containing much of metal and plastic stuff; hence, mixed-waste composting may not be a sound practice to apply. In mixed-waste composting, mechanical pre-processing and separation systems have to operate effectively. But these have failed to produce either a clean stream of compostables or marketable recyclables. In developing countries, the waste stream contains high levels of organic wastes, since the main non-compostables are not thrown out or are picked out prior to final disposal. Since the resulting waste stream is highly compostable, using even a low-technology can be more practical. Moreover, urban agriculture provides a strong demand for the resulting compost.

Wastewater sludge and human fecal matter: Wastewater sludge, septage, and human fecal matter are high-nitrogen materials that can be aerobically composted under certain conditions. They are high in moisture, sometimes actually liquid, and composting can only work if they are combined with carbon sources such as wood or paper, and bulking agents such as chipped wood or rubber which are dry, and maintain air spaces in the compost piles. Anaerobic digestion of these materials can also work, and is a sound practice on farms in industrialized countries. It is operationally more difficult than aerobic composting. Alternatively, sludges can be left to dry into cakes which may be used as fertilizer. The main problem with composting of wastewater sludge is that the wastewater usually includes both industrial and residential discharges. In most countries, wastewater from urban sources may include metals and contaminants that affect the quality and safety of the resulting compost. Such contaminated compost has to be landfilled at a high cost. Sound practice in composting these materials always includes an industrial wastewater pre-treatment practice and a careful monitoring of the compost. Adequate care has to be taken to ensure the health and safety of the workers and surrounding environment in handling these materials.

Manures and animal wastes: Manures and animal wastes have been composted for centuries. Manures are high in nitrogen, but most bedding materials are carbon sources, so manures bedded on straw, wood, plant wastes, or paper are easy to compost. Most manure composting begins with a hot aerobic phase, followed by a slow vermicomposting (worm culture) phase. Manure composting produces excellent fertilizer that is important for sustainable agriculture.

5.5 Composting Challenges

Centralized composting is a successful, cost-effective, environmentally sound waste management approach in Europe and increasingly in North America. In developing countries, however, it has not been very successful. The reasons are varied and are discussed in later pages. The composting industry all over the world faces several problems such as lack of consistent product quality, market research and planning, investment, accepted national compost specifications, so on. There are many technical and aesthetic problems to be solved in the composting of waste. For example, an important privately run vermiculture experiment in Indonesia failed when toxics in the MSW killed the earthworms. The role of Government agencies in the promotion of compost products is also limited. The compost uses range from city and county landscaping to niche markets such as soil remediation. New technologies allow compost companies to tailor their products to specific end-uses, increasing the market value of the material. In fact, more and more compost producers are producing multiple innovative compost products for applications such as bioremediation of contaminated soil, erosion control at construction sites and so on. Some companies use compost to control odours through new process technologies such as bio filters. Compost is also used as a filter in water treatment system.

5.6 Composting in Developing Countries

The processes of composting are the same all over the world. The main differences in developing countries relate to the nature of waste stream to be composted, the agricultural practices relating to production and use of compost, and the physical infrastructure of the built and natural environment.

There are three scales at which composting has been implemented: the residential level, the decentralized community level, and the centralized large-scale (municipality-wide) level; the larger the undertaking, the more capital investment is required. Most developing countries have found success with composting when implemented at the household level, with some projects doing well at the community level as well. At the municipal level, certainly overall cost and functionality are the primary reasons for the success of a given process; the financial commitment required as well as the effort required to maintain equipment satisfactorily to keep a large scale operation running, has resulted in widespread failures in India, Brazil, and elsewhere (Hoornweg, et al 1996). In developing countries, most of the city-based large mixed-waste compost plants which are designed by outside consultants, have either failed or operate at less than 30% of capacity. In many urban places, unreliable collection systems contribute to the inefficient running of the composting facilities. In India and China, small- and medium-scale composting facilities operate successfully while many failures are reported in other countries. Large cities - Bangkok, Hanoi, Shanghai, and Tokyo - earlier installed imported mechanical composting plants; some of them are now defunct and the remaining ones are not operating at full capacity for various reasons. Two industrial composting plants operated in Dakar (Senegal) and Abidjan (Ivory Coast) during the 1970s. These were financially unproductive and beset by mechanical problems and had to be closed. In the suburbs of South African cities, Durban, Johannesburg, and Pretoria, there are community composting centers where the garden waste dropped by residents is used. The compost is sold for household gardens. In the suburbs of urban centers of Africa, NGOs, community based organizations, and economic interest enterprises also promote composting of MSW. These projects are generally highly labor-intensive with a low capital investment. The compost produced is largely for self consumption or for sale to households or hotels in the city. By and large, compost systems fail or operate poorly for 'economic' and 'technical' reasons. The economic reasons relate to (i) the ability to secure waste and (ii) the need to market the compost that is produced. In many parts of Asia, where there is a long tradition of successful composting, the inexpensive disposal of waste in dumps or landfills does not seem to obstruct composting. In most of Latin American and African countries, efforts to setup composting have failed because sufficient waste is not available. The technical failures relate to: (i) failure of the mechanical systems that control waste streams (i.e., pre-processing), and (ii) failure to create the right environment for the biological process to occur successfully. Pre-processing methods based on manual separation have consistently produced the best compost in developing countries, as well as in industrialized ones. There are small-scale bio waste composting facilities in both industrialized and developing countries that are successfully operating because of the high degree of manual pre-processing. But in the larger facilities, it is difficult to depend on mechanical separation because of the diversity in the waste.

Composting and digestion of bones is carried out as a small scale industry in some developing countries. It can produce ingredients in the manufacture of fertilizer, animal feed, and glues. The traditional methods of sun-drying, breaking up bones manually, composting in pits (sometimes with the addition of household organics), and steam digestion carry various health risks, and cannot be considered a sound practice. Small-scale aerobic composting of hide scrapings, and tannery and slaughterhouse wastes can also produce fertilizers, but carry some pathogenic risks. These types of small enterprises though profitable and provide subsistence income, they are associated with poor working conditions and risks to workers' health, due to generation of leachate and associated bad odours. Introducing technical and health improvements rather than eradicating the activities themselves could make the practice sound.

In developing countries, the high animal and vegetable waste content of the waste stream (sometimes as high as 90%), combined with materials recovery by waste picking, source separation, or pre-processing indicates that the mixed waste stream is highly compostable to produce good compost at a small or medium scale.

In developing countries, *backyard composting* is also practiced. But the presence of rodents and vectors are a concern in cities with high pest populations, for example, Bangalore. That is why, municipal authorities frequently advise against backyard composting, and in some places it is prohibited by the health code. In African countries, Backyard composting is limited. Some NGOs promote the practice in Benin, Cameroon, Egypt, Ghana, Kenya, Nigeria, South Africa, and Zimbabwe but the practice does not have a significant impact on MSWM at the city level. Backyard composting is used in rural areas and in poorer areas of cities in Latin American countries and the compost is used in households and vegetable plots. The open-air windrow process is used in some countries, especially China. In many Chinese cities and towns the wastes are delivered directly by collection vehicles to the farmers who are instructed to compost the waste in windrows or pits for a prescribed period of time, but they often do not do this.

Since the organic fraction of the waste stream is high in most places in South and West Asia, there is considerable interest in composting of MSW in the region, and a long history of experiments with composting. But, large-scale centralized composting (as distinct from neighborhood composting) has had little success in this region. Centralized composting refers to composting of animal and plant wastes transported from multiple sources to a facility that can receive 10 to 200 tons per day. This is also referred to as Municipal-scale composting plants. These operations call for technical and environmental assistance, pre-processing system, and marketing structure for the finished compost. Centralized regional composting facilities generally have a capacity of more than 50 tons per day, and as much as 1,000 tons per day. Centralized composting has not been successful even in Latin America and the Caribbean, mostly due to high operating costs; most of the municipalities are unable to subsidize. For example, the plants installed in Brazil (Sao Polo city) and in Mexico had to be closed.

Decentralized composting or small scale composting is done using the wastes of a number of households, shops, or institutions, on unused land or in parks. These sites usually process less than five tons of waste per day and generally reduce the need for movement of compostable materials. This can be extended to village or community scale where the facility can handle about 2 to 50 tons/ day, depending on the size of the community and the amount of compostables in the waste stream. Here, more turning, processing, screening, and storage of the compost may be required.

Fig. 5.10 Composting Plant in Bangkok, Thailand

Decentralized Community-based Composting through Public-Private-Community partnership is practiced in Bangladesh as discussed here. *Decentralized Community Based Composting in Dhaka through Public-Private-Community Partnerships:* A research based organization, Waste Concern, initiated a community based decentralized composting project in Dhaka in 1995, in order to recover the value from organic portion of waste (Fig.5.11). The prime goal of this project was to explore technical and commercial feasibility of labor intensive aerobic decentralized composting technique and to promote the principle of 4Rs (Reduce, Re-use, Recycle, and Recovery of waste) in urban areas of Bangladesh.

Fig. 5.11 Decentralized composting Project in Bangladesh
(source: 'Waste Concern'

The project activity includes not only house-to-house waste collection, composting of the collected waste in a decentralized manner and marketing of compost, but also recycling of materials, mostly plastics. During 2004-05, Dhaka has generated 50,214 tons of plastic waste, of which 51% is recycled (see the Figs).

Barrel Type Composting Project for the Urban Poor: The Barrel Type Composting model invented by the SEVANATHA of Sri Lanka inspired 'Waste Concern' to implement the concept in the slums of Dhaka. With some modification and changes, Waste Concern with the support from Local Initiatives Facility for Environment (LIFE) of UNDP launched the barrel type composting units in two slums of Dhaka. After successful results, this concept is being replicated in a number of slums of Dhaka as well as other cities of Bangladesh. A specially designed 200 liter 3 bottomless perforated green barrel with a lid was supplied to the slum. One green barrel is provided to a group of six households and placed on a raised base with concrete ring. The cost of each specially designed barrel along with the civil work was around TK. 1800 (US$ 30) (Source: SAARC, 2004 and Waste Concern).

In Latin American and Caribbean countries, Colombia, Peru and Cuba, vermi composting (worm culture), which produces humus, appears to be more successful due to shorter production times (days vs. months) and the product having wider appeal and market. Successful vermiculture is not only done at a very small scale, typically with five or six persons, but also benefits from the public perception that its product comes from "clean" vegetable waste (market and agricultural wastes), whereas compost comes from "garbage." Moreover, humus is richer in nutrients than is compost, and compost suffers from worse quality control problems.

Field composting and using compost from dumps: The most widespread form of composting and use of compost from urban wastes today is the delivery of fresh garbage to farms by collection groups, the removal of compost from dumps by nearby farmers, and the conversion of old dump land into farms. The best known example of garbage farming is at CalcuttaÕs Dhapa dump, where the municipality leases out dump land for vegetable farming. The combination of dirt, dust, organics, human and animal feces, and ash in CalcuttaÕs garbage produces a fertile growing medium that requires no additives. The dump is in a wetland and there are numerous ponds between the ridges of garbage that provide water. At BeijingÕs main dump, the authority has provided sifting machines to encourage farmers to remove compost and thereby extend the life of the dump. In Yangon, Myanmar, the City Development Corporation allows small enterprises to mine the oldest inner-city dump (now closed) for metals (materiel dating from World War II) and screen the compost. Both Ho Chi Minh City and Medan allow the mining of compost from dump sites for fees. This way, the hill of compost is gradually removed so that land can be available for redevelopment or farming. These practices which are largely undocumented and informal in the developing countries use valuable organic matter, and help in waste reduction. However, they carry a risk of bacterial, glass, or chemical contamination which can cause health hazards during the work of gardening or in the consumption of the crops in some places. At old dumps in cities with low levels of industry, the subsurface compost probably contains few heavy metals; the compost therefore needs to be tested for use in farming. If the land, compost, and crops are monitored and are found safe for use, these traditional practices of natural composting and garbage farming could be regarded as sound practices for many places. Unfortunately, these countries hardly do the testing, because they need assistance for testing and long-term monitoring for the quality and contaminants. Also, advice on crops that may be safely grown on old garbage dumps is lacking.

In most developing countries, it is very difficult, if not impossible, to promote source separation of compostable materials due to lack of economic and environmental incentives. Nevertheless, since the success of composting systems and the quality of the compost depend heavily on the nature of waste that is composted, a separate collection system for compostable materials would help the production of high-quality compost. Community collection bins for compostable materials are one possibility and may, under certain situation, be easier to implement than household collection of compostable materials.

5.7 Composting in Developed Countries

Centralized composting programs in America and Canada were insignificant prior to the mid-1980s. Currently more than 5% of the MSW stream is managed by composting process because around 30 to 60% of a community's waste stream contains compostable portion. Composting programmes have been designed for a variety of organic waste streams, including yard wastes (grass trimmings, leaves, or tree prunings), food wastes, agricultural wastes, and wastewater treatment sludge. Mixed waste composting is another process which has been used on a limited basis in these countries. Mixed waste processing facilities accept unsorted MSW in the same form as it would be received at a landfill or a waste-to-energy facility, and separate recyclable materials (These methods will be discussed in the later pages).The relatively small community of Guelph, Ontario, has been operating such a facility successfully since mid-1990s. Large number of waste composting facilities has been in operation in the US, some of them are relatively small-scale. If the residents are encouraged to start their own composting activities, the waste quantity reaching centralized composting as well as the costs of such programmes would be reduced. To support this plan, the City of Toronto in mid-1990s, have brought in legislative changes that do not allow lawn clippings to be landfilled or taken to centralized composting. Residents can either compost them or leave them on the lawn. About half of the states in US have enacted similar bans resulting in the growth of yard waste composting programmes in the US. Canada currently has over 160 composting projects throughout the country (UNEP). Since Backyard composting and mulching is a source reduction activity that saves money for both the municipality and the resident, many communities in North America have developed programmes to encourage backyard composting, by offering educational materials and by distributing composting bins. A number of communities buy the bins in bulk and distribute them free of charge or for a nominal fee. In Vancouver, Canada, however, they charge $25 for the bin. The City of Seattle in US funds a backyard composting education programme run by a local organization of urban gardeners that trains volunteers to be proficient at composting. Toronto operates a similar programme through the Recycling Council of Ontario. In 1989, 5% of both Seattle's and Toronto's total waste stream was composted by residents. Seattle has also implemented a sophisticated multi-point composting programme that uses a combination of curbside pickup, drop-off, and backyard composting elements.

In North European countries, the collection of compostable materials for transport to centralized composting facilities is vital. Collection of compostable portion, referred to as 'bio-waste' or 'green waste' (garden and kitchen organics) from households is generally performed using 120-liter green rolling carts, and in some urban areas, using smaller, 35-liter pails or paper bags. The bio waste is collected alternate weeks bringing the cost and energy usage within acceptable budgetary and environmental levels in many Northern European countries. Centralized composting has a long tradition in Western Europe in particular, where some plants still in operation. Most European compost installations are aerobic systems, with the compost having a short residency time in a reactor or pre-composter and a longer time in aerated static piles. Windrow composting exists but is less common.

Centralized composting installations are designed to compost mixed-waste. Pre-processing and separation machinery are included to remove the non-compostable materials before mixing and composting and the recovered non-compostable materials are sent for recycling. These plants are under operation in Greece and Spain. Because the waste arrives at the facility not only mixed but compacted, both compostable materials and recyclables are exceedingly contaminated, requiring frequent modifications. Two other approaches - Wet-dry approach and source-separated collection of compostables approach – to centralized composting are also tried; the later one is favoured in Denmark, Germany, The Netherlands and other European countries. The current

Centralized composting facilities can be divided into pre-processing stages (that include removal of non-compostables by magnets, eddy currents, ballistics separators, and/or vibrating screens); size reduction; mixing and/or pre-composting; composting; curing; post-processing (usually consisting primarily of screening); packaging; and marketing. For the mixing and composting stages, the use of a vessel, usually a large horizontal drum or tunnel reactor is commonly used. Centralized composting facilities typically have a design capacity of 50 to 200 tons per day, and a 100-ton-per-day facility will produce only 30-50 tons of compost per day because of various steps involved.

Backyard composting is a longstanding tradition in countries like Australia, Japan, and New Zealand, especially in rural towns. It is now being promoted by local governments by supplying inexpensive compost bins. In Europe, even high quality compost has limitation to market because it is not usually considered as a fertilizer; rather, it is a soil conditioner, useful for its water-holding capacity (it limits evaporation and erosion, and functions as a mulch), its slow release of the nutrients nitrogen, phosphorus, and potassium, and its ability to return organic matter to depleted, excessively mineralized soils. It is especially useful on slopes, or for reclaiming land degraded by erosion or through mining, quarrying, or rapid construction and development.

CHAPTER 6

Energy from Waste

6.1 Introduction

Recovering recyclable materials and reusing them in the original form or recycling into new products is one of the economic benefits that the community could derive from municipal solid waste. Waste processing offers other benefits depending on the choice of technology. We have seen that *'composting'* the biodegradable component of the municipal waste provides a product which can be used for enriching the soil and/or as a fertilizer. The most important benefit is to produce *energy* from waste in the form of 'heat' or 'electricity'.

Municipal solid waste-to-energy plants can generate large amounts of electricity and syngas, and are a great alternative to other waste disposal methods as well as an alternative to fossil fuel usage which contribute to global warming and pollution. Waste-to-energy is a clean technology, and can be a vital component of the future of energy. This method of disposing of MSW that is generated will become even more important as the population around the world continues to rise, and even more energy will be needed and more waste will be generated. Several Waste-to-energy technologies are available which have potential to produce clean energy (electricity). Specially designed power plants fitted with pollution control equipment to clean unwanted emitted gases are in operation globally. Among renewable energy sources, waste is all the more attractive since its valorization enables both production of useful energy and disposal of waste in environmentally acceptable way. Waste-to-energy (WTE) has been recognized by the US EPA as a clean, reliable, renewable source of energy. Worldwide, about 130 million tons of MSW are combusted annually in over 600 WTE facilities that produce electricity and steam for district heating and recovered metals for recycling (Themelis 2003). The recovery of energy from wastes also offers a few additional benefits: (i) The total quantity of waste gets reduced by nearly 60% to over 90%, depending upon the waste composition and the adopted technology; (ii) Demand for land, which is already scarce in cities, for landfilling is reduced; (iii) Cost of transporting waste to far-away landfill sites also gets reduced; and (iv) Reduction in environmental pollution and thereby global warming.

Modern WTE technologies can be broadly classified (Fig. 6.1) as thermal, biochemical, and chemical processes. These will be individually discussed in the following pages. Most conversion technologies have three distinct components: (i) pre-processing to separate recyclables and to prepare the MSW, (ii) conversion stage where the processed material is treated by an appropriate technology, and (iii) energy production system for producing electricity, heat or other useful chemicals. Waste-to-energy has gained significance mostly because it helps to avoid greenhouse gases, and thus contributes to the reduction of global warming.

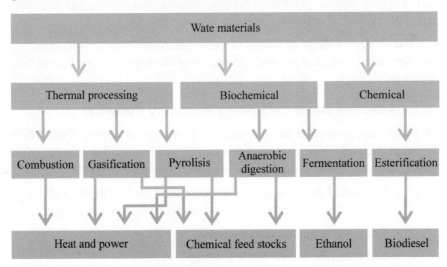

Fig. 6.1 Pathways which waste can be converted to energy or energy related products (source: The Australian Business Council for Sustainable Energy; Taken from Wagner 2007)

The energy recovery from waste not only depends on the chosen pathway but critically on certain physical and chemical parameters of waste.

The important *physical* parameters requiring consideration are:

(i) Size of constituents (smaller size of the constituents aids in faster decomposition of the waste),

(ii) Density (wastes of the high density reflect a high proportion of biodegradable organic matter and moisture whereas low density wastes indicate a high proportion of paper, plastics and other combustibles), and

(iii) Moisture content (high moisture content causes biodegradable waste fractions to decompose more rapidly than in dry conditions; also makes the waste rather unsuitable for thermal conversion – incineration, pyrolysis/gasification – for energy recovery because the waste must be supplied to remove moisture).

The important *chemical* parameters to be considered for determining the energy recovery potential and the suitability of waste treatment through Bio chemical or Thermal conversion technologies include (a) Volatile solids, (b) Fixed carbon content, (c) Inerts, (d) Calorific value, (e) Carbon/Nitrogen ratio, and (f) toxicity.

The desirable range of values of these parameters has to be maintained for technical viability of energy recovery through different treatment methods.

6.1.1 Assessment of the Energy Recovery Potential

A rough assessment of the potential of recovery of energy from MSW through different treatment methods can be made from knowledge of its calorific value and organic fraction, as under (chap 15 Efw.pdf):

(i) In thermal processing, all of the organic matter, biodegradable as well as non-biodegradable, contributes to the energy output:

Total waste quantity: W tonnes

Net Calorific Value: NCV k-cal/kg.

Energy recovery potential (kWh) = NCV × W × 1000/860

= 1.16 × NCV × W

Power generation potential (kW) = $1.16 \times NCV \times W/24$

$$= 0.048 \times NCV \times W$$

Conversion Efficiency = 25%

Net power generation potential (kW) = $0.012 \times NCV \times W$

If NCV = 1200 k-cal/kg., then

Net power generation potential (kW) = $14.4 \times W$

(ii) In bio-chemical processing, only the biodegradable fraction of the organic matter can contribute to the energy output:

Total waste quantity: W (tonnes)

Total Organic / Volatile Solids: VS = 50 %, say

Organic bio-degradable fraction: approx. 66% of VS = 0.33 x W

Typical digestion efficiency = 60 %

Typical bio-gas yield: B (m^3) = $0.80 \, m^3$ / kg. of VS destroyed

$$= 0.80 \times 0.60 \times 0.33 \times W \times 1000$$
$$= 158.4 \times W$$

Calorific Value of bio-gas = $5000 \, kcal/m^3$ (typical)

Energy recovery potential (kWh) = $B \times 5000 / 860 = 921 \times W$

Power generation potential (kW) = $921 \times W/24 = 38.4 \times W$

Typical Conversion Efficiency = 30%

Net power generation potential (kW) = $11.5 \times W$

In general, 100 tonnes of raw MSW with 50-60% organic matter can generate about 1-1.5 MW power, depending upon the waste characteristics.

Electricity from Municipal Solid Waste: Organic wastes come from a range of agricultural or industrial activities and include crop residues, forest and wood process residues, animal wastes including human sewage, municipal solid waste (excluding plastics and non-organic material), and food processing wastes. Numerous processes have been developed to produce energy – heat and electricity – from these wastes (Fig. 6.1).

Some waste streams produce energy carriers almost directly. For instance, biogas production by anaerobic digestion occurs naturally in landfills which can be recovered directly. Similarly, waste oil from the food industries can be reused as biofuels with some degree of processing.

6.1.2 Environmental Impacts of the Technologies

Although MSW power plants are regulated by laws to protect human health and the environment, there is a wide range of environmental impacts associated with power generation technologies.

(a) *Air emissions impacts*: Burning MSW produces nitrogen oxides and sulfur dioxide as well as trace amounts of toxic pollutants such as mercury compounds and dioxins. Although MSW power plants do emit the major greenhouse gas, carbon dioxide, the biomass-derived portion is considered to be part of the Earth's natural carbon cycle. The plants and trees that make up the paper, food, and other biogenic waste remove carbon dioxide from the air while they are growing, which is returned to the air when this material is burned. In contrast, when fossil fuels (or products derived from them such as plastics) are burned, they release carbon dioxide that has not been part of the Earth's atmosphere for a very long time (i.e., within a human time scale).

The variation in the composition of MSW affects the emissions' impact. For example, if MSW containing batteries and tyres are burned, toxic materials can be released into the air. A variety of air pollution control technologies are used to reduce toxic air pollutants from MSW power plants. For example, the average air emission rates in the United States from municipal solid waste-fired generation are: 2988 lbs/MWh of carbon dioxide (it is estimated that the fossil fuel-derived portion of carbon dioxide emissions represent approximately one-third of the total carbon emissions), 0.8 lbs/MWh of sulfur dioxide, and 5.4 lbs/MWh of nitrogen oxides.

(b) *Water discharge impacts*: Power plants that burn MSW are normally smaller than fossil fuel power plants but typically require a similar amount of water per unit of electricity generated. When the required water is removed from a lake or river, fish and other aquatic life can be killed, affecting those animals and people who depend on these resources. These power plants discharge water after usage which contains pollutants. In addition, the cooling water is considerably warmer when it is discharged than when it was taken. These water pollutants and the higher temperature of the discharged water can adversely affect water quality and aquatic life.

(c) *Solid residue impact*: While the combustion of MSW reduces the waste streams, it creates a solid waste called *ash* that contains any of the elements that were originally present in the waste. MSW power plants reduce the need for large landfill capacity because disposal of ash requires less land area than does unprocessed MSW. However, the ash and other residues from MSW treatment may contain toxic materials; hence the power plant wastes must be tested regularly to ensure that the wastes are safely disposed to prevent toxic substances from getting into the ground and water supplies. Hazardous ash must be disposed of as hazardous waste. Depending on state and local restrictions, non-hazardous ash may be disposed of in a MSW landfill or recycled for use in the laying of roads and/or parking lots. It can also be used for daily covering for sanitary landfills.

The environmental pollution control methods are briefly mentioned in later pages.

6.2 Thermal Processing

6.2.1 Combustion/Incineration

A controlled burning process called *combustion* or *incineration* is primarily implemented to reduce the volume of waste. In addition to reducing volume, the combustors, when properly equipped, can convert water into steam to fuel heating systems or generate electricity. Incineration facilities can also remove materials for recycling.

The basic technology for modern waste-to-energy *combustion* was developed in Europe during the 1960s and 1970s. This technology has been modified and improved since its development, and has been widely implemented in Western Europe and the US. Despite the fact that incineration of solid waste can decrease its volume ninefold and improve the final waste disposal into landfills, the full potential of utilizing solid waste for energy production has not been or is being realized because of widespread fears regarding environmental pollution. Modern WTE combustion facility with adequate environmental safeguards and careful monitoring, however, has been shown to be safe and cost-effective technology that is likely to gain importance during the coming decades.

Municipal solid waste combustion to the very highest standard is thus possible with today's combustion and emission control technologies. MSW can be considered as a biofuel that helps conserve energy sources, and provides heat and/or electric power at reduced CO_2 emission levels compared with conventional fossil-fuel technologies. The potential resources saved through EfW combustion processes are given in Table 6.1.

Table 6.1 Energy equivalence of 1 ton of MSW

1 t of MSW is equivalent to	2.5 t of steam (400 °C, 40 bar)
1 t of MSW is equivalent to	30 t of hot wat (at 180-130 °C)
1 t of MSW is equivalent to	200 kg of oil
1 t of MSW is equivalent to	500 kWh electricity

Combustion of organic materials and/or substances is the basic principle involved in incineration/ combustion technology (Knox 2005). Combustion is the rapid oxidation of combustible substances with release of heat. Oxygen is the sole supporter of combustion. Carbon and hydrogen are by far the most important of the combustible substances. These two elements occur either in a free or combined state in all fuels, whether solid, liquid or gaseous. Sulfur is the only other element considered to be combustible. In combustion of MSW, sulfur is a minor constituent with regard to heating value. However, its presence is a concern, and has to be considered in the design of the air pollution control equipment. The only source of oxygen considered here will be the oxygen in the air around us.

Table 6.2 shows the elements and compounds that participate in the combustion process. Water occurs as vapour in atmospheric air, and in the products of combustion, and as a liquid or vapor constituent of MSW fuel.

Table 6.2 Elements and compounds encountered in Combustion
(*Source*: Velzy and Grillo 2007)

Substance	Molecular Symbol	Molecular Weight	Form	Density (1b/ft^3)
Carbon	C	12.0	Solid	----
Hydrogen	H_2	2.0	Gas	0.0053
Sulfur	S	32.1	Solid	----
Carbon monoxide	CO	28.0	Gas	0.0780
Oxygen	O_2	32.0	Gas	0.0846
Nitrogen	N_2	28.0	Gas	0.0744
Nitrogen atmos.	$N_{2\,atm}$	28.2	Gas	0.0748
Dry air		29.0	Gas	0.0766
Carbon dioxide	CO_2	44.0	Gas	0.1170
Water	H_2O	18.0	Gas/liquid	0.0476
Sulfur dioxide	SO_2	64.1	Gas	0.1733
Oxides of nitrogen	NO_x	----	Gas	----
Hydrogen chloride	HCl	36.5	Gas	0.1016

The chemical reactions of combustion are shown in Table 6.3. These reactions result in complete combustion; that is, the elements and compounds which are capable of reacting chemically with oxygen connect with all the oxygen. In reality, combustion is a more complex process in which heat in the combustion chamber causes intermediate reactions leading up to complete combustion. Some of the resulting components are pollutants. These are important for the purpose of establishing air pollution control requirements in the incinerators.

Table 6.3 Chemical reactions of Combustion (Velzy and Grillo 2007)

Combustible	Reaction
Carbon	$C + O_2 = CO_2$
Hydrogen	$2H_2 + O_2 = 2H_2O$
Sulfur	$S + O_2 = SO_2$
Carbon monoxide	$2CO + O_2 = 2CO_2$
Nitrogen	$N_2 + O_2 = 2NO$
Nitrogen	$N_2 + 2O_2 = 2NO_2$
Nitrogen	$N_2 + 3O_2 = 2NO_3$
Chlorine	$4Cl + 2H_2O = 4HCl + O_2$

Source: From Hecklinger, R. S. 1996. *The Engineering Handbook*, CRC Press, Inc., Boca Raton, FL.

The combustion/incineration method is most suitable for high calorific value waste with a large component of paper, plastic, packaging materials, certain hazardous and medical wastes etc. As already seen, incinerators reduce the mass of the original waste by 80% to 85 % and the volume (already compressed somewhat in garbage trucks) by 95% to 96 %, depending upon composition and extent of recovery of materials such as metals from the ash for recycling (Ramboll 2006). In many countries, garbage trucks often reduce the volume of waste in a built-in compressor before delivery to the incinerator.

Incineration generally entails burning garbage to boil water to convert to steam that drives generators to produce electricity. Incineration is carried out both on a small scale and a large scale. It is considered as a realistic method of disposing certain hazardous, non-metallic organic wastes and medical wastes because the high temperature breaks down bacteria and viruses, in addition to being the most common process in WTE activity. This process is relatively sterile, noiseless and odourless; the land requirements are minimal.

Incineration may also be implemented without energy and materials recovery. Incinerators which do not include a materials separation to remove hazardous, bulky or recyclable materials before combustion cause great risk to the health of the plant workers and the local environment. Most of these facilities do not generate electricity.

The major concern with the incineration is the adverse environmental and public health effects due to the emission of fine particulate matter, heavy metals, nitrogen oxides, sulphur dioxide, carbon monoxide, and acid gases, in addition to dioxins and furans (see Table 6.3). But the more modern incinerators are equipped with the pollution control apparatus that include, in most cases, flue gas cleaners in the form of acid gas scrubbers, together with either electrostatic precipitators or bag house filters. Acid gases, SOx, and NOx are removed in flue gas cleaning systems, which usually consist of either wet or dry scrubbers, or the use of lime scrubbers on smokestacks. The limestone mineral used in these scrubbers has a pH of about 8 indicating that it is a base. By passing the smoke through the lime scrubbers, any acids that may be in the smoke are neutralized preventing the acid from reaching the atmosphere and adversely affecting the environment. Heavy metals are more likely to be removed in post-scrubbing filters, or via the injection of sodium sulfate in an electrostatic precipitator. This type of pollution control equipment can also remove dioxins and furans. The other important concern is the management of toxic fly ash and incinerator bottom ash (IBA).

The designers and engineers, therefore, have to incorporate designs and systems that provide environmentally sound sustainable incineration technology reducing and preventing air, ground and water pollution. The incineration technology is *highly expensive* requiring large capital as well as substantial operation and maintenance costs.

6.2.1.1 Incineration System types

The most widely used and technically proven incineration technologies are mass-burn incineration, and Refuse-derived-fuel Incineration. Modular incineration and Fluidized-bed incineration have been employed to a lesser extent and is expanding recently.

Some facilities have also experimented with pyrolysis, gasification, and other related processes (will be discussed later) that convert solid waste to gaseous, liquid, or solid fuel through thermal processing. For example, in the past, some MSW facilities in Japan have used a two-stage process where pyrolysis is followed by thermal combustion. Majority of attempts to use these technologies have been unsuccessful and ceased.

Currently, they are not considered as a reliable and cost-effective alternative, especially for developing countries.

6.2.1.1.1 Mass-Burn Technology (MBT)

Mass-burn systems are the predominant type of MSW incineration. The *unprocessed* or minimally processed waste is combusted in a mass-burn system. This feature makes mass-burn facility convenient and flexible. However, it is desirable to separate household hazardous wastes such as cleaners and pesticides and also to sort out and recover materials (example, iron scrap etc.,) to make certain that incineration is environmentally-sound. It further helps resource conservation.

In the 20th century, the major advance in waste incineration was the development of moving grates, which allow waste to be fed continuously into a furnace, initially either by gravity or mechanical means. The moving-grate unit has been the heart of the so called mass-burn system, where 'as received' waste is processed at the plant. The grate and furnace technologies of today are ideally suited to the combustion of black bag waste as well as the residual waste stream following extensive source separation for recyclate recovery.

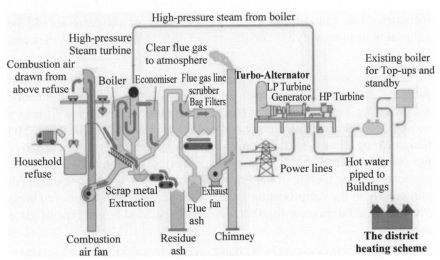

Fig. 6.2 Flow diagram of a Energy-from-Waste Plant, incorporating District Heating System (source: CIWM 2003)

Fig. 6.2 shows the main operating zones and parameters of furnace-boiler assembly. The major components of a mass burn facility are: (a) waste receiving, handling and storage system; (b) the combustion and steam generation system (boiler); (c) a flue gas cleaning system; (d) the power generation equipment (steam turbine and generator); (e) a condenser cooling water system; and (f) a residue handling and storage system. Mass burn combustion system includes a water wall furnace or water-cooled rotary combustion furnace or controlled air furnace (Kumar Sudhir 2000).

The combustion grate and the furnace is the critical part of a WTE plant. The first task of the combustion system is (i) to ensure the destruction of all the organic elements and pollutants contained in the waste by providing the necessary high-temperature profiles through the system as well as the required burn-out residence times, and (ii) to minimise the entrained fly ash to prevent the formation of pollutants such as dioxins and furans. For the plant to achieve the required levels of performance, fully engineered integration of grate, furnace and boiler are critical (CIWM 2003). The objective of a mechanical grate in a mass-burn furnace is to convey the refuse from the point of feed through the burning zone to the point of residue discharge with a proper depth of fuel and sufficient retention time to achieve complete combustion. The refuse bed should be agitated so as to enhance combustion. However, the agitation should not be so distinct that particulate emissions are unduly increased. The rate of movement of the grate or its parts should be adjustable to meet varying conditions or needs in the furnace (Velzy and Grillo 2007).

Several types of mechanical grates have been used in continuous feed furnaces burning 'as-received' (unprocessed) MSW. These include reciprocating grates, rocking grates, roller grates, and water wall rotary combustors for mass-burn units and traveling grates for RDF units. The reciprocating grates, rocking grates, and roller grates agitate and move the refuse material through the furnace by the movement of the grate elements and the incline of the grate bed. Additional agitation is obtained, particularly in the reciprocating grate, by drops in elevation between grate sections. The furnace configuration is largely decided by the type of grate used.

The most common grate technology developed by Martin GmbH, Munich, Germany (Fig. 6.3) in the year 2000, has an annual installed capacity of about 59 million metric tons, worldwide. A second very popular mass burning technology is provided by Von Roll Inova Corp (Switzerland) with an installed worldwide capacity of 32 million tons (Themelis 2003). Both the Von Roll and the Martin grates use a reciprocating motion to push the refuse material through the furnace.

However, in the Martin grate, the grate surface slopes steeply down from the feed end of the furnace to the ash discharge end and the grate sections push the refuse uphill against the flow of waste, causing a gentle tumbling and agitation of the fuel bed (Velzy and Grillo 2007).

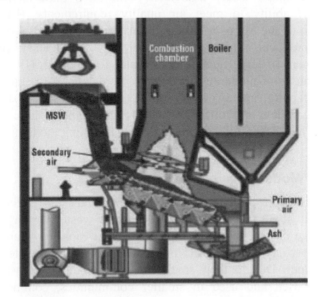

Fig. 6.3 The Martin Grate Combustion System – Mass Burn
(taken from Estevez 2003; originally from Brescia Plant in Italy)

The operation is as follows:

The trucks bring and deposit waste into pits and the cranes mix the refuse and recover bulky and large non-combustible materials, if any. The refuse storage area is maintained under pressure less than atmospheric in order to prevent odours from escaping. The cranes move the refuse to the combustor charging hopper to feed the boiler. The heat from the combustion process is used to convert water into steam, which is then used to run a steam turbine-generator for power generation. The steam is then condensed via wet-cooling towers and sent back to the boiler. The residues produced are 'bottom ash' (which falls to the bottom of the combustion chamber), and fly ash which leaves the combustion chamber with the flue gas. The combined ash and air pollution control residue typically is about 20 to 25% by weight of the waste processed. The composition of the MSW decides whether the ash is hazardous or not.

The incoming residue can be controlled or the ash can be treated to avoid the hazardous nature of the ash. Non-hazardous ash is mixed with soils and used for landfill cover and other applications.

There are major issues associated with mass burn facilities. For example, (a) ability to meet air quality requirements, (b) possible classification of the ash as a hazardous material, (c) disposal of ash and other by-products, (d) disturbance to biological resources, (e) requirement of large quantities of water for cooling, (f) impacts of transportation from numerous truck trips from the waste source to the mass burn facility, and (g) likely public opposition due to fears over health, safety, odour and traffic impacts etc., are some of them. Many of these issues are successfully addressed while installing modern mass burn facilities.

6.2.1.1.2 Modular incinerators

Modular incinerator units are usually prefabricated with relatively small capacities of about 5 to 120 tonnes of solid waste per day. Typical facilities have one to four units for a total plant capacity of about 15 to 400 tonnes per day. The majority of modular units generate steam as the energy product. Due to their small capacity, modular incinerators are generally chosen and used in smaller communities, commercial and industrial operations. The modular facilities can be built in short times due to their prefabricated design. On average, capital costs per tonne of capacity are lower for modular units compared to other incineration options.

Modular incinerators utilize a slightly different process than mass-burn incinerators, typically involving two combustion chambers. Gases generated in the primary chamber flow to an afterburner, which ensures more complete combustion. It often serves as the most important means of pollution control. Smaller-scale plants with capacity less than 50 tonnes per day sometimes operate in a batch process, operating only 8 to 16 hours per day rather than continuously. However, the modular incineration option is not very common, partly due to concerns over the reliability and inadequacy of air pollution controls.

6.2.1.1.3 Fluidized-bed incinerators

A fluidised bed is a bed of solid particles with gas flowing through the bed to give an expanded, suspended mass that behaves like a liquid. The fluidised bed therefore exhibits a zero angle of repose. It seeks its own level and assumes the shape of the containing vessel.

In a fluidized-bed incinerator, the stoker grate is replaced by a bed of limestone or sand that can withstand high temperatures which is fed by an air distribution system. When the bed is heated and the air velocities are increased the bed starts bubbling. Two types of fluidized-bed technologies - a bubbling bed and a circulating bed - are in operation. The differences are reflected in the relationship between air flow and bed material, and have implications for the type of wastes that can be burned as well as the heat transfer to the energy recovery system. The major parts of a fluidised-bed system can be identified as follows (CIWM 2003):

Fluidisation vessel - comprising a gas distributor (plate or nozzles) below the bed, a fluidised-bed section and a freeboard section above the bed for disengaging the bed particles from the flue gases,

Fuel feeder - charging MSW fuel into the top of the bed. The fuel feeder may also be used to feed additives into the bed such as new sand and limestone/lime,

Solids discharge - drawing solid material from below the bed. Material is constantly drawn from the bottom of the bed and classified to recover the sand. The smaller particles are returned to the bed with the larger items, including glass, ceramics and metals rejected as bottom ash.

The fluidized-bed incinerators require fuel preparation. Source separation is also essential as the presence of glass and metals are undesirable in these systems. Also, fluidized-bed systems can successfully burn wastes of widely varying moisture and heat content. Hence the inclusion of recyclable and burnable materials like paper and wood is not a critical factor in the operation of these systems. The paper therefore can be extracted for higher-value recycling. Fluidized-bed technologies are more compatible with high-recovery recycling systems, since there might be less competition for waste streams that are both burnable and recyclable. For this reason, fluidized-bed technology may be a sound choice for cities with high recycling facilities in developing countries when they first introduce incineration.

Fluidized-bed systems are more consistent in their operation than mass-burn and can be controlled more effectively to achieve higher energy conversion efficiency, less residual ash, and lesser air emissions. Despite unclear cost-wise comparison with mass-burn facility, fluidized-bed incinerators appear to operate efficiently on smaller scales than do mass-burn incinerators. This aspect may make them attractive in some situations.

Fluidized-bed incineration of MSW has been most extensively used in Japan. The Japanese plants are typically of medium scale, processing around 50 to 150 tonnes per day. The market share of Fluidized-bed incineration is increasing in the European MSW incineration market, although mass-burn technology still dominates. Overall, the experience with fluidized-bed incineration is less compared to mass burn. With the installation of more incinerators in Europe over the next few years, and as more experience is gained in Japan, fluidized-bed incineration of MSW could turn out to be a wholly proven technology commercially.

6.2.1.1.4 Incineration with Refuse-derived Fuel (RDF)

In contrast to 'mass burn facility' where the 'unprocessed' MSW is introduced into the combustion chamber, the RDF facilities are equipped to recover recyclables (e.g., metals/ metal products, iron scrap, glass) before converting the combustible fraction into a fluff or pellets for incineration. Using raw unprocessed MSW as a fuel is problematic due to the heterogeneous nature of the material, which varies with the country and the season. It also has a low heat value and high ash and moisture content. This makes it difficult for plant operators to always provide acceptable pollution-free levels of combustion. Processing of the waste to 'refuse derived fuel'(RDF) partially overcomes these problems and the fuel can then be used more successfully in either chain grate water-tube boilers or in circulating fluidised beds.

The solid waste that has been mechanically processed to produce a more homogeneous fuel for combustion is called 'refuse-derived fuel'. The RDF technology processes the MSW not only for incineration but as a supplementary fuel source by basically altering the physical characteristics of the material.

RDF systems perform two functions: RDF production and RDF combustion. RDF production facilities prepare RDF in different forms – fluff or pellets or bricks – after separating non-combustible materials, glass, metals, grit etc., using mechanical means, and size reduction. Waste with high organic (carbon) content is suitable for briquetting and pelletizing after non-combustible and recyclable materials have been separated. Although RDF processing has the advantage of removing recyclables and contaminants from the combustion stream, the complexity of this processing has increased the operating costs and reduced the reliability of the facilities. The capital costs per ton of capacity for incineration units that use RDF, on average, are higher than for other incineration types.

RDF production plants, like mass burn incinerators, typically have an indoor tipping floor. In the production plant the waste is typically fed onto a conveyor. The loader doing the feeding sometimes will separate corrugated and bulky items, like carpets. Once on the conveyor, the waste travels through a number of processing stages, beginning with magnetic separation (Fig. 6.4).

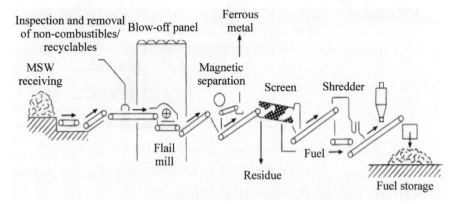

Fig. 6.4 Processes in RDF manufacturing

(Source: Wagner 2007)

The processing steps are customized to the desired products (pellets or bricks) that include (a) one or more screening stages, using trommel or vibrating screens, (b) shredding or hammer-milling of waste with additional screening steps, (c) wet separation, drying, and pressing, and (c) pelletizing or baling. Sometimes, manual separation is also performed. RDF is also extracted from MSW using other methods such as mechanical heat treatment, mechanical biological treatment or waste autoclaves. The permitting issues mentioned for mass burn facilities apply to RDF combustion systems also.

The SEMASS facility (RDF-type process) in Rochester, Massachusetts, US (Fig. 6.5(a) and (b)) was developed in 1989 by Energy Answers Corp. In 1996, it was taken over and operated by American Ref-Fuel till Coventa acquired it in 2005. It has a capacity of 0.9 million tons/year and is one of the most successful *RDF-type* processes. The schematic representation of the SEMASS facility is shown in Fig. 6.5(a).

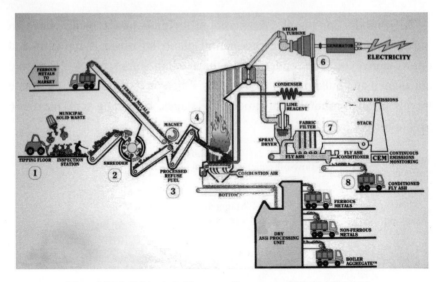

Fig. 6.5(a) Schematic Process diagram at SEMASS facility
(Source: Energy Answer's Brochure, 2008)

The process steps are as follows (Fig. 6.5(a)):

1. Trash is delivered to the tipping floor where it is inspected, and pushed onto conveyors which feed hammermill shredders. Material which should not be shredded is removed and processed by more appropriate systems,

2. Trash is shredded to 6 inches or less in size, then passed under magnets which remove ferrous metals for recycling,

3. This shredded material is called Processed Refuse Fuel (PRF). It would take 72 gallons of fuel or about one-third ton of coal to create as much heat as one ton of PRF,

4. The PRF is blown into specially-designed boilers. Light materials burn in suspension, while heavy portions of the fuel burn on a *traveling grate* at the bottom of the boiler,

5. Dry bottom ash is conveyed to the processing facility. Here, the metals are recovered and recycled through scrap dealers. Boiler Aggregate TM is also produced, which can be used in asphalt road construction and in the production of concrete and concrete blocks,

6. Steam produced in the boiler can be used in manufacturing industry as process heat and to produce electricity. The excess steam is converted back into water by condensers for re-use in the boilers, and

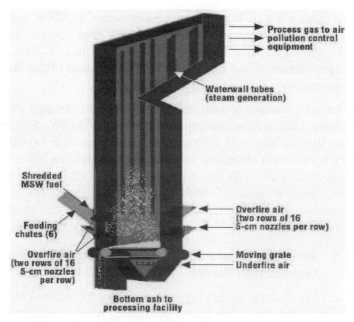

Fig. 6.5(b) Schematic process of SEMASS facility (stage 4 in Fig. 6.4(a))
(Source: Estevez 2003)

7. The combustion gases are passed through scrubbers to neutralize acids, and through other equipment to capture fine particles; afterwards, the Continuous Emissions Monitoring (CEM) system measures levels of compounds remaining. Fly ash, collected by the state-of-the-art air pollution control system, is kept separate from the bottom ash and conditioned for re-use or disposal (EnergyAnswers Brochure 2008).

SEMASS provides efficient total combustion of the waste, and complete recovery of usable or recyclable materials from the ash. The main objective of the design is to accomplish "zero landfill". When compared with other municipal solid waste combustion systems in operation in US, the outcome from SEMASS is excellent because of highest energy recovery, lowest percentage of ash, highest ferrous recovery rate, lowest residues requiring disposal, *lowest capital costs per ton, and lowest tipping fees.*

This facility was considered to be among the 10 finalists for the Waste-to-Energy Research and Technology Council (WTERT) 2006 Industrial Award. It was thus considered to be among the best in the world on the basis of the following: energy recovery in terms of kWh of electricity plus kWh of heat recovered per tonne of MSW and as the percentage of thermal energy input in the MSW feed, level of emissions achieved, optimal resource recovery and beneficial use of WTE ash, the aesthetic appearance of the facility, and the acceptance of the facility by the host community (Psomopoulos et al 2009).

Substantial quantities of heat energy have to be recovered during the thermal destruction of the combustible portions of MSW. Systems that have been successfully used to recover this energy include (a) mass-fired refractory combustion chambers followed by a convection boiler section; (b) a mass-fired water wall unit where the water wall furnace enclosure forms an integral part of the boiler system, and (c) an RDF semisuspension-fired spreader-stoker/ boiler unit. Each system has perceptible advantages as well as disadvantages (Velzy and Grillo 2007).

6.2.1.1.5 Environmental Pollution Control Measures for Incineration Plants

A number of pollutants, in varying concentration, like carbon monoxide, sulfur dioxide, and particulate matter containing heavy metal compounds and dioxins can be found in flue gas produced by Incinerators burning MSW. Many of these pollutants are formed as a result of incomplete/ partial combustion. The generation of these pollutants and their release into the atmosphere can be effectively reduced or prevented by incorporating a number of air pollution control devices and by proper operation of the WTE facility. Concentrations of heavy metals in particulates, particularly lead, zinc, mercury and cadmium, may be significant and care must be exercised in their removal and disposal. The most important of flue gas pollutants are sulphur dioxide (SO_2) and hydrogen chloride (HCl) which cause acid rain. These may be eliminated by wet scrubbers. Hydrogen fluoride and NO_x are produced in low concentrations but are not generally a problem. The emission of combustible, carbon-containing pollutants – dioxins and furans – is also of serious concern. These can be controlled by optimizing the combustion process.

Other concerns related to incineration include the disposal of the liquid wastes from floor drainage; quench water, and scrubber effluents, and the problem of ash disposal in landfills because of heavy metal residues.

The following gaseous emission control devices are currently used to remove pollutants from the incinerator load:

Dry Scrubbers: The particulate matter and gases from the air are 'washed' by passing them through a liquid. The scrubber removes acid gases by injecting a lime slurry (a watery mixture) into a reaction tower through which the gases flow. A dry powder containing salts is produced and collected along with the fly ash in an electrostatic precipitator or in filters, and discharged into the ash residue. The lime also causes small particles to stick together, forming larger particles that are easier to remove. Ash is stabilized by the addition of lime which enhances its natural alkalinity.

Electrostatic Precipitators (ESP): They use high voltage to negatively charge incoming dust particles, and then these charged particles are collected on positively charged plates, ESPs. These are very commonly used as WTE air pollution control devices. Nearly 43% of all existing facilities use this method to control air pollution.

Fabric Filters (Bag houses): These consist of hundreds of long fabric bags made of heat-resistant material suspended in an enclosed housing which filters particles from the gas stream. Fabric filters are able to trap fine, inhalable particles (<10 microns) and can capture 99% of the particulates in the gas flow coming out of the scrubber, including condensed toxic organic and heavy metal compounds.

Stack Height: Stack height is a safety measure to ensure that any remaining pollutants will not reach the ground in a concentrated area. When the gases enter the stack they are quite clean due to the controls explained above. Presently, Stacks built have a height of 200-300 feet (60-90 m) or more, nearly twice as high as the stacks used on older municipal incinerators. Stack heights should be determined by calculating quantity of fuel used and considering local weather conditions. Standard equations could be used for determining stack heights.

Dioxins and Furans: In recent years, Polychlorinated dibenzofurans (PCDFs), commonly called dioxins and furans, are of serious concern due their toxicity, carcinogeecity and possible mutagenicity. These compounds are found in many foods including fish, poultry and eggs, and occur in such common products as wood pulp and paper. About 75 different forms have been identified, of which five dioxins and seven furans are considered to be most toxic. But these compounds can be virtually eliminated by maintaining very high temperatures during the combustion process. Also, a combination of scrubbers and fabric filtration systems can remove up to 99 percent of these large molecules. Activated carbon injection before the flue gas treatment has also proved to be effective. Activated carbon reactor and catalytic rectors can be used for advanced processing.

However, the mechanism of their production and their removal methods are not completely understood and established.

The liquid wastes of incineration – floor drainage, scrubber effluents and quench water – have to be treated before discharging them into the surface waters and aquifers to avoid polluting them.

Dry ash in the bottom residue and fly ash captured from flue gases in electrostatic precipitations or bag filters contain heavy metals and will pollute the land unless treated or disposed of at special hazardous waste landfills.

6.2.1.2 Incineration in Developed countries

Modern MSW incineration plants operate quite well in cities of industrialized countries, recovering energy in the form of steam for heating and for electricity generation. Waste combustion is particularly popular in countries such as Japan where land is scarce. Energy produced by incinerators in large cities of Japan is widely used for heating community swimming pools or for air-conditioning.

European countries vary widely in their dependence on incineration. In Western European countries, at least 35% and in some cases as much as 80% of the residential waste stream is disposed of through incineration. Until relatively recently, mass-burn technology is relied upon, but there is increasing interest in and growing positive experience with fluidized-bed technologies. In 2005, waste incineration produced 4.8 % of the electricity consumption and 13.7 % of the total domestic heat consumption in Denmark (Danish Energy Authority 2007). Denmark and Sweden are highly reliant on mass-burn incineration, coupled with energy generation, accompanied by state of the art pollution control equipment. Other European Countries, particularly Luxembourg, The Netherlands, Germany and France (Ramboll 2006) rely heavily on incineration for handling municipal waste. Another reason for relying on incineration in Europe is that the energy generated by European WTE plants goes to supply steam to district heating loops.

In Europe, many RDF production units have come up in early 1970s. They have been producing refuse-derived fuel in the form of pellets or baled paper and plastic, though their ability to produce marketable recyclables has been limited. Producing RDF is another type of energy recovery system in Europe. RDF has been found useful in power generating stations that use fluidized-bed (boiler) technology. RDF is also used in Europe for generating heat required in industrial processes, particularly paper making, and in the cement kiln industry. The European Union is enforcing severe emissions standards for all types of incinerators, along with rigid rules for protecting the health and safety of workers. Many European countries are phasing out earlier generation of non-energy-generating incinerators, because these do not comply with emissions limitations in national and European Community law. In some cases, the older incinerators are being upgraded and retrofitted with pollution control equipment. In Eastern Europe, incinerators are older ones that usually do not have adequate environmental controls, and incineration is not economically attractive. Of the two byproducts of incineration, fly ash is often used in bonded asphalt and other road products, and the bottom ash and slag are used as an aggregate in road construction or in the production of brick materials. This practice is more common in a few countries, for example, The Netherlands. In European countries, where these materials are not put to these uses, they are generally land filled.

MSW incineration systems currently working in North America incorporate energy recovery in the form of steam used either to drive a turbine for generation of electricity or directly for heating or cooling. In the process, the volume of solid waste is reduced by up to 90% and its weight by up to 75%. About 10-15% of the MSW stream in North America is managed by waste-to-energy (WTE) incineration. The amount of solid waste processed in WTE facilities varies significantly by region. The northeastern US currently incinerates and recovers energy from over 40% of its solid waste, while many other states incinerate less than 2% of the solid waste generated. There are more WTE facilities in the US compared to Canada.

The most widely used and technically proven (a) mass-burn combustion, (b) modular combustion, and (c) refuse-derived-fuel production and combustion are used in North America. Over time, local governments have largely favoured mass-burn systems that recover electricity over other technologies. Mass-burn systems generally consist of either two or three combustion units ranging in capacity from 50 to 1,000 tons per day; thus, facility capacity ranges from about 100 to 3,000 tons per day. About 90% of operating mass-burn facilities generates electricity. These facilities can accept refuse with no preprocessing other than the removal of oversized items. This versatility makes mass-burn facilities convenient and flexible; however, local programmes to separate household hazardous wastes (e.g., cleaners and pesticides) and recover certain recyclables are necessary to help ensure environmentally responsible incineration and resource conservation. Modular combustors are generally used in smaller communities or for commercial and industrial operations because of their small capacity. On average, capital costs per ton of capacity are lower for modular units than for mass- burn and refuse-derived fuel plants. In US, this technology is used to generate steam. The vast majority of RDF combustion facilities in US generate electricity.

6.2.1.3 Incineration in Developing Countries

Incineration is not viable for many developing countries except those with fast growing economies, due to (a) the high capital and operating costs involved relative to national income levels, and (b) the availability of comparatively low cost sanitary landfilling. It is also difficult to incinerate wastes in many developing countries due to their high moisture and low energy content. Fig. 6.6 shows the calorific values of the waste generated in a few developing countries. The minimum calorific value required for sustainable combustion ranges between 5024-5861 kJ/Kg of MSW (www.no-burn.org).

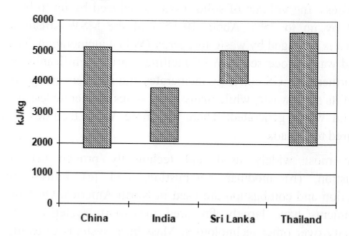

Fig. 6.6 Calorific values of waste in some developing countries
(Source: ARPEET 2004)

In addition, the technical infrastructure required to maintain incineration facilities such as pollution control equipment, is generally unavailable in developing countries. The other factors are lack of highly trained personnel and technologically advanced testing and repair facilities. Therefore, incineration with or without energy recovery does not appear to be a sound option in developing countries. That is why; there are few examples of successful MSW incineration in the developing countries and several examples of premature attempts to adopt this technology. For example, Buenos Aires (Brazil), Mexico City (Mexico), New Delhi (India), and Sao Paulo (Argentina) have to shut down incinerator facilities due to high costs or environmental considerations.

Virtually no incinerators operate in Latin America or the Caribbean because they are economically not viable. Barbados has one tiny (one ton/day) incinerator for processing wastes originating in the port with private financing. The refuse derived fuel (RDF) incinerators that operate in Mumbai and Hyderabad are questionable regarding their efficiency and activities. In the African continent, Incineration and WTE presently do not play significant roles in MSWM. High costs relative to other MSWM options, a limited infrastructure of human, mechanical and institutional resources, and the very composition of the waste stream itself, suggest that incineration is an inappropriate technology for Africa now. However, some medical waste incinerators, especially in major hospitals of cities in South Africa are operating. Some experiments have also been tried in some countries, e.g., Tanzania, Senegal so on.

However, some developing countries – Singapore, South Korea, Taiwan and Hong Kong in South East Asia and Pacific ring – do have considerable technical expertise and the capital necessary to install and operate incinerators. Some advanced developing countries like Singapore operates three MSW incinerators that handle about 90% of the MSW generated. South Korea also has many incinerators. Thailand has two MSW incinerators of 250 and 70 tons/day capacity at Phuket and Samui Island municipalities respectively. These incinerators have been in operation since 1998 and process less than 1% of the total waste generated in Thailand. Incineration will remain popular in cities like Singapore, Hong Kong, Taipei, and Tokyo as there is lack of landfill sites. In Sri Lanka, as of now there are no MSW incineration plants though some locals burn the waste in enclosures, and technical incineration remains vague to some local bodies. In the cities of Makkah and Medina in Saudi Arabia, several incinerators are operating as other disposal options are not available; the plant in Jeddah is abandoned.

In China, the first modern incineration plant with treating capacity of 300 tons per day was installed and operated in Shengzhen in 1989 by importing Japanese technology and equipments. Since the domestic manufacturing companies were unable to meet the equipment demand, incineration plants were constructed using the technologies from America, Europe and Japan. So far, there are about 36 incineration plants in the whole country for about 2% of the MSW and this figure would increase in future. As of now Shanghai (Fig.6.7), Xiamen, Zhuhai, Nanhai, Beihai, Ningbo, Guangzhou and Beijing have constructed incineration plants to generate electricity. It has also been observed that some of the existing operational incineration plants fitted with domestic equipments are not properly functioning due to poor combustion, inconvenience in operation with no standard unit for flue gas purification and so on (ARRPET 2004).

Fig. 6.7 Incineration plant in Shanghai, China
(Source: ARRPET 2004)

Incineration technologies are not fully functional globally as the most attractive ones; there are arguments 'for' and 'against' utilizing incineration process for energy production.

6.2.1.4 Reasons for/against Incineration

(a) *Reasons for utilizing incineration:* (Source: Wikipedia 2010)

(i) The progress in emission control equipment and designs and stringent governmental regulations, at least in some countries, have largely reduced the emissions of dioxins and furans.

Consequently the concerns over the health effects of dioxin and furan emissions have been significantly lessened.

(ii) The U.K. *Health Protection Agency* declared in 2009 that "modern, well managed incinerators" emit low concentrations of air pollutants locally. Such small additions could possibly have an impact on health, if they exist, but are likely to be very little and undetectable (UK-HPA 2009).

(iii) Electricity and heat generated by Incineration plants can substitute the power plants run by other fuels at the regional electric and district heating grid, and can supply steam for industrial customers. Incinerators and other waste-to-energy plants generate at least partially biomass-based renewable energy that offsets greenhouse gas pollution from coal- and oil-fired power plants. The EU considers energy generated from biogenic waste (waste with biological origin) incineration as non-fossil renewable energy under its emissions caps. These greenhouse gas reductions are in addition to those generated by the avoidance of landfill methane.

(iv) The bottom ash (remaining residue after combustion) is a non-hazardous waste that can be safely put into landfills or can be used as construction material. Samples are tested for ecotoxic metals (Abbott et al 2003). See later section.

(v) Incineration is desirable because finding space for additional landfills in densely populated cities has become increasingly difficult.

(vi) Fine particles can be efficiently removed from the flue gases with baghouse filters. Even though approximately 40 % of the incinerated waste in Denmark was incinerated at plants with no baghouse filters, measurements by the Danish Environmental Research Institute in 2006 showed that incinerators were only responsible for about 0.3 % of the total domestic emissions of particulate, $PM_{2.5}$, to the atmosphere (Ministry of Env., Denmark 2006; Nielsen et al 2006).

(vii) Incineration of MSW rather than landfill avoids the release of methane. Every ton of MSW incinerated, prevents about one ton of carbon dioxide equivalents from being released to the atmosphere (Themelis 2003).

(viii) Incineration of medical waste and sewage sludge produces an end-product ash that is sterile and non-hazardous.

(ix) Most municipalities that operate incineration facilities have higher recycling rates than others who do not send their waste to incinerators (EIA website). This is in part due to enhanced recovery of ceramic materials reused in construction, as well as ferrous and, in some cases, non-ferrous metals that can be recovered from combustion residue.

(x) Volume of combusted waste is reduced by approximately 90%, increasing the life of landfills. Ash from modern incinerators is vitrified at 1000°-1100°C (1,830°F-2,010°F), reducing the leachability and toxicity of residue. As a result, special landfills are generally no longer required for incinerator ash from municipal waste, and the life of existing landfills can be considerably increased by combusting waste. This reduces the need to search for site to construct new landfills (US EPA website).

(b) *Reasons against utilizing incineration*:

(i) Report by Scottish Environmental Protection Agency (SEPA, 2009) on health effects did not specify any conclusive evidence of non-occupational health effects from incinerators; however, admitted that any small significant effects were virtually impossible to detect. The report highlighted epidemiological deficiencies in previous UK health studies and suggested areas for future studies (Health Protection Scotland 2009).

(ii) The highly toxic fly ash requires safe disposal which involves additional transport of waste and the need for special type of toxic waste landfill. The local residents may be concerned otherwise about the impact (van Steenis 2005).

(iii) People are still concerned about the health effects of dioxin and furan emissions into the atmosphere from old incinerators.

(iv) Incinerators emit varying levels of heavy metals such as vanadium, nickel, manganese, chromium, arsenic, mercury, lead, and cadmium, which can be toxic at very minute levels.

(v) Incinerator Bottom Ash (IBA) has high levels of heavy metals with ecotoxicity concerns if not reused properly. Some consider that IBA reuse is still in its infancy and is not considered to be a desirable product, despite additional engineering treatments. Following several construction and demolition explosions in 2010, the UK Health and Safety Executive has expressed concerns about IBA use in foam concrete. IBA is currently banned from use by the UK

Highway Authority in concrete work until these incidents are investigated (Highways Agency 2009).

(vi) Alternative technologies such as Mechanical Biological Treatment (MBT), Anaerobic Digestion (AD), Autoclaving or Mechanical Heat Treatment (MHT) using steam or plasma arc gasification, or combinations of these treatments are available or in development. Installation of incinerators competes with the development and introduction of other emerging technologies. For instance, WRAP(UK) report, August 2008, found that median incinerator costs per ton were generally higher than those for MBT treatments by £18 per metric ton; and £27 per metric ton for most modern (post 2000) incinerators.

(vii) Building and operating waste processing plants such as incinerators requires long contract periods to recover initial investment costs, causing a long term lock-in. Incinerator lifetimes normally range 25–30 years.

(viii) Incinerators produce fine particles in the furnace. Even with modern particle filtering of the flue gases, a small part of these is emitted to the atmosphere. $PM_{2.5}$ is not separately regulated in the European Waste Incineration Directive, even though they are repeatedly correlated spatially to infant mortality in the UK. In June 2008, several European doctors' associations, physicians, environmental chemists and toxicologists represented to the European Parliament citing widespread concerns on incinerator particulate emissions and the absence of monitoring of specific fine and ultrafine particle size, or in depth industry/government epidemiological studies of these minute and invisible incinerator particle size emissions.

(ix) Local communities are often opposed to the idea of locating incinerators in their vicinity. Studies in Andover, Massachusetts, USA strongly correlated property devaluations by 10% with close incinerator proximity (Shi-Ling Hsu 1999).

(x) Prevention, waste minimization, reuse and recycling of waste should all be preferred to incineration according to the waste hierarchy. Supporters of zero waste consider incinerators and other waste treatment technologies as barriers to recycling and separation beyond particular levels, and that waste resources are sacrificed for energy production (Connett 2006; Friends of Earth 2007).

(xi) A 2008 Report found that under certain conditions and assumptions, incineration causes less CO_2 reduction than other emerging Energy-from-Waste and CHP technology combinations for treating residual mixed waste (Hogg et al 2008). CHP incinerator technology without waste recycling was found to be ranked 19 out of 24 combinations (where all alternatives to incineration were combined with advanced waste recycling plants); being 228% less efficient than the ranked 1 Advanced MBT maturation technology; or 211% less efficient than plasma gasification/autoclaving combination ranked 2.

(xii) Some incinerators are visually undesirable. In many countries they require a visually intrusive chimney stack.

(xiii) If reusable waste fractions are handled in incinerators in developing countries, it would reduce practical work for local economies. It is estimated that there are 1 million people making a livelihood through collecting waste (Medina 2000).

During 2001 to 2007, the WTE capacity increased by about four million metric tonnes per annum. Japan and China built several plants that were based on *direct smelting* or on *fluid bed combustion* of solid waste. Japan is the largest user in thermal treatment of MSW in the world with 40 million tonnes. Some of the newest plants use *stoker technology* and others use the *advanced oxygen enrichment* technology. There are also over one hundred thermal treatment plants using relatively novel processes such as direct smelting, the Ebara fluidization process and the Thermo- select -JFE gasification and melting technology process. A Greek company in Patras has developed and tested a system that shows potential. It generates 25 kwatts of electricity and 25 kwatts of heat from waste water.

6.2.1.5 *Management of Residues*

The incineration of MSW gives rise to several forms of residues created directly by the process i.e., heat energy in the form of high-temperature flue gas and solid residues (IBA), and the residue(s) created by the associated flue gas treatment system (fly ash and acid gases).

(i) Incineration Bottom Ash (IBA) is the principal residue stream, accounting for 80–95% of the total weight of residues generated, with boiler ash, flyash and FGT residue(s) making up the balance. IBA is a heterogeneous mix of slag, ferrous and non-ferrous metals, ceramics, glass, minerals and other materials. This mix can contain up to 20% by weight, of oversized (>10 cm) items, e.g., metal objects, construction materials, large pieces of slag. When these materials are excluded, the particle size distribution is similar in classification to sand and gravel. The moisture content of bottom ash which is important in compaction of IBA and, hence, its properties for utilisation can vary widely and is generally considered to be a function of the type of quench tank/discharge system employed. Compacted IBA has good load-bearing capacities with acceptable durability for a number of commercial applications. The chemical characteristics of IBA largely influence its interactive behaviour. IBA is alkaline in nature with pH values ranging from 9.5 to 11.5. The major elements comprising IBA (accounting for ~ 80% by weight) are oxygen, silicon, iron, calcium, aluminium and sodium. The concentration of certain trace metals found in IBA is influenced by the type of combustor used and the quality of combustion. The degree of burnout can affect the quantity of calcium and silicon found in the IBA, with poor burnout resulting in higher concentrations of volatile elements such as cadmium, arsenic and organic carbon in the ash. For treating IBA, chemical and thermal procedures are also employed. Forced carbonisation of IBA reduces alkalinity and can significantly reduce release of some trace metals. Similarly, treatment with selective chemicals can also achieve reductions in trace metal releases. Thermal treatment of WTE residues can be carried out to achieve sintering, i.e., melting to form a slag or a vitrified material. For the utilisation of IBA, pre-processing of IBA by removal of oversized material, including ferrous and non-ferrous metal items and screening is important. The aging process of IBA happens when it is stockpiled for a suitable period of 3 months and above when a number of beneficial reactions take place. IBA has been used commercially for different purposes such as landfill cover (daily and final), road foundations, wind and sound barriers, lightweight concrete masonry, structural fill, aggregate in asphalt, shore-line protection and marine reefs (CIWM 2003).

(ii) *Fly ash*: If separately removed from the flue gases, fly ash offers several applications:

    (a) Washing of fly ash yields an alkaline water stream that can be used in a wet scrubber to remove acid gases and provide make-up water for the water evaporated in quenching the flue gases;

    (b) Removal of the soluble constituents, by controlled washing, benefits the subsequent management of the fly ash, for example, treated fly ash mixed with IBA produces several civil engineering products; (c) The processing of fly ash in order to recover selected metal constituents is also being practised in certain locations; and (d) Fly ash is vitrified in some instances.

(iii) *Acid gases*: The principal acid gases found in MSW incinerator flue gas are, nitrogen oxides ($NO + NO_2 = NOx$), hydrogen chloride (HCl), and sulphur dioxide (SO_2). In addition, there are minor concentrations of hydrogen fluoride (HF) and sulphur trioxide (SO_3) present. A major pollutant, NOx is not considered for selective separation here. Separate extraction of HCl and SOx ($SO_2 + SO_3$) offers residue streams for the production of commercial grade products (Table 6.4).

Table 6.4 Commercial products and markets for acid gases recovered from MSW incinerator flue gases

Acid gas	Product	Market
HCl	NaCl	Road treatment (winter) Secondary aluminium smelting Animal hide treatment Water softening
HCl	$CaCl_2$	Road treatment (winter A/C dessicant Soil stabilization Lightweight concrete Coal washing
HCl	Hydrochloric acid	Regeneration of cationic exchangers for boiler feedwater Electroplating
SO_2	Gypsum	Building products Blender in cement manufacture

(*Source*: CIWM 2003)

To obtain a salt product, a dedicated wet scrubber circuit, using either a calcium or sodium-based reagent is used. A number of European MSW incineration plants process the hydrogen chloride to produce hydrochloric acid. The efficiency of the selective separation of SO_2, by a dedicated wet scrubber circuit using lime or limestone as reagent to produce gypsum, requires the circuit liquor to be at an optimum pH.

(iv) *Heat recovery*: The energy content of the exit flue gases from the (conventional) heat recovery plant used on MSW grate and furnace WTE plant, is of the order of 3.2-3.5 × 106 kJ per ton of MSW combusted; the flue gas energy ('FGT energy') is of the order of 20% of the calorific value of typical MSW. Grate and furnace WTE plant that operate on a combined heat and power energy concept, recover a substantial proportion of this 'FGT Energy' by using a hybrid flue gas treatment system that incorporates a wet scrubber. Wet scrubber operation can convert sensible heat of gases into latent (water vapour) heat for the recovery of 'FGT energy'; an additional closed-loop water circuit, linked to a heat pump, is installed downstream of the wet scrubber. Plants with this configuration recover around 1.95 × 106 kJ per ton of MSW, i.e., around 50% of the 'lost energy'. A by-product of this supplementary energy recovery concept is that around 300-400 kg per ton of water is produced through condensation of flue gas water vapour. In some regions, this water by-product could be important (CIWM 2003).

6.2.2 Pyrolysis

Thermal gasification and *Pyrolysis* are considered advanced thermal technologies, however, they are yet to be utilised on a large scale like incineration technologies. These processes involve the thermal breakdown of solid materials into a synthetic gas or syngas, and in some cases a solid char and/or liquid (oil). The process energy is provided in a reactor. One advantage of these technologies is that the syngas can be utilized both in boilers or low profile reciprocating engines for generating clean electricity more efficiently compared to conventional incinerators. Since net energy recovery and proper destruction of the waste are ensured in these processes, they have an edge over incineration.

These two technologies are related and are established for homogenous organic matter like wood, pulp, and so on. They are now being offered as an option for disposal of MSW. Gasification and pyrolysis have now reached thermal conversion efficiencies of up to 75%; however a complete combustion is superior in terms of fuel conversion efficiency. Some pyrolysis processes need an outside heat source which may be supplied by the gasification process, making the combined process self sustaining.

Reported energy outputs for gasification pyrolysis systems are in the range 300 to 750 kWhe/ ton of processed feedstock (i.e., sorted and/or unsorted MSW). This compares to a typical output of 550 kWhe/ ton of MSW for conventional combustion systems using unsorted waste. Higher efficiencies should not be expected unless and until proven by long-term (several years) continuous operation and making due allowance for the higher calorific value of processed feedstock (CIWM 2003).

In Pyrolysis, *thermal decomposition* of organic substances occurs at elevated temperatures in *the absence of air or oxygen*. It is a special case of thermolysis related to the chemical process of charring, and is most commonly used for organic materials. It occurs spontaneously at high temperatures, above 300°C for wood. It produces gas and liquid products (fuels) and a carbon-rich solid residue (Fig. 6.8). The process does not involve reactions with oxygen or any other reagents but takes place in their presence. The liquid products (fuels) typically comprises of 15-25% water with the remainder (organic) fraction comprising acetic and other acids, sugars and minor quantities of aldehydes, ketones and alcohols.

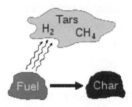

Fig. 6.8 Chemistry of pyrolysis (Source: Wikipedia)

Pyrolysis plants typically have a chamber or reactor that is sealed to prevent air entry. In practice, complete elimination of air is difficult to achieve and some oxidation is likely to occur. The relative proportions of produces gas, liquid and char that are produced in the process depend on the operational temperature, exposure time and type of feedstock. The production of char (charcoal is produced in this way) will be maximised with long exposure (hours) at low temperatures, 400–500°C. Short exposure (< 1 s) with high temperatures, 500–1000°C, referred to as 'flash' pyrolysis, will give a higher proportion of gas or liquid. If a liquid fuel (pyrolysis oil) is desired, rapid quenching of the gaseous product is necessary. The liquid fuel can be stored and transported easily for beneficial use.

Pyrolysis is used to turn waste into safely disposable substances. Pyrolysis is the basis of *several methods* that are being developed for producing fuel from biomass, which may include either biological waste products from industries or crops grown for the purpose. In many industrial applications, the process takes place under pressure and at temperatures above 430°C. For agricultural waste, typical temperatures are 450 to 550 °C. In vacuum pyrolysis, organic material is heated in a vacuum in order to decrease boiling point and avoid adverse chemical reactions.

Plasma Pyrolysis Vitrification (PPV) is a relatively new technology for disposal of hazardous wastes, radioactive wastes etc. Toxic materials get encapsulated in vitreous mass, which is relatively much safer to handle than incineration/gasifier ash. Pyrolysis can be integrated with other processes such as mechanical biological treatment (MBT) and anaerobic digestion.

6.2.3 Gasification

Gasification, unlike pyrolysis, is the process of conversion of carbonaceous materials (coal, petroleum, biofuel, biomass, or solid waste) into a combustible gas *in the presence of a controlled amount of oxygen or air.* The resulting combustible gas is called synthesis gas (or syngas), and consists of carbon monoxide and hydrogen, carbon dioxide and methane, hydrocarbon oils, char and ash. Syngas is itself a fuel. Gasification is an endothermic thermal conversion technology for extracting energy from different types of organic materials. The emissions from a gasification plant include nitrogen oxides, sulphur dioxide, particulate matter, hydrogen chloride, hydrogen fluoride, ammonia, heavy metals, dioxins and furans.

The advantage of gasification is that using the syngas is potentially more efficient (than direct combustion of the original fuel) because it can be combusted at higher temperatures, so that the thermodynamic upper limit to the efficiency defined by Carnot's rule is higher or not applicable.

Syngas may be burned directly in internal combustion engines (IC engines) or used to produce methanol and hydrogen, or converted via the Fischer-Tropsch process into synthetic fuel. Materials that are not otherwise useful fuels, such as organic waste can be used in the gasification.

Gasification relies on chemical processes at elevated temperatures >700°C, which distinguishes it from biological process like anaerobic digestion that produces biogas.

Process: In a gasifier, the material undergoes several different processes. The *pyrolysis* (or devolatilization) process occurs as the carbonaceous particle heats up. Volatiles are released and char is produced, resulting in weight loss up to 70% for coal. The process is dependent on the properties of the material and determines the structure and composition of the char, which will then undergo gasification reactions (Fig.6.9a).

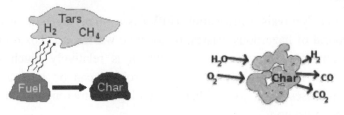

Fig. 6.9 (a) Pyrolysis of carbonaceous fuels **(b)** Gasification of char

(Source: Wikipedia)

The *combustion* process occurs as the volatile products and some of the char reacts with oxygen to form carbon dioxide and carbon monoxide, which provides heat for the subsequent gasification reactions. The basic reaction here (Fig. 6.8b) is $C + \frac{1}{2} O_2 \rightarrow CO$ where C represents the carbon-contained organic compound.

(i) The *gasification* process occurs as the char reacts with carbon dioxide and steam to produce carbon monoxide and hydrogen, via the reaction

$$C + H_2O \rightarrow H_2 + CO \quad \text{(Fig. 6.9b)}$$

(ii) In addition, 'the reversible gas phase water gas shift reaction' reaches equilibrium very fast at the temperatures in a gasifier. This balances the concentrations of carbon monoxide, steam, carbon dioxide and hydrogen.

$$CO + H_2O \rightarrow CO_2 + H_2 \quad \text{(Fig. 6.9b)}$$

In essence, a limited amount of oxygen or air is introduced into the reactor to allow some of the organic material to be 'burned' to produce carbon monoxide and energy, which drives a second reaction that converts further organic material to hydrogen and additional carbon dioxide. Table 6.5 shows the typical major components in fuel gas derived from MSW.

Table 6.5 Typical composition of fuel-gas from processed MSW
(% volume – dry gas)

	Fluidised bed gasification (air-blown)	Pyrolysis
CO	10	34
CO_2	16	19
H_2	7	33
CH_4	5	6
C_xH_y	4	3
N_2	54	5
Gross CV ($MJ\ m^{-3}$)	5-7.5	11

Source: NERL 1996: IEA/EU 1997

In addition to these major components, the fuel gas will also contain trace contaminants such as particulates, acid gases and heavy metals. These have the potential to cause operational problems in any downstream energy conversion process (CIWM 2003).

The gasification process is not new; it was originally developed in the 1800s to produce town gas for lighting and cooking, and the first four-stroke engine was run on producer gas in 1876. Electricity and natural gas later replaced town gas for these applications, but the gasification process has been utilized for the production of synthetic chemicals and fuels since 1920s. Wood gas generators, called Gasogene or *Gazogène*, were used to power motor vehicles in Europe when there was fuel shortage during World War II (Fig.6.10). The application to wastes, and in particular to MSW or products derived from MSW is, however, relatively new and still in the commercial demonstration stage.

Fig. 6.10 Adler Diplomat 3 with gas generator (1941)
(Source: Wikipedia, Free Encyclopedia)

Four types of gasifiers are currently available for commercial use: (1) counter-current fixed bed (2) co-current fixed bed, (3) fluidized bed, and (4) entrained flow (Beychok M R 1975, 1974a, 1974b). The working principles of these gasifiers are not covered in this book.

Applications: Industrial-scale gasification is mostly used to produce electricity from fossil fuels such as coal, where the syngas is burned in a gas turbine. In applications where the wood source is sustainable, 250-1000 kWe capacity new zero-carbon biomass gasification plants have been installed in Europe. These plants produce tar-free syngas and burn it in the reciprocation engines connected to a generator with heat recovery. This type of plant is often referred to as 'a wood biomass CHP unit' in which seven different processes are involved: biomass processing, fuel delivery, gasification, gas cleaning, waste disposal, electricity generation and heat recovery (Wikipedia).

Gasification is also used in the industry to produce electricity, ammonia and liquid fuels (oil) using Integrated Gasification Combined Cycles (IGCC), with the possibility of producing methane and hydrogen for fuel cells.

Gasification technologies have been developed in recent years that use plastic-rich waste as a feed.

Syngas can be used for heat production and for generation of mechanical and electrical power. Like other gaseous fuels, producer gas gives greater control over power levels when compared to solid fuels, leading to more efficient and cleaner operation.

Gasifiers offer a flexible option for thermal applications, as they can be retrofitted into existing gas fueled devices such as ovens, furnaces, boilers, etc., where syngas can replace fossil fuels.

Diesel engines can be operated on dual fuel mode using producer gas. Diesel substitution of over 80% at high loads and 70-80% under normal load variations can easily be achieved (Wikipedia). Spark ignition engines and SOFC fuel cells can operate on 100% gasification gas (Ahrenfeldt 2007, Hofmann et al 2007). Mechanical energy from the engines may be used for driving water pumps for irrigation or for coupling with an alternator for electrical power generation.

Small-scale rural biomass gasifiers have been extensively applied in India especially in Tamil-Nadu in South India. Most of the applications are 9 kWe systems used for water pumping and street lighting operated by the local panchayats. Despite their technical soundness, the systems have to face financial and maintenance problems locally. Most of the systems are not running after 1 to 3 years.

6.2.3.1 Waste Gasification

Several gasification processes for *thermal treatment of waste* are under development as an attractive alternative to incineration. Waste gasification has several advantages over incineration: (1) the cleaning may be confined to syngas instead of the much larger volume of flue gas after combustion, (2) electric power may be generated in *engines* and *gas turbines*, which are much cheaper and more efficient than the steam cycle used in incineration, (3) even fuel cells may potentially be used, though they demand severe requirements regarding the purity of the gas, (4) chemical processing of the syngas may produce other synthetic fuels instead of electricity, (5) some gasification processes treat ash containing heavy metals at very high temperatures so that it is released in a glassy and chemically stable form.

A major challenge for waste gasification technologies is to reach an acceptable electric efficiency. Significant power consumption in the waste preprocessing, consumption of large amounts of pure oxygen (often used as gasification agent), and gas cleaning are the factors that affect the efficiency of converting syngas to electric power.

Several waste gasification *processes* have been proposed, but very few have been built and tested. Based on the *Thermoselect process,* several commercial plants are constructed in Japan at Chiba (1999), Tokushiki Yoshino (2003), Matsu (2003), Sainokuni city (2004), Kyokuto (2005), Kurashiki city (2005) and Isahaya (2005) which are in operation. Plants started at Fondotoca (Italy) in 1992 and at Karlsruhe (Germany) in 2000 are shut down.The strengths and weaknesses of Thermoselect process are the following (Neissen 2007): By using raw MSW as its feed material, Thermoselect avoids all of the RDF preparation problems that are intrinsic in all other gasification technologies. The environmental benefits of gasification can thus be obtained without the implicit risks of RDF. Also, Thermoselect was conceived to discharge only products and not secondary wastes. In principle, the system is very well attuned to 'Zero Waste' concept. The disadvantages of Thermoselect include its high operating cost relative to more conventional technology and the absence of either operating plants or experienced technical support operations. However, if the processes' solid byproducts can be sold or, at least disposed at no net cost for use as an aggregate or fill (possible because of the binding of heavy metals into vitrified, non-leachable particles), the operating economics of the Thermoselect process are greatly improved. Fig. 6.11 is a schematic display of the 'high temperature conversion of waste'. This is one of several proposed waste gasification processes. According to sales management consultants KBI Group, a pilot plant Arnstadt implementing this process has completed initial tests. (*Ref:* HTCW commercial webpage)

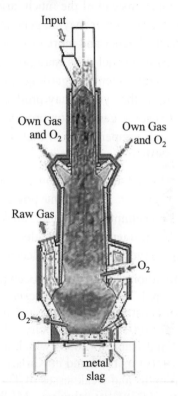

Fig. 6.11 High Temperature Conversion of Waste reactor
(Source: Wikipedia, Free Encyclopedia).

Waste gasification is *a high potential process for producing renewable energy*. Gasification can be utilized with any organic material, biomass or solid waste including plastic waste. The resulting syngas, if it is clean enough, may be used for power production in gas engines, gas turbines or even fuel cells. The syngas can also be converted efficiently to dimethyl ether (DME) by methanol dehydration, or to methane via the Sabatier reaction, or diesel-like synthetic fuel via the Fischer-Tropsch process.

Greenhouse gases such as carbon dioxide are not directly emitted either during the gasification process or the subsequent process. But, due to significant power consumption in the gasification and syngas conversion processes, there may be indirect CO_2 emissions. In slagging and plasma gasification, the electricity consumption may even exceed any power production from the syngas. Combustion of syngas or derived fuels emits exactly the same amount of carbon dioxide as would have been emitted from combustion of the initial fuel.

6.2.4 Plasma Arc Gasification

It is a waste treatment technology that uses electrical energy and the high temperatures created by an electrical arc in a closed vessel/reactor. The radiant energy of the plasma arc is so powerful that it breaks down waste into elemental gas and slag. The process is capable of breaking down almost *any* waste including toxic, except nuclear waste.

Inside a sealed stainless steel vessel filled with an inert gas such as nitrogen or air, a high voltage, high current electricity is passed between two electrodes spaced apart, and creating an electrical arc which converts the gas into plasma. Current flows continuously through this plasma creating a field of intense energy (high temperatures). This energy disintegrates most types of waste into basic elemental gaseous components, and the complex molecules are separated into individual atoms. This vessel is referred to as Plasma Reactor. The by-products are a glass-like material and 'syngas' – primarily, a mixture of hydrogen and carbon monoxide. The glassy material has many applications, and the syngas may be refined into various fuels such as natural gas, ethanol and hydrogen. A portion of the syngas may be used to generate steam that drives turbines to produce electricity; this electric power can be used to run the Plasma reactor, and the excess can be utilized for heating on the site or sell to the utility grid. This is an excellent technology to derive energy from all kinds of waste, and large cities have opportunities to make money from the garbage.

Syngas is produced exclusively from organic materials with a conversion rate of *greater than 99%* using plasma gasification. Other inorganic materials in the waste stream that are not broken down go through a phase change (solid to liquid) and add to the volume of slag with minimal energy recovery and increased cost for refining.

There are *several plasma arc facilities* working all over the world. For example,

(a) PEAT International constructed a plasma arc waste disposal facility at National Cheng Kung University (NCKU) in Tainan City, Taiwan, which handles 3–5 metric tons of waste per day from a variety of waste streams, including incinerator fly ash, medical waste, organic industrial process waste and inorganic sludges. It can also process waste consumer batteries and other materials, including heavy metal sludges, and refinery catalysts (waste streams that would generate valuable metal alloys). It has been in operation from 2005;

(b) Three similar small plants are in operation in Japan — a 166-short-ton (151,000 kg) per day "pilot" plant in Yoshii, co-developed by Hitachi Metals Ltd. and Westinghouse Plasma, which was certified after a demonstration period in 1999–2000; a 165-short-ton (150,000 kg) per day plant in Utashinai City, completed in 2002; and a 28-short-ton (25,000 kg) per day plant commissioned by the twin cities of Mihama and Mikata in 2002;

(c) Advanced Plasma Power (APP) has developed 'Gas plasma', a patented modular process, based on proven gasification and plasma conversion technology which uses refuse-derived fuel feedstock to produce a hydrogen rich syngas, energy (electric power) and vitrified gravel called Plasmarok. APP which originally had a test facility in Faringdon, Oxfordshire has moved to Swindon, Wiltshire, in 2007 where they operate a scale pilot plant. The plant runs in a building under a light vacuum and contains all odours. The entire process occurs within the building. A full scale plant will treat 100,000 short tons (91,000 t) per annum of municipal waste and produce (1) enough power for 10,000 homes, (2) enough heat for around 700 homes, (3) over 99% landfill diversion of feedstock with minimal residues and emissions, (4) increase recycling rates by over 20%, (5) high performance, high-value aggregate glass (trademark Plasmarok), (6) novel combination of three existing and proven technologies (termed Gasplasma) and (7) negative carbon footprint and lowest environmental impact plant and building (Wikipedia 2010);

Fig. 6.12 Startech's trash converter (Photo courtesy: Kevin Hand, Source: Popular Science,' The Prophet of Garbage', 2007)

(d) A US based company, Startech Environmental Corporation, Bristol, CT, has developed a Plasma Converter based on plasma gasification. Startech's trash converter uses superheated plasma – an electrically conductive mass of charged particles (ions and electrons) generated from ordinary air – to reduce garbage to its molecular components. First the trash is fed into an auger that shreds it into small pieces. Then the mulch is delivered into the plasma chamber, where the superheated plasma converts it into two by-products. One is a syngas composed mostly of hydrogen and carbon monoxide, which is fed into the adjacent Starcell system to be converted into fuel. The other is molten glass that can be sold for use in household tiles or road asphalt (Fig.6.12). This Converter can handle 2000 tons of waste daily, an amount generated by a city of one million people in a day. This 15-ft-tall machine which costs roughly $250 million can consume any type of waste, from diapers to chemical weapons, and annihilates toxic materials. Considering the *national* average tipping fee of $35/ton that an American city pays to the operator of landfills in large cities, and additional costs that include transporting garbage to landfill sites, the capturing of leaky methane from the decomposing site etc., this Startech machine could pay for itself in about *ten* years, even without considering the money recovered by selling excess electric power and syngas. It is all profit after break-even point (Popular Science – Popsci.com, March 2007).

More facilities that are *planned* include: (a) Utilizing technology licensed from Europlasma, the plasma arc facility proposed for lands in the vicinity of Wesleyville in Port Hope, Ontario (approximately 45 minutes east of Toronto) will handle 400 short tons (360 t) per day of Municipal Solid Waste (MSW) and Tire Derived Fuel (TDF). Sunbay Energy is currently obtaining the required approvals from Provincial authorities and intends to have the facility operational during the 4th Quarter of 2009; (b) The city of Tallahassee, Florida has signed the largest plasma arc waste to energy contract (35 MW) to date with Green Power Systems to process 1,000 short tons (910 t) daily from the city and several surrounding counties. Completion of the project is scheduled for October 2010; (c) EnviroParks Ltd plan (31/9/07) a consortium to build an Organic Park in Tower Colliery at Hirwaun, South Wales. This includes a plasma gasification plant combined with advanced anaerobic digestion to divert municipal solid waste from the landfill. Enviroparks are currently collaborating with Europlasma of Bordeaux to provide the plasma gasification unit to the park. The Hirwaun site is large enough for the processing of over 250,000 metric tons (280,000 short tons) of non-hazardous waste a year. Initially, though, an anaerobic digestion plant will be designed to handle 50,000 metric tons (55,000 short tons) of organic wastes a year; (d) PR Power Co. plans to open a plant south of Atlanta, near Jackson, Georgia, that will use a "plasma torch" to vaporize tires down to their natural elements — mainly hydrocarbons and scrap steel. The gases will be converted to electricity for sale to electric utilities and the scrap steel will be sold at an estimated $50 a ton.; (e) Plasco Energy group Inc., is preparing to start construction on a commercial-scale facility in Red Deer, Alberta in the Summer of 2009. This facility, the company's first commercial plant, is expected to be completed by the end of 2010; (f) SMS Infrastructures Limited (SMSIL), India's largest civil engineering and infrastructure development company, constructed 68 tonne-per-day hazardous waste-to-energy plants, located in Pune, India, that will use Westinghouse Plasma Corporation's (WPC) plasma technology and reactor vessel design. Each plant will provide comprehensive disposal services for a wide variety of hazardous waste, and will produce up to 1.6 MW (net) of electricity. The facilities will be the largest plasma gasification WTE plants in the world processing hazardous waste; (g) Plasco announced in April 2010 that it is planning a joint venture with Beijing Environmental Sanitation Engineering Group Co. to construct a 200 tonne per day demonstration facility in Beijing. If successful, it is intended to construct a larger 1,000 tonne per day facility to help dispose of the city's 18,000 tonnes of municipal waste generated per day (Wikipedia 2010).

6.3 Biochemical Process

6.3.1 Anaerobic Digestion (Biomethanation)

6.3.1.1 *Introduction*

Organic waste such as cattle manure is treated using Anaerobic Digestion (AD). The process is also called 'Biomethanation'. It is a process whereby wet residues, for instance waste from dairy industries or cattle manure is broken down in a controlled oxygen-free environment, by bacteria that naturally occur in the waste material. A gas called 'biogas' is produced. Biogas is a mixture of methane (CH_4) and carbon dioxide (CO_2). Hydrogen sulphide, water, and numerous trace gasses are also present in smaller amounts. The biogas production ranges from 50-150 m^3/tonne of wastes, depending upon the composition of waste.

The biogas can be utilised either for cooking/ heating applications, or through dual fuel or gas engines or gas/steam turbines for generating electricity. Depending on the nature of waste input and the system design, biogas is typically 55 to 75% pure methane. State-of-the-systems producing biogas with more than 95% methane are reported. Biomethanation is a fast growing process and is being utilized in many parts of the world.

The merits of the AD process are: (a) most of the materials that are currently sent to landfill can be utilized, (b) natural methane emissions are reduced, and (c) conventional generation with its associated carbon emissions is removed. The residual digestate which is nutrient-rich can be used as fertiliser reducing the need for chemical fertilisers.

Scientific interest in the production of biogas by the natural decomposition of organic matter was first reported in the 17th century by Robert Boyle and Stephen Hale, who noted that flammable gas was released by disturbing the sediment of streams and lakes (Fergusen and Mah 2006). In 1808, Sir Humphry Davy found that methane was present in the gases produced by cattle manure (Cruazon 2007). The first anaerobic digester was built in Bombay, India in 1859. Biogas Plants of different designs and capacities have been installed in the rural India to help farmers to generate biogas, a source of energy, and fertilizer using the animal dung generated by their livestock. In 1895 the technology was developed in Exeter, England, where a septic tank was used to generate gas for the sewer gas destructor lamp, a type of gas lighting. In 1904, the first dual purpose tank for both sedimentation and sludge treatment was installed in Hampton, England. In 1907, a patent was issued for the Imhoff tank, an early type of digester in Germany. Anaerobic digestion gained academic recognition in the 1930s through scientific research that led to the discovery of anaerobic bacteria, the microorganisms that facilitate the process. Investigations related to the conditions under which methanogenic bacteria were able to grow and reproduce (Humanik et al 2007) were carried out. This work was developed during World War II, when there was an increase in the application of anaerobic digestion for the treatment of manure in both Germany and France (Wikipedia).

6.3.1.2 Process

Anaerobic digestion is a complex process that involves interaction between many different microorganisms, so-called *consortia*. Each consortium lives optimally at a desirable set of chemical and physical conditions.

A number of macro- and micronutrients are required in order to facilitate the biological conversion and growth processes. Ten macronutrients, namely, carbon, hydrogen, oxygen, nitrogen, sulphur, phosphorous, calcium, potassium, iron, and magnesium, should be present in concentrations exceeding 10^{-4} M. Among important micronutrients, nickel and cobalt should be present in concentrations below 10^{-4} M (ProBiogas 2007).

The *degradation processes* can be divided into four major phases: *hydrolysis, acidogenesis, acetogenesis, and methanogenesis.*

Each of the four steps relies on certain microbial consortia to perform the conversion processes. Some consortia are highly tolerant and can utilise multiple substrates, while others are very sensitive towards environmental changes. In addition, they are only capable of utilising a single substrate. Hence, in order for the four degradation processes to be in balance, the overall chemical environment in the biogas reactor has to satisfy the needs of all consortia all the time.

1. The digestion process starts with bacterial *hydrolysis* of the feed stock (plant and animal matter). In this phase, hydrolytic and fermentative microorganisms excrete hydrolytic enzymes that convert biopolymers (the insoluble organic polymers) such as carbohydrates into soluble compounds, to make them available for other bacteria.

Hydrolysis is an important step enabling fermentation and subsequently biogas formation.

2. In the acidogenesis, the second phase, the products from the hydrolysis are converted by the fermentative microbial consortia into methanogenic substrates, which include volatile fatty acids, alcohols, carbon dioxide, ammonia, and hydrogen. Volatile fatty acids constitute the most frequently encountered intermediate products in anaerobic digesters. Approximately 30 % of the hydrolysis products will be converted into volatile fatty acids and alcohols. In case of an imbalanced process, the concentrations of volatile fatty acids will continue to rise, affect the chemical environment including the pH, and eventually lead to process failure. Under a given set of operating conditions, the acidogenic microorganisms chose the thermodynamically most favourable metabolism. Hence the product formation depends on the current conditions in the biogas reactor.

3. During the third phase of acetogenesis, products from the acidogenesis are converted onto methanogenic substrates (acetate, carbon dioxide, and hydrogen), since not all fermentation products can be converted to methane by the methanogenic microbial consortia. For instance, volatile fatty acids with carbon chains longer than two units and alcohols with carbon chains longer than one unit need to be oxidised into acetate and hydrogen. This operation is performed by the acetogenic consortia during acetogenesis.

4. Finally, two distinct types of methanogenic bacteria are utilised. The first type reduces carbon dioxide to methane and the second decarboxylates acetic acid to methane and carbon dioxide. i.e.,methanogens convert the products in step 2, to biogas (a mixture of methane and carbon dioxide). The operating conditions have severe influence on the methanogenesis. Composition of the feedstocks, feeding rate, temperature, and pH are the parameters that affect the methanogenesis.

As already mentioned, several microorganisms, are, thus, involved in the process of anaerobic digestion: acidogenic bacteria, acetogenic bacteria (acetogens), and methane-forming archaea (methanogens).

Factors that influence the AD process are temperature, pH, nutrient concentration, loading rate, toxic compounds and mixing.

Since oxygen-free environment is an essential factor for the anaerobic process to occur, oxygen is prevented from entering the system by proper control in sealed tanks. When the oxygen source is derived from the organic material itself, the resulting 'intermediate' end products are primarily alcohols, aldehydes, organic acids and carbon dioxide. In the presence of specialised methanogens, these intermediates are converted to the 'final' end products, methane and carbon dioxide with traces of hydrogen sulfide (Beychok 1967).

In the anaerobic process, most of the chemical energy contained within the starting material is released by methanogenic bacteria as methane. Since anaerobic microorganisms normally take a significant period of time to become fully effective, 'seeding' the digester is done by adding sewage sludge or cattle slurry. ('Seeding' is introducing anaerobic microorganisms from materials with existing populations). A temperature of about 35-38^{0}C is generally considered optimal in mesophilic zone (20-45^{0}C), and higher gas production can be obtained under thermophillic temperature in the range of 55-60^{0}C. Methanogens come from the primitive group of archaea. These species are more resistant to heat and can operate at thermophilic temperatures. Because of these variations, close monitoring of temperatures is very vital. A simplified generic chemical equation for the overall processes can be represented as:

$$C_6H_{12}O_6 \rightarrow 3CO_2 + 3CH_4$$

6.3.1.3 Design Parameters

(a) Digesters typically can accept any biodegradable material; however, if the objective is to produce biogas, the level of putrescibility is very vital, because the more putrescible the material, the higher the possibility of gas yields from the system.

Substrate composition is a major factor in determining the methane yield and its production rates from the digestion of biomass. Techniques are available to determine the compositional characteristics of the feedstock (Jerger and Tsao 2006). Anaerobes can breakdown material to varying degrees of success; for example, short chain hydrocarbons such as sugars can break down readily, whereas cellulose and hemicelluloses take longer periods of time. Anaerobic microorganisms are not capable of breaking down long chain woody molecules such as lignin.

Early anaerobic digesters were designed for treating sewage sludge and manures which are *not the ones with the most potential for anaerobic digestion* as the biodegradable material has already had the energy content taken out by the animal that has produced it.

(b) The moisture content of the feedstock is another important parameter. If the material is wetter, standard pumps are more suited to handle instead of concrete pumps which are energy intensive. Also the wetter the material, the more volume and area it takes up relative to the levels of gas that are produced. The moisture content of the feedstock will also affect the type of system utilised for its treatment. To utilise a high solids anaerobic digester for dilute feedstocks, bulking agents such as compost should be applied to increase the solid content of the input material.

(c) *pH:* The methanogenic consortia exert the highest *intolerance* towards fluctuations in the pH. The recommended pH range is 6.5 to 8, which is quite narrow. Below pH 6.6, due to slow growth of the methanogens, there is a risk of being washed out of the biogas reactor. Monitoring of pH, however, does not provide the correct state of the process, and pH has to be compared with the buffer capacity, total alkalinity and bicarbonate alkalinity to know the correct state of the process (ProBiogas 2007).

(d) The carbon/nitrogen ratio of the input material which balances the food required by a microbe to grow is another key parameter. The optimal C/N ratio for the 'food' a microbe needs is 20/30:1. Excess nitrogen can lead to ammonia inhibition of digestion (Richards 1991).

(e) Ammonium: Degradation of manure- and protein-rich feedstock causes ammonium to be released in the reactor which is an important nutrient in many of the microbial processes. However, depending on the chemical environment, ammonium can be toxic and inhibit the process. It has been proposed that the species responsible for inhibition is free ammonia, NH_3.

The concentration of free ammonium increases with temperature. i.e., processes operated at thermophilic temperatures are more vulnerable towards ammonia inhibition than processes run at mesophilic temperatures.

(f) The level of contamination of the feedstock is another factor. If the feedstock has significant levels of plastic, glass or metals, pre-processing has to be done so that the digesters function efficiently without getting blocked. This aspect is taken care of in the designing of mechanical biological treatment plants. A higher level of pre-treatment of feedstock requires more processing machinery resulting in higher capital costs.

After sorting or screening to remove any physical contaminants, such as metals and plastics from the feedstock, the material is often shredded, minced and pulped (mechanically or hydraulically) to increase the surface area available to microbes in the digesters which in turn increase the speed of digestion. The feedstock material is then fed into the airtight digester where anaerobic process occurs.

6.3.1.4 Design Types

Different process configurations can be used in several ways to design anaerobic digesters to operate: Batch or continuous; Based on temperature (Mesophilic or thermophilic); Based on Solids content (High solids or low solids); Single stage or multistage.

(i) *Batch or continuous*: A batch system is the simplest form of digestion. Biomass is added to the reactor at the start of the process in a batch and is sealed for the duration of the process. Batch reactors suffer from odour issues which can pose a severe problem when they are emptied. The designing of batch digestion system is simple and requires less equipment and hence cheaper. In *continuous* digestion processes, organic matter is added to the reactor constantly or in stages. The end products are regularly removed, resulting in regular production of biogas. Examples of this form of anaerobic digestion include, continuous stirred-tank reactors (CSTRs), Upflow anaerobic sludge blanket (UASB), Expanded granular sludge bed (EGSB) and Internal Circulation reactors (IC). These systems have been developed since the late eighties principally for the organic fraction of municipal solid waste but have also been extended to other industrial, market and agricultural wastes. The digestion occurs at solid content of 16% to 40%. These systems are referred to as 'Dry Digestion' or Anaerobic Composting when the solid concentration is in the range of 25-40% and free watercontent is low. Systems in this category vary widely in design and include both completely mixed and plug-flow systems.

Continuous plants are more suitable for rural houses because the operation fits better into their daily routine. Gas production is constant and slightly higher than in batch plants. If straw and dung are to be digested together, a plant can be operated in a semibatch basis (Ludwig Sasse 1988).

(ii) *Temperature-based*: AD processes can be run at different temperatures. Usually, the different operating ranges are divided into three groups: psychrophilic (below 25°), mesophilic (25°C-45°C), and thermophilic (45°C-70°C). The later two are conventional operating temperature zones. Mesophiles have comparatively more tolerence to changes in environmental conditions; that is why, mesophilic systems are considered to be more stable than thermophilic digestion systems. However, thermophilic temperatures are desirable because highest growth rate is obtained at these temperatures. However, there are a number of disadvantages such as (a) elevated risk of ammonia inhibition, (b) requiring relatively more energy to maintain high temperatures, and (c) higher degree of instability. But, several advantages such as (a) effective pathogen reduction, (b) increased organic load, (c) reduced retention time enabling higher substrate throughput, (d) better degradation of solid substrate, i.e., better substrate utilization, (e) facilitating greater sterilisation of the end digestate, and (f) higher biogas yield due to increased solubility of hydrolysis products can outweigh the disadvantages, if the process is kept within *safe operating conditions.*

The thermophilic temperatures result in faster chemical reaction rates, higher solubility, and lower viscosity. Consequently, the substrate is better utilised compared to mesophilic conditions. Therefore, demand for more process energy can be justified by the higher biogas yield.

It is essential that the process temperature is kept constant to maintain a sound microbial environment. Otherwise the biogas production will drop until the bacteria have adapted to the new temperature. Temperature fluctuations will negatively affect the biogas production and thus the overall economy of the plant.

(iii) *Based on Solid Content:* Typically there are two different operational parameters associated with the solids content of the feedstock. Digesters can either be designed to operate in 'high solids' content, with a total suspended solids (TSS) concentration greater than ~ 20%, or 'low solids' concentration less than ~ 15% (Jewell et al 1993). High-solids digesters process thick slurry that requires more energy input to move and process the feedstock. The thickness of the material may also lead to associated problems with abrasion. High-solids digesters require less land due to the lower volumes associated with the moisture. Low-solids digesters can transport material through the system using standard pumps that need significantly lower energy input. Low-solids digesters require larger stretch of land than high-solids due to the increased volumes associated with the increased liquid-to-feedstock ratio of the digesters. Operation in a liquid environment has several benefits as it enables more thorough circulation of materials and contact between the bacteria and their food. This enables the bacteria to more readily access the substances they are feeding off and increases the speed of gas yields.

A large number of systems presently available worldwide for digestion of solid wastes are for low (< 10%) or medium (10-16%) solid concentrations. Some of these systems, when applied to MSW or market waste, require the use of water, sewage sludge or manure.

(iv) *Single stage and multistage systems*: Digestion systems can be configured with single-stage or multistage. In a single-stage digestion system, all of the biological reactions occur within a single sealed reactor or holding tank. Utilising a single stage reduces construction costs; however, it facilitates less control of the reactions occurring within the system. Acidogenic bacteria, through the production of acids, reduce the pH of the tank. Since methanogenic bacteria operate in a strictly defined pH range, the biological reactions of the different species in a single stage reactor can be in direct competition with each other.

Another one-stage reaction system is an anaerobic lagoon. These are earthen basins, pond-shape, used for the treatment and long-term storage of manures. Here the anaerobic reactions are contained within the natural anaerobic sludge contained in the pool.

In a two-stage or multi-stage digestion system, different digestion vessels are optimised to get maximum control over the bacterial groups existing within the digesters. Hydrolysis, acetogenesis and acidogenesis typically occur within the first reaction vessel. The organic material is then heated to the required operational temperature (either mesophilic or thermophilic) before pumping into a methanogenic reactor. The initial hydrolysis or acidogenesis tanks prior to the methanogenic reactor can provide a buffer to the rate at which feedstock is added. If an amount of higher heat treatment to kill harmful bacteria in the input waste is required, there may be a pasteurisation or sterilisation stage prior to digestion or between the two digestion tanks. It is not possible to completely isolate the different reaction phases. There is often some biogas that is produced in the hydrolysis or acidogenesis tanks. The residence time in a digester varies with the quantity and type of feed material, and the configuration of the digestion system, whether one-stage or two-stage. In the case of single-stage thermophilic digestion, residence times may be around 14 days, which is relatively fast compared to mesophilic digestion. The plug-flow nature of some of these systems will mean that the full degradation of the material may not have been realised in this timescale. In such an event digestate will be darker in colour and will typically have more odour.

In two-stage mesophilic digestion, residence time may vary between 15 and 40 days. Continuous digesters have mechanical or hydraulic devices, depending on the level of solids in the material, to mix the contents enabling the bacteria and the food to be in contact. They also allow excess material to be constantly extracted to maintain a reasonably constant volume within the digestion tanks. At the end, three principal products – biogas, digestate and water - result in anaerobic digestion. However, requirement of two reactors and more process controls may lead to higher capital costs and system complications.

6.3.1.5 End products

Biogas: Biogas is the main final product with a composition of methane 50-75%, carbon dioxide 25-50%, nitrogen 0-10%, hydrogen 0-1%, hydrogen sulphide 0-3%, and oxygen 0-2%. 'As-produced' biogas may also contain water vapor depending on biogas temperature (Richards et al 1991). Most of the biogas is produced during the middle of the digestion and is normally stored on top of the digester in an inflatable gas bubble or extracted and stored in a gas holder next to the facility. The Calorific Value of biogas is about 5000 kcal/m^3 and depends upon the methane percentage. The biogas, by virtue of its high calorific value, has tremendous potential to be used as fuel for power generation through either IC Engines or Gas Turbines. Heat energy also can be derived from a Co-generation arrangement.

The simplest and most cost-effective option for use of biogas (even landfill gas) is utilization locally. This option requires that the gas be transported, typically by a dedicated pipeline, from the point of collection to the point(s) of gas use. If possible, a single point of use is preferred so that pipeline construction and operation costs can be minimized. Prior to transporting the gas to the user, the gas must be cleaned to some extent. Condensate and particulates are removed through a series of filters and/or driers. With this minimal level of gas cleaning, gas quality of 35 to 50 percent methane is typically produced. This level of methane concentration is generally acceptable for use in a wide variety of equipment, including boilers and engines.

The equipment usually designed to handle natural gas can be adjusted easily to handle the biogas gas with the lower methane content. For use in electricity generation, the gas requires more cleaning.

The engines to be fuelled by biogas can tolerate H_2S content of up to 1000 ppm, beyond which the H_2S can cause rapid corrosion. Although biogas generated from MSW is generally not expected to contain high percentage of H_2S, adequate arrangements for cleaning of the gas have to be made in case it is beyond 1000 ppm. Hydrogen sulfide is a toxic product formed from sulfates in the feedstock. Gas scrubbing and cleaning equipment (such as amine gas treating) are needed to process the biogas to within accepted levels. An alternative method is, adding ferrous chloride ($FeCl_2$) to the digestion tanks to inhibit hydrogen sulfide production.

Volatile siloxanes which are frequently found in household waste and wastewater can also contaminate the biogas. In digestion facilities accepting these materials, the low molecular weight siloxanes volatilize into biogas. When this gas is combusted in a gas engine, turbine or boiler, siloxanes are converted into silicon dioxide (SiO_2) which gets deposited internally in the machine, increasing wear and tear (Wheles and Pierece 2004). Practical and cost-effective technologies to remove siloxanes and other biogas contaminants are currently available (Tower et al 2006). In certain applications, *in situ* treatment can be used to increase the methane purity by reducing the carbon dioxide content (Richards 1994). In Switzerland, Germany and Sweden, the methane in the biogas is concentrated in order to use as a transportation fuel.

Electricity generation: Electricity can be generated for on-site or for pumping into local electric power grid. Internal combustion engines (ICs) and Gas turbines are the most commonly used for biogas-to-power generation projects.

ICs are stationary engines, similar to conventional automobile engines that can use medium quality gas to generate electricity. While they can range from 30 to 2000 kW, IC engines associated with landfill gas typically have several hundred kW capacities.

IC engines are a proven, reliable and cost-effective technology. Their flexibility, especially for small generating capacities, makes them the only electricity generating option. Some IC engines also produce significant NO_x emissions, although designs exist to reduce NO_x emissions.

Gas turbines can use medium quality gas to generate power of sale to nearby users or electricity supply companies, or for on-site use. Gas turbines typically require higher gas flows than IC engines in order to be economically attractive. Also, gas turbines have significant parasitic loads; when idle, gas turbines consume roughly the same amount of fuel as when generating power. Additionally, the gas must be compressed prior to use in the turbine.

Steam turbines can be used for power generation where extremely large gas flows are available.

Biogas does not contribute to atmospheric CO_2 concentrations because the gas is not released directly into the atmosphere.

Digestate: The solid residue of the original input material to the digesters that the microbes cannot use is referred to as 'digestate'. It also consists of the mineralised remains of the dead bacteria from within the digesters. Digestate can come in three forms; fibrous, liquor or a sludge-based combination of the two fractions. In two-stage systems the different forms of digestate come from different digestion tanks. In single stage digestion systems the two fractions will be combined and if necessary, separated by further processing.

The second by-product (acidogenic digestate) is a stable organic material comprised largely of lignin and cellulose, but also of a variety of mineral components in a matrix of dead bacterial cells. Some plastic may also be present. The material resembles domestic compost and can be used as compost or to make low grade building products such as fibreboard.

The third by-product is a liquid (methanogenic digestate) that is rich in nutrients and can be used as a fertiliser depending on the quality of the material being digested. In industrial waste, the levels of potentially toxic elements (PTEs) may be higher which needs to be considered when deciding a suitable end use for the material.

Wastewater: The final output from anaerobic digestion systems is water. This water originates both from the moisture content of the original waste that was treated, and water produced during the microbial reactions in the digestion systems. This water may be released from the dewatering of the digestate or may be implicitly separated from the digestate.

This wastewater will typically have high levels of biochemical oxygen demand (BOD) and chemical oxygen demand (COD) which indicate an ability to pollute. Some of this material is termed 'hard COD', i.e., it cannot be accessed by the anaerobic bacteria for conversion into biogas. If this effluent is released into watercourses it would negatively affect them by causing eutrophication. As such further treatment of the wastewater is often required. This treatment will typically be an oxidation stage where air is passed through the water in a sequencing batch reactors or reverse osmosis unit.

6.3.1.6 Valorga Process (Anaerobic digestion)

In Europe, the French company Valorga developed a process to treat mixed solid waste based on anaerobic digestion. Their initial facility was set up in Amiens France in 1987 to process 165 tons/day; expanded to 256 tons/day in 1995. The fuel gas generated in their facilities is used for process heat (warming the digesters consumes about 5% of the product gas) and the rest is sold for application to district heating and electrical generation. The detention time in the digester approximates 18 to 21 days or more. As an example of the digester volume-to-throughput ratio, in the Cadiz, Spain plant, 590 tons of MSW that is received would become 350 tons/day as prepared feed material, after mechanical and manual sorting, to the four 4,000 m^3 digestion vessels. Biogas is produced at a nominal rate of 4,650 sft^3 per short ton processed.

These facilities show a 100% pathogen kill and have not experienced odour problems since the process is contained in closed digesters. There are, however, a number of post-digestion steps involving dewatering of the digested mass, stabilization and drying of the compost and disposal of the non-compost reject streams. The off-gas from these steps can contain odour.

Buildings are maintained under draft and odour control is effected using a biofilter and scrubber. No net odour problems have been reported by the process developers.

The refined compost produced following digestion has been shown to be favorable to soil improvement so as to be marketable to agriculture. Heavy metal content is within acceptable limits. It achieves a pH of 8 and a carbon-to-nitrogen ratio of 18. There are problems, however, in the feeding of the residual combustible refuse fraction to either combustors or gasifiers. Specifically, the free moisture in the material from the digesters is undoubtedly high, probably in excess of 50%. This presents a problem in achieving combustion completeness to combustors and has a profound effect on the thermal efficiency of gasification systems.

Further, the product gas from digestion is, realistically, only useful as a boiler fuel (replacing natural gas or other fossil fuel) whereas the synthesis gas from gasification has broader use; therefore, there may little benefit associated with the digestion alternative. This technology has been operating successfully for many years in Europe, and a few are listed in Table 6.6.

Table 6.6 Installation of Valorga Process Plants
(Source: Niessen 2007)

Location	Start year	Capacity Mg/day	Capacity Tons/day	Waste Type
Amiens, France	1987	230	253	MSW
Tilburg, Netherlands	1993	140	157	Veg, Garden
Engelskirchen, Germany	1997	96	105	Food waste, Garden
Freiburg, Germany	1998	100	110	Food waste, Garden
Mons, Belgium	1998	160	175	Food waste, MSW
Geneva, Switzerland	1999	27	30	Food waste, Garden
Cadiz, Spain	1999	590	650	MSW
La Coruna, Spain	1999	500	550	MSW
Varennes-Jarcy, France	2000	390	430	Food waste, MSW
Hanover, Germany	2000	275	300	MSW
Bassano, Italy	2000	145	160	Food waste, MSW
Barceloan, Spain	2000	820	900	MSW
Calais, France	2000	74	81	Food waste, Garden
Shanghai, PRC	005	735	810	Food waste, MSW
Beijing, PRC	2005	290	320	Sorted MSW
Marseilles, France	2007	1095	1200	MSW
Zarragoza, Spain	2007	760	690	MSW
Tondela, Protugal	2007	82	90	Food waste, Garden
Madrid #1, Spain	2007	550	600	Sorted MSW
Madrid #2, Spain	2007	820	900	Sorted MSW

The maximum strength of the Valorga process relate to the expected good emissions characteristics and the high heat content of the product gas (500-600 Btu/cubic foot). The gas can be used directly in a gas turbine as part of a combined cycle generation system. Also, display of the basic features of the process in large plants in Europe confidently supports the view that the technology is transferable to commercial MSW processing. However, one must remember that European (mostly in Spain and Italy) and Chinese MSW tend to be high in food waste which is not a favoured feedstock to digestion processes.

There is, therefore, necessity to use separate collection of food waste to enhance both gas production and processing rate from Valorga.

6.3.1.7 Applications

Anaerobic digestion is particularly suited to wet organic material and is commonly used for effluent and sewage treatment. Almost any bio degradable waste materials such as *waste paper, grass clippings, food waste, sewage* and *animal waste* can be processed with anaerobic digestion. Woody wastes are largely unaffected by digestion as most anaerobes are unable to degrade lignin. Anaerobic digesters can also be fed with specially grown energy crops such as silage for dedicated biogas production.

In developed countries, the application of anaerobic digestion has increased among Solid waste disposal methods, as a process for reducing waste volumes and generating useful by-products.

Anaerobic digestion may either be used to process the source separated fraction of municipal waste, or alternatively combined with mechanical sorting systems, to process residual mixed municipal waste. These facilities are called mechanical biological treatment plants, described later in this book.

Utilising anaerobic digestion technologies can *help to reduce the emission of greenhouse gases* in a number of ways: (a) replacement of fossil fuels, (b) reducing methane emission from landfills, (c) replacing industrially-produced chemical fertilizers, (d) reducing vehicle movements, and (e) reducing electrical grid transportation losses.

Digestate (residue) can be used as a fertilizer providing vital nutrients to soils. The solid, fibrous component of digestate can be used as a soil conditioner. The sludge can be used as a substitute for chemical fertilizers which are not only carbon-intensive but require large amounts of energy to produce and transport. This solid digestate can be used to boost the organic content of soils. In countries, such as Spain, where there are many depleted soils, the markets for the digestate are high like the biogas.

In countries that collect household waste, the utilization of local anaerobic digestion facilities can help to reduce the amount of waste that requires transportation to centralized landfill sites or incineration facilities. If localized anaerobic digestion facilities are integrated into an electrical distribution network, they can help reduce the electrical losses that are associated with transmitting electricity over a national grid.

6.3.1.8 Environmental Pollution Control Measures for AD Plants

The main points of concern relating to Anaerobic Digestion plants include:

(a) Biogas emissions/ leakage posing environmental and fire hazards,

(b) Gaseous exhaust from the power generating units which must be duly cleaned to meet specified standards for air emissions,

(c) Disposal of large quantities of water and of liquid sludge which can pose potential water pollution problem. While the liquid sludge can be used as rich organic manure, either directly or after drying, its quality needs to be duly ascertained for particular application. In case of use for food crops it needs to ensure that it is not contaminated by heavy metals/ toxic substances beyond permissible levels.

6.3.1.9 Status in Developed countries

The power potential from sewage works is limited; in UK there are about 80 MW total of such generation, with potential to increase to 150 MW, which is insignificant compared to the average power demand in the UK of about 35,000 MW. The *scope for biogas generation from non-sewage waste biological matter* – energy crops, food waste, abattoir waste etc., is *much higher*, estimated to be capable of about 3,000 MW. Farm biogas plants using animal waste and energy crops are expected to contribute to reducing CO_2 emissions and support the grid, while providing farmers with additional income. The number of biomass methanisation units is rapidly increasing in the European Union (EU). The growth of such technologies should be facilitated by their relatively low price compared to other renewable energies.

The EU adopted a Biomass Action Plan in December 2005 to enhance the use of renewable energies. This action plan outlines measures to increase the development of biomass energy from wood, wastes, and agricultural crops, by creating market-based incentives to its use and removing barriers to the development of the market. The plan expects to double the biomass use by 2010. The largest biogas producing countries in Europe are Germany and the UK, other countries being way below in terms of biogas primary energy production.(1923 ktoe in Germany, 1696 ktoe in the UK, and 353 ktoe in the third ranked country, Italy). One ktoe is kiloton oil equivalent. In the UK, the biogas is directly recovered from landfills. According to Cardiff University Waste Research Station, anaerobic digestion has not taken off as a waste treatment and disposal option in the UK mainly due to the lack of market for the produced soil conditioner. In Germany, biogas production is mainly due to electricity production from small agricultural methanisation units operating in combined power and heat production. Six hundred such units were installed in 2005 and 800 in 2006 to reach a total 3500 by the end of 2006.

Biogas from sewage is also used to run a gas engine to produce electricity which can be used to power the sewage works. Some waste heat from the engine, generally enough to heat the digester to the required temperatures, is then used to heat the digester. The Anaerobic Lagoon at the Cal Poly, SLO Dairy in USA is shown in Fig.6.13. The biogas from this facility is used to fire a 25-kW power plant.

Fig. 6.13 Anaerobic lagoon and generators at the Cal Poly Dairy, USA 2003; 25 kW Power plant is fired using the biogas
(photo by Kjkolb, From Wikipedia, Free encyclopedia)

Centralised co-digestion of manure and suitable organic wastes is today a mature technology, economically sustainable and a cost efficient tool for reducing the emissions of green house gases and environmental improvement. The technology provides economic and environmental benefits by renewable electricity and heat production, improved manure management and increased waste recycling. It reduces the nutrient losses to water systems, the emissions of methane and nitrous oxide, the odours and flies nuisance from manure storage and application, and increases the veterinary safety by sanitation. The experience from Denmark proves that biogas from centralised co-digestion is a multifunctional concept, providing quantifiable environmental and economic benefits for agriculture, industry, energy and the overall society and could be an important tool in controlling GHG emissions from agriculture and the waste management. The details of the technology under utilization in Denmark are given in the Annexure.

6.3.1.10 Status in Developing countries

Biogas plants are appropriate to the technical abilities and economic capacity of farmers in developing countries. Home and farm-based anaerobic digestion systems offer the potential for low-cost energy for cooking and lighting (FOE 2004, Cardiff Univ. 2005). Biogas technology is progressive and extremely appropriate to the ecological and economic demands of the future. Anaerobic digestion facilities have been recognized by the UNDP as one of the most useful decentralized sources of energy supply (UNDP Report 1997). Government backed schemes for adaptation of small biogas plants for use in the household for cooking and lighting are launched on a large-scale both in China and India from 1975. A great deal of experience with biomethanation systems exists in India, but a large part of this is related to farm-scale biogas plants and industrial effluents. There is little experience in the treatment of solid organic waste, except sewage sludge and animal manure. Some details on Biogas plants used in India are given in the later pages. Presently, projects for anaerobic digestion in the developing world can gain financial support through the Clean Development Mechanism if they are able to show they provide reduced carbon emissions (Irrd.org). Biogas and compost production from organic waste fractions has been widely accepted in Africa as a best practice, and progress is being made in developing and implementing specific projects in various countries.

Biogas technology is well known in smaller towns in rural areas in China through government policies of comprehensive utilization of wastes. However, these biogas digesters use human and animal feces as the main feedstock. Some agricultural wastes may be added. The number of biogas digesters in rural towns is declining with the breakup of communes. Household-level digesters have not proved practical. The potential remains for anaerobic digestion of wastes, but organizational problems need to be overcome. The Indian subcontinent has wide experience with anaerobic digestion of cattle dung, and it was assumed that similar digesters could be adapted to ferment MSW, but producing animal-dung-like slurries from urban organic waste proved energy-intensive and the product was poor. A major problem with anaerobic digestion is that MSW used as feedstock tends to float. A number of design changes have to be achieved to produce small-scale digesters. Despite several advantages that the systems offer to farmers, a biogas plant never meets the owner's need for status and recognition, and biogas technology unfortunately has a poor image!

6.3.2 Mechanical Biological Treatment

A Mechanical Biological Treatment **(MBT)** system is a form of waste processing facility in which a sorting facility is combined with a form of biological treatment such as *composting* or anaerobic digestion. MBT plants are designed to process mixed household waste as well as commercial and industrial wastes. The terms 'mechanical biological treatment (MBT)' or 'mechanical biological pre-treatment (MBP)' relate to a group of solid waste treatment systems. Initially the technique was applied to pretreatment and hence its acronym is MBP (Ludwig *et al.,* 2003). The sorting component or materials recovery facility of the plant is either configured to recover the individual elements of the waste or produce a refuse-derived fuel that can be used for the power generation.

Fig. 6.14 Anaerobic digestion and air processing components of Lübeck mechanical biological treatment plant in Germany (From Wikipedia, Free encyclopedia)

Approach (different processors) to MBT is schematically shown in Fig. 6.15. Mechanical sorting: The 'mechanical' element is usually an automated mechanical sorting stage which either removes recyclable elements from a mixed waste stream (such as metals, plastics, glass and paper) or processes them.

It typically involves factory style conveyors, industrial magnets, eddy current separators, trommels, shredders and other tailor made systems. Manual sorting is also done. The mechanical element has a number of similarities to a materials recovery facility (MRF) (Sita 2004).

Some systems integrate a wet MRF to recover and wash the recyclable elements of the waste in a form that can be sent for recycling. MBT can alternatively process the waste to produce a high calorific fuel, RDF. RDF can be used, as mentioned earlier, in cement kilns or power plants and is generally made up from plastics and biodegradable organic waste. Systems which are configured to produce RDF include the Herhof and Ecodeco Processes. It is a common misconception that all MBT processes produce RDF. This is not the case and depends strictly on system configuration and suitable local markets for MBT outputs. Biological processing refers to anaerobic digestion, or composting, or biodrying. The biogas produced in anaerobic digestion can be used to generate electricity and heat. In composting, the organic component is treated with aerobic microorganisms. They break down the waste into carbon dioxide and compost. There is no green energy produced by systems employing only composting treatment for the biodegradable waste (see 'composting'). In the case of biodrying, the waste material undergoes a period of rapid heating through the action of aerobic microbes. During this partial composting stage the heat generated by the microbes result in rapid drying of the waste. These systems are often configured to produce a refuse-derived fuel where a dry, light material is advantageous for later transport combustion.

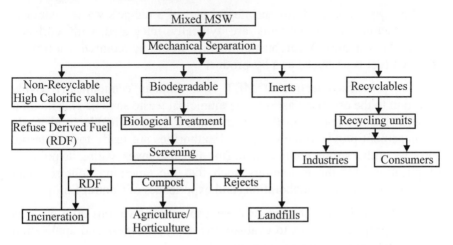

Fig. 6.15 Approach to MBT process (Source: ARRPET 2004)

Some systems incorporate both anaerobic digestion and composting. This may either take the form of a full anaerobic digestion phase, followed by the maturation (composting) of the digestate. Alternatively a partial anaerobic digestion phase can be induced on water that is percolated through the raw waste, dissolving the readily available sugars; the remaining material is sent to a windrow composting facility. By processing the biodegradable waste either by anaerobic digestion or by composting, MBT technologies help to reduce the contribution of greenhouse gases that cause global warming.

Municipal solid waste and Sewage sludge are the ones that are used in this system. The final products derived from this system are (i) recyclable materials such as metals, paper, plastics, glass etc., (ii) organic fertilizer (separate collection of organic waste), (iii) unusable materials prepared for their harmless final disposal (compaction > 1.3 t/m³), (iv) revenues from Carbon credits, and (v) additional revenues from high calorific by product, refuse derived fuel (RDF). Other advantages are: (i) rapid improvement of SWM and the finally deposited waste is inert and harmless; (ii) reduction of the waste volume to be deposited to at least a half (density > 1.3 t/m³), thereby increasing the lifetime of the landfill by at least twice as long as usually; (iii) utilization of the leachate in the process; (iv) no additional facilities for the collection and combustion of biogas required as there is no biogas; (v) daily covering not necessary; (vi) benefits to human health by decreased emission of pollutants; and (vi) aftercare not required for 3 to 5 years. The only disadvantage is the risk of climatic effects.

MBT systems can form an integral part of a region's waste treatment infrastructure. These systems are typically integrated with curbside collection schemes. A combustion facility would be required if a refuse-derived fuel is produced as a by-product.

The primary objective of MBT is to find an economically viable solution to the organic wastes in the municipal waste stream. This method would find application in the Asian context for sustainable solid waste management provided the concerned authorities are ready to initiate the process with a strong political will. It is aimed at investigating its possible applications for the technology such that the emerging countries can benefit from it with collaborations and experiences of Central Europe.

GTZ and German enterprises have implemented various pilot projects in their project countries to evaluate the appropriateness and application of MBWT. Three of the projects are:

1. Pilot project in São Sebastião, Brazil by FABER-AMBRA® process
2. Pilot project in Phitsanulok, Thailand by FABER-AMBRA® process
3. Scale-model MBWT trial in Al-Salamieh, Syria by GORE®

Various other projects have been supported for field-testing of the system in:

1. City of Atlacomulco, Mexico for composting; sorting of recyclables and management of a micro-enterprise; and treatment of waste inputs according to the MBWT process using the informal sector, and

2. Armenia, Columbia with a view to compile the experience gained and makes it available to interested parties across South America via the internet (www.foro-z.com). Detailed information at *http://www.gtz.de/mba/english/pilotprojekte.html*

6.3.3 Fermentation

This is a WTE technology which creates ethanol from biomass, using waste cellulosic or organic material. In the fermentation process, the sugar in the waste is changed to carbon dioxide and alcohol, as in the case of general process used to make wine. Normally fermentation occurs in the absence of air. A research group at Centurymarc has discovered that the cellulose materials can be processed to create a highly efficient clean burning fuel very similar to e-85 Ethanol. This process can convert one ton of waste into ethanol in 24 hours, about 7 times faster than the current methods of producing ethanol from corn. The process is not only faster and more efficient but is also considered to be much less harmful to the environment than ethanol derived from corn. According to Argonne National Laboratory (USA), cellulose based ethanol emits 80 percent less emissions into the atmosphere than standard gasoline, where as corn based ethanol emits only 20 to 30 percent less than gasoline. Also cellulose based ethanol does not require fossil fuels when being distilled, which further lowers the amount of greenhouse gas emissions. The reader may refer to literature for more details.

6.4 Chemical Processing: Esterification

Introduction: Biodiesel can be produced from waste vegetable oil by esterification. This chemical treatment is suitable for processing waste cooking oils into a renewable, biodegradable, eco-friendly and non-toxic fuel called biodiesel (Marchetti et al 2007a; Nie et al, 2006; Shibasaki-Kitakawa et al 2007; Kim et al 2004).

Biodiesel has gained prominence as a substitute for petroleum based diesel due to environmental benefits and depletion of oil sources. Biodiesel can be used with little or no modifications to existing diesel engines. In addition, biodiesel has a low emission profile (Biodiesel/Now 2006).

Biodiesel is not yet cost competitive with petroleum diesel (Marchetti et al., 2007b; Loreto et al 2005). To make the production more competitive, the production costs have to be lowered. The cost of raw materials – oil, alcohol, a catalyst – consists as much as 70-95 percent of the total production cost of biodiesel (Zhang et al, 2003). To reduce the costs, it is beneficial to use 'waste cooking oil' which is available at low cost rather than virgin oil. But waste cooking oil presents problems compared to virgin oil because waste oil contains dirt, charred food, and other material including water. However, the potential of waste cooking oil as a feedstock has been investigated (Wang et al 2007; Felizardo et al., 2006; Halim et al., 2009; Canakci 2007; Gui et al., 2008; Pahn and Pahn 2008; Haas and Foglia 2005).

Currently, most of biodiesel production processes use *the chemical* approach, which involves an acid catalyst, a base (alkali) catalyst or both. There are two major sources of biodiesel: (a) the oil (triglyceride) and (b) the free fatty acids (FFA) contained with oil. Both of them require alcohol to convert to biodiesel. The difference lies in the co-product of the reaction. For oil, biodiesel is produced with glycerol as a co-product in a *transesterification* reaction. For FFA, biodiesel is produced with water as a co-product in an *esterification* reaction. See (1) and (2) below.

Transesterification: Oil + Alcohol <-----→ Biodiesel + Glycerol   (1)

Esterification: FFAs + Alcohol <------→ Biodiesel + H_2O (2)

The most common method of biodiesel production is the transesterification of vegetable oils and animal fats. Transesterification is the process of exchanging the organic group of an ester with the organic group of an alcohol. These reactions are often catalysed by the addition of an acid or base. Enzymes catalyst is also utilized. i.e.,

alcohol + ester -----> different alcohol + different ester

Transesterification is influenced by the molar ratio of alcohol to triglycerides, the catalyst used, reaction temperature, reaction time, and the free fatty acids and water content of the cooking oil used (Ma, Hanna 1999).

Therefore, the three main catalysts to biodiesel production from waste oils are: (1) Base catalysed transesterification of the oil; (2) acid catalysed transesterification of the oil, and (3) enzyme catalyzed transesterification of oil. Comparatively, enzyme catalyzed transesterification consumes more reaction time, and also more costly. In the biofuel industry, however, the enzymatic process is theoretically known to be better than the chemical process because it can operate under mild temperature and pressure and co-produce glycerol with higher purity.

However, people have come to believe that enzymes are expensive and have short lifespans. In addition, an overwhelming amount of research papers overlook the negative effects of the second liquid phase that forms during the reaction which further contributes to misconceptions about the enzymatic process. This is why, to date, enzymes have rarely had commercial use in biodiesel production (Sunho Corp. 2010).

(a) *Acid catalyzed Reaction*: This type of transesterification reaction can be catalyzed by sulphuric or phosphoric or hydrochloric or organic sulfonic acids (Fukuda et al 2001). Since sulphuric acid is not sensitive to the free fatty acid content of the oil, typically sulphuric acid is preferred. i.e., a pretreatment is not necessary in the transesterification of waste cooking oil to biodiesel.

However, an acid catalyst is normally applied only for esterification. The main drawback is that due to corrosive nature of sulphuric acid, the cost of special materials to construct a biodiesel reactor will be high. Further, the reaction is too slow requiring increased reaction time and uses a much higher amount of alcohol to drive the reaction.

In addition, neutralization remains unavoidable for the purpose of removing the homogeneous acid catalyst. The acid catalyzed reaction is, therefore, not considered practical.

(b) *Alkali (Base) Catalyzed Reaction*: This type of reaction has received more attention (Wang et al 2007). In this reaction, either sodium or potassium hydroxide is normally used. The limitation of a base catalyst is that it can be applied only for *transesterification*. The main problem is that the alkali catalyzed reaction is sensitive to the content of free fatty acids (FFAs) in the waste oil; consequently, it is limited to feedstock with low water and FFA levels. When a liquid type alkali catalyst is used, the resulting soap formation due to presence of water not only consumes the catalyst but also makes the separation of glycerol and biodiesel difficult. It also involves wastewater disposal problems and results in low quality. Hence, a pre-treatment step is necessary to reduce free fatty acid content in the oil to less than 0.5%. This problem is more in waste oils with free fatty acid content more than 20%.

The other drawbacks are (a) the process is energy intensive and (b) recovery of glycerol is difficult because of formation of soap when the product was washed to remove alkaline catalyst (Wan Omar et al., 2009).

Therefore, a two-step process is followed. The first step is the pre-treatment *esterification* which is an acid catalyzed conversion of free fatty acids to esters using methanol. This reaction (shown below) would decrease the free fatty acid content of the oil to the desired level. This reaction, however, produces water which needs to be removed to prevent saponification.

$$\text{Free fatty acid + alcohol} \xrightarrow[\text{catalyst}]{\text{acid}} \text{ester + water}$$

The pre-treatment step has converted free fatty acids into esters. What is generally done to make the process cost-effective and time efficient is as follows (Canaki and Gerpen 2006): The acid catalyst and the methanol are added, allowed reacting and then settling. The methanol/ water mixture can then be removed. Again, add more acid catalyst and methanol, allow reacting and then settling. The methanol/ water mixture is then removed, and the transesterification step is performed. It is highly desirable to carry out this reaction in at least two stages as the addition of water to the transesterification reaction causes the reaction to slow down. It is essential to ensure that there is no water present to avoid gels and emulsions forming with biodiesel. Although the esterification reaction in pre-treatment step changed the FFAs in waste oils into corresponding Fatty Acid Methyl Ester (FAME) or known as biodiesel, un-converted FFAs and triglycerides still remained in the pre-treated oil. Consequently, transesterification reaction with alkali was performed to complete the reaction (second step).

The second step is transesterification of the pretreated product by using alkali catalyst. i.e, after the pretreatment, sodium hydroxide (catalyst) and methanol are added and the transesterification reaction begins. After several hours, biodiesel is formed along with glycerol. The next step is glycerol separation. A successful transesterification reaction is signified by the separation of the ester and glycerol layers. The heavier co-product, glycerol, settles out and may be sold as it is. Or it may be purified for use in other industries, e.g. the pharmaceutical, cosmetics, etc.

Then, methanol is distilled from biodiesel and glycerol phases. Finally, biodiesel is washed with warm water to remove any excess catalyst or soap from it. The biodiesel is then stored or used (Beth Knight et al 2006).

This two-step catalystic biodiesel production was also carried out using ferric sulphate as solid acid and calcium oxide (CaO) as solid 'base'. In this study, relationships between reaction temperature, reaction time, and molar ratio of methanol to oil, and the optimum conditions in the pre-treatment step are investigated. This study concludes that optimum condition for pre-treatment step is estimated to be 3 hours for reaction time, 60^0C for reaction temperature, and 7:1 for molar ratio of methanol to oil to produce maximum total FAME (biodiesel) yield of 81.3% (Wan Omar et al 2009). Several studies have been undertaken for the production of biodiesel from waste cooking oil (e.g, Prafulla Patil et al., 2010; Wang et al 2010; Saifuddin et al., 2009).

The process flow diagram proposed for the production of biodiesel from waste vegetable oil at Oregon State University is shown (Beth Knight et al 2006).

The cost effectiveness of esterification will depend on the feedstock being used, and the other relevant factors such as transportation distance, amount of oil present in the feedstock and others.

The economic analysis of all types of options has shown that the alkali catalyzed reaction minimizes capital cost and will provide a comparable biodiesel product to one produced from virgin vegetable oil (Beth Knight et al 2006). It is the most economical process.

Enzymatic approach: This approach is considered to be costly. But, from the technology point of view, the main advantage of the enzymatic approach is that the reaction can be performed in mild conditions, and can handle both transesterification and esterification simultaneously. Initially, applications of the enzymatic approach did not make use of an inert solvent. As such, the reaction time was deemed too long and the biocatalyst was eventually deactivated by glycerol or water. Only a batch operation was possible and the overall operating cost was determined to be high. Even then, product quality is unpredictable as the immobilized lipase deactivates after several runs. Sunho Corporation has developed 'Biodiesel's Enzymatic Transesterification Process (ET Process)' which makes use of an inert solvent that protects the lipase so that it can have a long lifespan. The reaction can be done at ambient temperature and pressure in a continuous, integrated process. Since there is no water washing involved after the reaction, biodiesel and glycerol can be recovered with high purity. The reaction time is also reduced to less than 30 minutes, hence allowing for better time and cost efficiency (Sunho Corp. 2010).

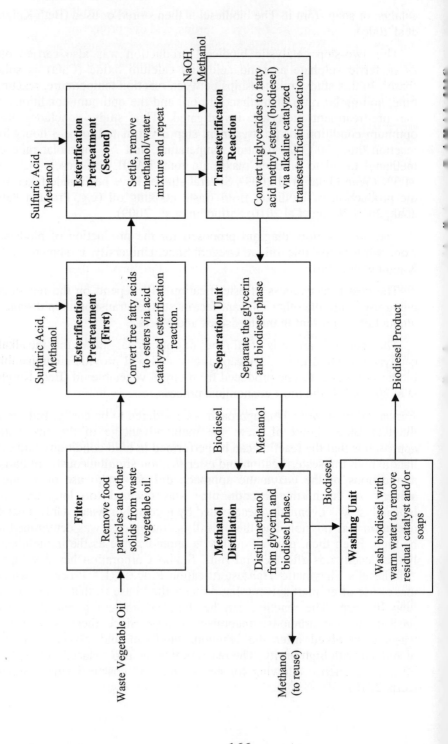

Safety aspects: The people working with biodiesel production have to take precautions and wear gloves, aprons, and eye protecting devices etc., for personal safety. Inhaling methanol vapours can cause blindness and even death. Sodium hydroxide is caustic and should be kept away from skin and clothes. Sulphuric acid is corrosive to skin and can cause severe chemical burns.

6.5 Recent Developments in WTE Technologies

Technology is moving fast in the WTE sector with a number of new approaches or renewed technologies. Some of the recent technology developments utilizing the principles explained already is presented (Wagner 2007):

EnerTech – SlurryCarb™ process: The EnerTech SlurryCarb™ process (www.enertech.com) is based on a pre-treatment of MSW in water slurry form to help the removal of recyclables. The slurry is then subjected to high pressure and temperature conditions and partial dewatering to turn it into a higher calorific value RDF suitable to gasification for combustion in a high-pressure steam boiler or to power a gas turbine. It is currently under demonstration in the US, and if succesful, this process, even though expensive, will have very low pollution levels and significantly higher thermal efficiency than mass burn technology.

EcoEnergy Oy – Wabio process: The *Wabio* process is bio-thermal waste treatment developed by EcoEnergy Oy, Espoo, Finland. Waste is pre-treated and divided into organic and combustion fractions. The organic fraction is degraded into biogas and compost matter. The RDF is burned in a specially designed fluidized bed unit. The temperature is kept below 900°C to avoid the formation of thermal NOx and of dangerous slagging compounds that could reduce the life of the boiler.

Centre Nationale de Recherche Scientifique (CNRS) – Valgora process: The *Valgora* process, developed in France and adopted by Babcock-Borsig Power, uses a similar approach as EcoEnergy. MSW is shredded and sorted mechanically to recover glass, metals, plastics, inerts such as sand and gravel, and remove sources of toxic compounds such as batteries. The remaining fractions are separated into dry RDF and fermentescibles; the RDF is directed to a rocking kiln for producing steam and base load power generation, and the fermentescibles are sent to a specially-designed, high solids (above 45% solids), high yield methane digester which is computer-controlled. The methane is used to produce peak load power. The organic residues are composted to produce a sterile high quality soil conditioner. A plant which has a processing capacity of 120,000 tons per year of fermentescibles could generate 31 GWh of power from the methane produced and 57,000 tons of soil conditioner.

Convertech Group – Convertech process: The *Convertech* technology is intended for the processing of biomass into valuable products, such as chemicals, reconstituted wood products like panel boards, heat and power. As such, it is not specifically designed to handle mixed waste. In the long run, in the field of waste management, its main application could be in the treatment of MSW to produce a dry, cleaner burning RDF.

Martin GmbH – SynCom process: The *SynCom* process, developed by Martin GmbH (www.martingmbh.de), involves oxygen enrichment of underfire air, recirculation of flue gas and a combustion control system using infrared thermography of the waste layer on the grate. At the demonstration plant in Coburg (Germany), operational reliability and plant availability using SynCom process could be proven under real disposal conditions with a waste throughput of 7 tons per hour. Oxygen enrichment of the underfire air promotes the destruction of pollutants due to the high oxygen partial pressures and temperatures. This results in very low residual amounts of organic combustion by-products in the bottom ash and flue gas from the *SynCom* unit.

Demonstration of a typical solid and liquid waste management (EcoSolutionsManila 2009):

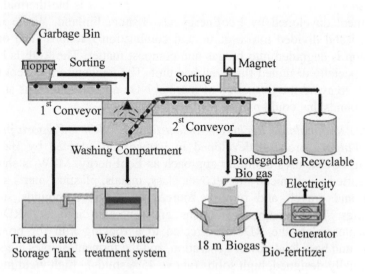

Fig. 6.16 A typical Solid and Liquid Waste management system (EcoSolutionsManila)

A typical Lagon and Vista group Solid and Liquid Waste management Technology Demonstration of EcoSolutions Manila is schematically presented. Fig.6.16 is self explanatory. The system is conveyor-linked and facilitates manual and automated sorting of waste. It also processes biodegradable wastes through a biodigester and treats the wastewater for continued water reuse within the system. Although the technology claims that the operations could be carried out 'without the use of dumping areas', it is clear that since the system lacks any provision for recycling residual wastes, final disposal in a landfill would still be required.

Although the whole system attempts to pursue an integrated approach to waste management, it tends to complicate (in a very costly way) the simple task of waste segregation which in fact should be carried out in the household level. It would be much better leaving the recovery of recyclables at the household level and managing only biodegradable wastes by having the biodigester, thus saving costs of conveyer belt system.

Another serious concern is the consumption of more time for the process to complete. The use of conveyors could be optimized and rendered cost-efficient only in large scale applications where the garbage is transported to longer distances, allowing more labourers to sort the waste at a faster rate, thereby increasing the system's overall processing capacity.

6.6 Planning and Execution of WTE Technologies

To plan a Waste to Energy facility and to select most appropriate, techno-economically viable technology, several important factors need to be considered.

1. *Cost of Collection and Transportation of Wastes:*

   Sufficient consideration should be given to the costs involved in the collection, segregation and transportation of waste. In the MSWM, collection and transportation costs are often highest, as high as 70%. This factor may rule out consideration of certain technologies like Sanitary Landfilling, if situated at faraway sites.

2. *Level of Treatment:*

   The waste quantity available/ to be processed is another major factor. Large scale treatment would be favourable where large waste quantities are discharged in limited area. But small scale treatment may be more suitable for low discharge density of wastes because they can be operated easily and quickly. However, collection and transportation costs in this case (involving wide area) are bound to be higher than in the latter case involving a limited area, and a trade-off will be necessary.

3. *Local Conditions/ Existing Waste Management Practices*

The viability of any WTE Project critically depends upon the availability of the required quantities and quality of the waste. The waste management practices generally vary with the local socioeconomic and physical conditions, rate of waste generation, and composition of waste. The last two factors also determine the potential for energy recovery within the over all structure of the SWM system. Therefore, there may be a need to improve the existing waste management practices/ local conditions to suit the selected technology in order to maximize energy recovery.

For example, some sources of wastes in an urban area have a very high percentage of organic matter and hence a high energy recovery potential. It is to be ensured that such wastes are collected and transported directly to the energy recovery facility and not allowed to get mixed up with other waste streams.

4. *Characteristics of the Waste:*

Proper evaluation of the fraction of bio-degradable/ combustible constituents/ moisture content of the waste, and its chemical composition is essential for selecting the most appropriate WTE technology. For e.g., Wastes from vegetable/ fruit yards and markets, agricultural and food processing units etc., contain high concentration of bio-degradable matter and are suitable for energy recovery through anaerobic de-composition. Solid wastes from offices, timber shops etc., having a high fraction of paper and wood products will be suitable for incineration.

The wastes in urban areas in India are characterized, generally, by low percentage of combustibles and high percentage of inorganic/inerts and moisture and are not very suited for incineration. The waste is generally rich in bio-degradable matter and moisture content and can be suitably treated in Sanitary Landfills or Anaerobic Digesters for energy recovery.

The Incineration/Gasification/Pyrolysis options can be usefully utilised where waste containing high percentage of combustibles and low percentage of inorganic/inerts and moisture, is either available or can be ensured.

Seasonal fluctuations in wastes quantity and quality must be considered because any imbalance between the availability of requisite quantity and quality of wastes and the energy demand/ utilization pattern may badly affect the project's viability.

5. *Treatment/Disposal of Residues*

Treating and disposal of the final residues/ effluents and the utility of the same should be planned in advance. For e.g., in anaerobic digestion, about 70% of the input is left as sediment (digested sludge); after being stabilized through aerobic treatment, it can be used as a good fertilizer. Similarly, the fly ash and incinerator bottom ash.

Secondly, the possibility of Toxic and Hazardous wastes being present in the MSW should be carefully examined and duly taken into consideration during their treatment/ processing and in the design of the WTE plants.

Plastic wastes may account for 1-10% of the total MSW. They are highly resistant to bio-degradation, and require special attention in waste management. Plastics have a high heating value making them very suitable for incineration. However, PVC when burnt, under certain conditions, may produce dioxin and acid gas, which calls for adequate safety measures as already discussed.

6. *Marketing Energy produced*

Effective marketability of end products (thermal energy/ power/ fuel oil/gas/pellets) depending on the WTE technology chosen will be a crucial factor determining the project's economic viability. In projects where electric power is generated, the availability of utility grid close to plant site would be necessary to enable wheeling of the generated power.

7. *Economics of the Project:*

The capital and recurring costs have to be assessed. The land area requirements, the auxiliary power/water requirements, the required infrastructure, and manpower with adequate expertise and skill for smooth operation and maintenance have to be analysed and the costs worked out.

8. *Environmental Impact*

The basic objective should be to promote environmentally sound waste disposal and treatment technologies, wherein energy recovery is an additional benefit. A solution in waste disposal should not lead to air or water pollution.

The ideal technology is generally considered the one which, per unit volume of the waste treated, (i) requires minimum space, (ii) requires the least initial capital investment, (iii) generates the minimum rejects requiring least treatment for further disposal or final usage on discharge, (iv) demands least O and M efforts in terms of both recurring expenditure and manpower, (v) has the best impact on minimizing environmental pollution, and (vi) recovers the maximum net energy. However, in actual situation, a trade-off between these aspects would have to be made, and the technology should be chosen based on techno-economic viability at the specific site subject to the local conditions and the available physical and financial resources.

Feasibility Studies: The Feasibility Studies are most essential for ascertaining the techno-economic viability of different waste treatment options. These studies should cover the following aspects:

1. *Quantity of Municipal Solid Waste Generated per day*: Per capita and total generation, Zone-wise quantity, Number of collection points along with quantity of waste available at each point has to be known clearly.

2. *Current Mechanism for Collection/ Transportation:* Existing mode of collection, Details of collection and dumping points, and waste quantities collected/ dumped per day at each point, and Site maps showing the location of collection and disposal sites need to be known.

3. *Physical and Chemical Characteristics of Collected Waste:* Data on Size of different constituents, density, moisture content, calorific value, ultimate/ proximate analysis, percentage of volatile solids and fixed carbon, etc., Sampling of waste over minimum period of 3 consequtive seasons, and Sampling procedure to be as per BIS norms are to be collected.

4. *Present Mode of Disposal:* Whether the waste is disposed by Burning/ composting/ other methods and the respective costs involved are required.

5. *Provisions in the Existing System:* In the existing waste management system, whether provision for Segregation of inert material, Recycling, Scientific disposal/ energy recovery *and* Revenue generation exist or not.

6. *Private public Partnership, if any*: Arrangements of Concerned Municipality with Private Parties regarding MSWM activities such as Waste Collection/ Disposal, if any, need to be known in detail.

7. *Details about the Proposed Scheme of Energy Recovery:* Details regarding Suitability of site with details, sizing of plant capacity, capacity of estimated waste processing/ treatment, estimated energy recovery potential/ other by-products, assessment of alternative options/ technology selection; quantity and quality of final rejects to be disposed off and their disposal Method.

8. Environmental Impact Assessment Analysis of the Selected WTE technology, the Energy End-Use and Revenue Generation

9. *Cost Estimates:* Capital cost, OandM costs including manpower, Revenue, Cost benefit analysis, etc.

6.7 Application of Important Industrial Wastes

Some industrial wastes have useful applications which are listed in the Table 6.7. Management of a few residues are described under 'Incineration' and 'Anerobic Digestion' in the earlier pages.

Table 6.7 Industrial wastes and their applications

S.No.	Waste	Areas of application
1	Fly ash	• Cement
		• Raw material in Ordinary Portland Cement (OPC) manufacture
		• Manufacture of oil well cement
		• Making sintered fly ash light-weight aggregates
		• Cement / silicate bonded fly ash/clay binding bricks and insulating bricks
		• Cellular concrete bricks and blocks, lime and cement fly ash concrete
		• Precast fly ash concrete building units
		• Structural fill for roads, construction on sites, land reclamation, etc.
		• As filler in mines, in bituminous concrete
		• As plasticiser
		• As water reducer in concrete and sulphate resisting concrete

Table Contd....

S.No.	Waste	Areas of application
2	Blast Furnace Slags	• Manufacture of slag cement, super sulphated cement, metallurgical cement • Non-portland cement • Making expansive cement, oil well, coloured cement and high early-strength cement • In refractory and in ceramic as sital • As a structural fill (air-cooled slag) • As aggregate in concrete
3	Ferro-alloy and other metallurgical slags	• As structural fill • In making pozzolona metallurgical cement
4	By product gypsum	• In making of gypsum plaster, plaster boards and slotted tiles • As set controller in the manufacture of portland cement • In the manufacture of expensive or non-shrinking cement, super sulphated and anhydrite cement • As mineraliser • Simultaneous manufacture of cement and sulphuric acid
5	Lime sludge (phos-phochalk paper and sugar sludges)	• As a sweetener for lime in cement manufacture • Manufacture of lime pozzolana bricks / binders • For recycling in parent industry • Manufacture of building lime • Manufacture of masonry cement
6	Chromium sludge	• As a raw material component in cement manufacture • Manufacture of coloured cement as a chromium – bearing material
7	Red mud	• As a corrective material • As a binder • Making construction blocks • As a cellular concrete additive· Coloured composition for concrete • Making heavy clay products and red mud bricks • In the formation of aggregate • In making floor and all tiles • Red mud polymer door
8	Pulp and paper	• Lignin

(Source: Manual on Municipal Solid Waste Management, CPHEEO, New Delhi)

CHAPTER 7

Landfilling

7.1 Introduction

Landfills are the ultimate storage area of a city's MSW after all other MSWM options have been exercised. In many cases, especially in developing countries, the landfill is the only MSWM option available and practiced. The safe and effective operation of landfills depends on sound planning, administration, and management of the total MSWM system. Landfills are one of the ways of treating solid waste by burying after separating recyclable materials from the collected waste. Landfills range in nature, from 'uncontrolled open dumps' to 'controlled open dumps' to 'sanitary landfills'. Uncontrolled open dumps are primitive and not a sound practice, but 'controlled dumps' and 'sanitary landfills' can provide effective disposal of a city's MSW in accordance with appropriate local health and environmental guidelines/ standards. Fig.7.1 shows the disposal methods of municipal solid waste practiced before 2000 in some South Asian countries (Visvanathan and Glowe 2006). It is clear that 'open dumping' has been the predominant practice for long in these countries, and controlled landfills is a recent concept. Same is the case in most of the developing countries, especially poor economies, in Africa and Latin America. These are further discussed under 'Landfills in Developing countries'.

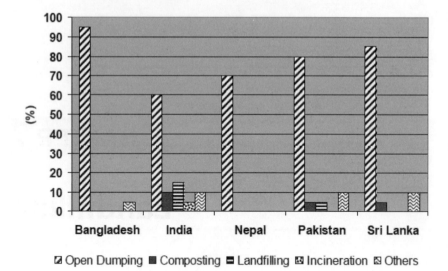

Fig. 7.1 Disposal methods in select South Asian countries
(Source: Visvanathan and Glowe 2006)

Landfill is a carefully designed structure built into or on top of the ground in which trash is dumped and isolated from the surrounding environment (ground water, air, rain) using a bottom liner and a covering of soil on top.

A Sanitary landfill uses a clay liner to isolate the trash from the environment, and a MSW landfill uses a synthetic liner. Sanitary landfills involve well-designed engineering methods to protect the environment from contamination by solid or liquid wastes. The main requirement in designing a sanitary landfill is the availability of vacant land that is accessible to the community and has the capacity to handle waste material for several years. The location must also be acceptable to the local community and in addition, soil must be available to cover the landfill.

Historically, landfills were built in a particular location more for convenience of access than for any environmental or geological reasons. Currently more care is taken in determining the location of new landfills to avoid pollution impact on the nearby residences.

Unlined unsanitary landfills and open dumps allow the precipitation (rain) to mix with degradable organic matter from MSW to form leachate (a liquid with contaminants) which percolates into the soil. The leachate eventually contaminates the surface and groundwater. For instance, in India, leachates of around 14.9×106 m^3/ year may be generated from degradable organic matters from open dumping/unlined landfills (Kumar et al., 2001).

7.2 Environmental Impact Study

Before construction of a landfill, an environmental impact study is inevitably undertaken on the proposed site to ascertain that

(a) sufficient land is available to build the landfill, and for other support areas/ activities such as runoff collection ponds, leachate collection ponds, drop-off stations, areas for borrowing soil and buffer areas,

(b) the sites are not located on faulted or highly permeable rock, but on sites with a water-tight or less-permeable foundation. Mines or quarries should be avoided because these structures frequently contact the ground-water supply,

(c) rivers, lakes, floodplains, wetlands, and ground-water recharge zones are avoided so that any potential leakage from the landfill does not enter the ground-water and that the excess water from the landfill is not allowed to flow into surrounding areas,

(d) the site has no historical or archeological value, and

(e) the landfills are located far away from airports to totally avoid the birds attracted by landfills, interfering with the movement of aircraft.

The environmental impact study must be scrupulously undertaken.

Care must also be taken that the initial location of a landfill will reduce the necessity for future clean-up and site rehabilitation. Due to these conditions and other factors, it is becoming increasingly difficult to find suitable locations for new landfills in many places. Easily accessible open space is becoming scarce especially in big towns/cities, and many residential colonies or communities are unwilling to accept the construction of a landfill within their boundaries.

7.3 Landfill Construction

The three basic procedures that are accepted a in sanitary landfill are: spreading the solid waste materials in layers; compacting the wastes as much as possible; and covering the material with dirt at the end of each day. This methodology reduces the breeding of rats and insects and the threat of spontaneous fires at the landfill, prevents uncontrolled settling of the materials, and ensures efficient use of the available land. Although this method does help control some of the pollution generated by the landfill, the fill dirt also occupies up to 20% of the landfill space, reducing its waste-holding capacity. Another important consideration for landfill design is the use of the site after it is filled. Some sites have become parks, housing projects or sites for agriculture. In the context of limited natural and financial resources available to them, municipalities have to plan the construction of landfills carefully to avoid some of the later costs of clean-up. The trash that is buried in the landfill is isolated from groundwater and air. It is kept dry to prevent decomposition as much as in the case of compost pile.

The critical elements in a secure landfill are: Bottom liner, Cells, Storm water drainage system, leachate collection system, and Cover.

(a) *Bottom liner system*: The bottom liner prevents the trash and subsequent leachate from coming in contact with the outside soil, particularly the ground water. The liner is usually clay or plastic or composite type. Natural clay is often fractured and cracked, and certain organic chemicals in the waste can degrade clay over time. Some type of durable, puncture-resistant synthetic plastic such as polyethylene, high-density polyethylene (HDPE), or polyvinylchloride (PVC) is used. But a few household chemicals such as moth balls degrade high-density polyethylene and soften or make it brittle and crack. Other chemicals such as vinegar, shoe polish, margarine etc., can cause to develop stress cracks also. Generally, a composite liner consisting of plastic liner and compacted clay soil is used. It is usually about 100 mils thick. If plastic liner is used, it may be surrounded on either side by a fabric mat (geotextile mat) that will prevent the plastic liner from tearing or puncturing from the nearby rock and gravel layers. For the ultimate safe disposal of MSW, the use of geosynthetic clay liners (GCL) as barrier to prevent leachate percolation both on the open-dumps and landfill sites should be adopted (Rachel et al., 2009). Studies, however, indicate that certain amount of permeation should be expected; a 10-acre landfill will have a leak rate between 0.2 and 10 gallons per day.

(b) *Cells*: The most precious component in a landfill is air space. The amount of space is directly related to the capacity and usable life of the landfill. If the air space is increased, the usable life of the landfill will be increased. To achieve this, the garbage is compacted into areas, called cells that contain only one day's garbage. Trench' and 'area' methods, and combination of both, are used in the operation of landfills. Both methods operate on the principle of a 'cell'. The trench method is appropriate in areas where there is relatively little waste, low ground water, and the soil is over 6 ft (1.8 m) deep. The area method is usually used to dispose of large amounts of solid waste.

In the trench method, a channel with a typical depth of 15 ft (4.6 m) is dug, and the excavated soil is saved for later use as a cover over the waste. Grading in the trench method is so arranged that the rain water is drained-off. The other consideration is the type of subsurface soil that exists under the topsoil. Clay is a good source of soil because it is nonporous. In the area method, the solid wastes and cover materials are compacted on top of the ground. This method can be used on flat ground, in abandoned strip mines, gullies, ravines, valleys, or any other suitable land. This method is useful when it is not possible to create a landfill below ground.

A combination method is called the 'progressive slope or ramp' method, where the depositing, covering, and compacting are performed on a slope. The covering soil is excavated in front of the daily cell. If there is no cover material at the site, it is brought from outside.

Fig. 7.2 A landfill compaction vehicle (source:Wikipedia)

(c) ***Storm water drainage system***: It collects rain water that falls on the landfill. It is essential to keep rain water out of landfill and keep it as dry as possible to reduce the amount of leachate production. To exclude liquids from the solid waste, the waste must be tested for liquids before entering the landfill. This is done by passing samples of the waste through standard paint filters. If no liquid comes through the sample after 10 minutes, then the waste is accepted into the landfill. Plastic drainage pipes and storm liners collect water from areas of the landfill and channel it to drainage ditches situated around the landfill's base. The ditches are either concrete or gravel-lined and carry water to collection ponds to the side of the landfill. In the collection ponds, suspended soil particles are allowed to settle and the water is tested for leachate chemicals. After the water has passed tests, it is pumped or allowed to flow off-site.

(d) ***Leachate collection system***: There is no perfect system to keep out water from getting into the landfill. The water percolates through the cells and soil in the landfill and as the water percolates through the garbage, it picks up contaminants such as organic and inorganic chemicals, metals, biological waste products of decomposition. This liquid called leachate is typically acidic. Perforated pipes are run throughout the landfill to collect and drain leachate into a leachate pipe, which carries the liquid to a leachate collection pond. Leachate is either pumped or flow by gravity to the collection pond. The leachate in the pond is tested for acceptable levels of various chemicals (biological and chemical oxygen demands, organic chemicals, pH, calcium, magnesium, iron, sulfate and chloride) and allowed to settle. Then it is treated at a wastewater treatment plant. The solids removed from the leachate during this step are returned to the landfill, or are sent to some other landfill. If leachate collection pipes clog up and leachate remains in the landfill, fluids can build up in the bathtub. The resulting liquid pressure may force the waste out of the bottom of the landfill when the bottom liner fails. Leachate collection systems can clog up in less than a decade due to silt or mud, or due to growth of microorganisms in the pipes, or due to a chemical reaction leading to the precipitation of minerals in the pipe. In course of time, the pipes may become weakened by chemical attack (solvents, acids, oxidizing agents, or corrosion) and get crushed by the weight of garbage piled on them.

(e) *Covering or cap*: It helps to seal off the top of the landfill to keep water out (to prevent leachate formation). It generally consists of several sloped layers of clay or membrane liner to prevent rain from intruding, overlain by a very permeable layer of sandy or rough soil to promote rain runoff, and by topsoil in which vegetation can start off to stabilize the underlying layers of the cover. This covering also seals the compacted garbage from the air and prevents pests such as birds, rats, mice, flying insects, etc., getting into the garbage. This soil is quite thick and takes up quite a bit of space. Many landfills are therefore experimenting with tarps or spray coverings of paper or cement or paper emulsions. These emulsions can effectively cover the garbage, but take up only a quarter of an inch instead of six inches! The vegetation generally consists of grass. Shrubs or plants with deep penetrating roots are not planted to avoid plant roots contact the underlying garbage and allow leachate out of the landfill.

Covers are often vulnerable to attack from various factors:

1. erosion by natural weather events (rain, hail, snow, freeze-thaw cycles, and wind);

2. vegetation, such as shrubs and trees that continually compete with grasses for available space, sending down roots that will try to penetrate the cover;

3. burrowing or soil-dwelling mammals (woodchucks, mice, moles, voles), reptiles (snakes, tortoises), insects (ants, beetles), and worms posing constant threat to the life of the cover;

4. sunlight (if a portion of the cover is punched for some reason) causing dryness of clay permitting cracks to develop or destroy membrane liners through the action of ultraviolet radiation;

5. an uneven collapse of the cap caused by settling of wastes or organic decay of wastes, or by loss of liquids from landfilled drums resulting in cracks in clay or tears in membrane liners, or resulting in ponding on the surface which can make a clay cap mushy or can subject the cap to freeze-thaw pressures;

6. rubber tires, which "float" upward in a landfill; and

7. human activities of several kinds.

When the leachate seeps through weak point in the covering and come out on to the surface, it appears black and bubbly, slowly staining the ground red. Leachate seepages are quickly repaired by excavating the area around the seepage and filling it with well-compacted soil so that the flow of leachate turns back into the landfill. A schematic diagram (Fig. 7.3) of a typical municipal solid waste landfill is shown (Ref: UNEP: Sound Practices – Landfills: available at http://www.unep. or.jp/ietc/estdir/pub/msw/sp/sp6/sp6_4.asp)

Typical schematic of a state-of-the-art landfill

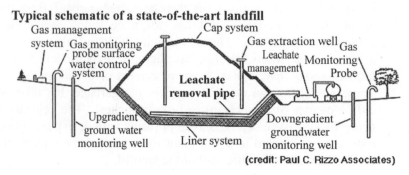

(credit: Paul C. Rizzo Associates)

Fig. 7.3 Typical schematic of a landfill

7.4 Decomposition in the Landfill

In a landfill, decomposition occurs in three stages.

The first one is an aerobic decomposition. The solid wastes that are biodegradable react with the oxygen in the landfill and begin to form carbon dioxide and water. Temperature during this stage of decomposition rises by about 30°F (16.7°C) over the surrounding air. Weak acid forms within the water and some of the minerals are then dissolved.

The next stage is anaerobic: microorganisms in the absence of oxygen break down the wastes into hydrogen, ammonia, carbon dioxide, and inorganic acids.

In the third stage of decomposition, methane gas is produced. Sufficient amounts of water and warm temperatures have to be present in the landfill for the microorganisms to form the gas. The gas produced during this stage will be carbon dioxide and methane in equal parts. *This gas is a usable energy source.*

Operational aspects: Landfills may appear simple; but they need to be operated carefully. Specific aspects such as where to start filling, wind direction, the type of equipment used, method of filling, roadways to and within the landfill, the angle of slope of each daily cell, complete isolation of the waste from groundwater, the handling of equipment at the landfill site have to be planned properly. The soil that is used as a daily cover which is usually 6 in (15.2 cm) thick, an intermediate cover of 1 ft (30.5 cm), and a final cover of 2 ft (61 cm) are important factors. The compacting of solid waste and soil is also vital for occurance of proper biological processes of decomposition.

Shredding of solid wastes helps to save space at landfills. Shredding also helps to perform compacting of waste more quickly, and to a greater density of compacting of materials extending the life of the landfill. Shredding proves to be advantageous in reducing the extent of cover and the danger of spontaneous fire. Landfills using shredded materials produce more organic decomposition than those disposing of unshredded solid wastes. Another method is baling of wastes. The advantages of baling are: an increase in landfill life due to an increase in waste density, and decrease in hauling times, litter, dust, odour, fires, traffic, noise, earth moving, and land settling. Less heavy equipment is needed for the cover operation and the amount of time it takes for the land to stabilize is reduced. Using biodegradable materials also helps save space in landfills because microorganisms can break down these materials more quickly. Garbage bags made of biodegradable materials are of particular use because microorganisms cause holes to form in the bags, enabling the material inside to break down more quickly. When the secure landfill reaches capacity, it is capped by a cover of clay, plastic, and soil, much like the bottom layers. Vegetation is planted to stabilize the surface and make the site nice-looking and visual.

Groundwater monitoring stations are installed around a landfill. These are pipes that are sunk into the groundwater; so water can be sampled and tested for the presence of leachate chemicals. The temperature of the ground water is measured because an increase in ground water temperature could indicate that leachate is seeping into the groundwater. The groundwater becoming acidic (as measured by pH) is also an indication of the seeping of leachate.

7.5 Benefits

The closed landfills have been used for different purposes. They include industrial parks, airport runways, recreational parks, ski slopes, ball fields, golf courses, playgrounds, and many others. When the bearing capacity of the landfill surface is found to be adequate, buildings can also be erected. Presently, the landfills are not considered as 'garbage dumps'; scientific methods are developed to engineer the establishment, maintain, close, and re-use of the area for benefit of the community.

7.6 Recovery and uses of Landfill Gas

Landfill gas containing about 45-55% methane can be recovered through a network of gas collection pipes and utilised as a source of energy. Typically, production of landfill gas starts within a few months after disposal of the wastes and generally lasts for about ten years or even more depending upon mainly the composition of wastes and availability/distribution of moisture. The MSW generated in major Indian cities is rich in organic matter and has the potential to generate about 15-25 l/kg of gas per year over its operative period whereas in full size sanitary Landfills in other countries, the production of gas ranges from 5 to 40 litre/ kilogram.

The proportion of various constituent gases changes with time since the onset of decomposition. The gas tends to escape through the cracks and crevices in the deposited material unless suitable outlet is provided. It also moves by diffusion (concentration gradient) and convection (pressure gradient) mechanisms. Such lateral migration poses danger to adjoining structures and vegetation.

Passive or active systems are used for controlling the production of methane gas. In a passive system, which is relatively inexpensive, the gas is vented into the atmosphere naturally, and may include venting trenches, cutoff walls, or gas vents to direct the gas. An active system employs a mechanical method to remove the methane gas and includes recovery wells, gas collection lines, and a gas burner. Both active and passive systems have monitoring devices to prevent explosions or fires.

The technical feasibility of recovering methane gas depends on several factors, the most important being the composition of the MSW. The production of methane gas depends *on a relatively high percentage of organic MSW as well as proper nutrients, bacteria, pH, and high moisture content.* The size of the landfill must be *large enough and contain enough MSW to produce economically recoverable quantities of methane.* Generally, landfills having capacity of at least one million tons should produce enough methane to support recovery operations. The age of the landfill is also important because it can take anywhere from several months to a few years after the disposal of MSW before sufficient methane is produced.

Early methane production can be enhanced by using uncompacted waste as the first layer of a landfill, thus allowing it to compost more quickly. The engineering aspects of a landfill may also increase the quantity of methane gas that can be recovered.

Landfill liners help keep methane from escaping from the landfill and help maintain the anaerobic conditions necessary for methane production. Similarly, a daily cover that keeps methane from escaping and also avoiding the introduction of air into the landfill can increase the rate of methane production. However, landfills that do not have engineered liners or covers, and landfills sited in porous soils can still produce significant quantities of methane.

In practice, not all landfill gas generated in the landfill can be collected; some of it will escape through the cover of even the most tightly constructed and collection system. Newer systems may be more efficient than the average system in operation. A reasonable assumption for the gas collection efficiency for a properly planned gas collection system is 70-85%.

Landfill gas has a calorific value of around 4500 Kilo calories per m^3. It can be used as a good source of energy, either for direct thermal applications or for power generation. There are three primary approaches to using the landfill gas as in the case of biogas: (a) direct use of the gas locally (either on-site or nearby); (b) generation of electricity and distribution through the power grid; and (c) injection into a gas distribution grid, if available.

7.7 Associated Activities

In developed countries, recycling centers where residents can drop off recyclable materials (aluminum cans, glass bottles, newspapers, blend paper, corrugated cardboard) are organised, and it is mandatory for the residents to use these centres for dropping the trash. This helps to reduce the amount of material in the landfill. The materials that can be recycled are banned from sending to landfills by law. For using the site, customers are charged tipping fees which is used to meet the operating costs of the landfills. Along the site, there are drop-off stations for materials that are not wanted or legally banned by the landfill. A multi-material drop-off station is used for tires, motor oil, lead-acid batteries and drywall. Some of these materials can be recycled and used.

Some more materials are also banned from disposal in solid waste landfills, that include common household items such as paints, cleaners/chemicals, motor oil, batteries, and pesticides. There is a household hazardous waste drop-off station for these chemicals. These chemicals are disposed by private companies. Some paints can be recycled and some organic chemicals can be burned in incinerators or power plants.

Other structures alongside the landfill are the 'borrowed area' that supplies the soil for the landfill, the runoff collection pond, leachate collection ponds, and methane station. Landfills appear as complicated structures when these associated facilities are included; but when properly designed and managed, serve an important purpose. In the coming years, new technologies like *bioreactors* are likely to become popular to speed the breakdown of garbage in landfills and produce more methane.

Modern landfills are well-engineered facilities that are located, designed, operated, and monitored in accordance with the regulations.

7.8 Health and Environmental Impacts

Landfills are associated with a wide range of health and social effects. Health and social impacts include odour nuisance; ozone formation (from reaction of NOx and non methane organic compounds with sunlight) that can cause pulmonary and central nervous system damage; fire and explosion hazards from build-up of methane; an increase in the number of pests, birds, rodents and insects which act as disease vectors; and ground and air pollution from leachate and landfill gases (Daskalopoulos et al. 1998, El-Fadel et al., 1997, EPA 1995a, Neal and Schubel 1987). Water contamination by leachate can transmit bacteria and diseases. Typhoid fever is a common problem for the people of developing countries.

There are also many environmental impacts of landfills. Ozone formation can cause decreases in crop yield and plant growth rate. Methane and carbon dioxide are greenhouse gases that contribute to global warming. Globally1.4 billion tonnes of MSW that is landfilled generate an estimated 62 million tonnes of methane, and less than 10% of it is captured presently (WTERT Brochure 2006). Methane is twenty times more effective at trapping heat than carbon dioxide, and more persistent in the environment (EPA 1995a, Jayarama Reddy 2011). Leachate from the landfill can enter ground water systems, leading to increases in nutrient levels that cause eutrophication (El-Fadel et al., 1997). Further, bioaccumulation of toxins and heavy metals can occur.

Operation of landfills has to be in tune with the concept of sustainable development; that is, pollution control problems need to be eliminated or minimised for the present and future generations. This can be achieved by taking the following actions: (a) sending only inert wastes to landfills, (b) carrying out pre-treatment of the mixed waste to a quality which is least harmful, and (c) managing the bioreactive waste material (biological municipal wastes) in such a way that the landfill degrades rapidly to come close to a stable, non-polluting state within the design life of the landfill system. These approaches can form a strategy complying with sustainable principles.

In Europe and other regions of the world, Hazardous Waste Landfills are being constructed and used. The concept of these hazardous waste landfills is not sustainable in the long-term as it leads to a situation where controlled release does not occur. Concentrating hazardous materials in hazardous waste landfills at discrete locations does not comply with sustainable development practices. It will lead to the uncontrolled escape of these substances which will simply remain entombed for the entire lifetime of the containment system.

7.9 An Example

North Wake County Landfill in Raleigh, North Carolina, USA, which has both a sanitary landfill (closed in 1997) and a working MSW landfill, is located on about 230 acres of land, but only 70 acres is dedicated to the actual landfill. The remaining land is for the support areas (runoff collection ponds, leachate collection ponds, drop-off stations, areas for borrowing soil and 50-100 foot buffer areas. The cost of its construction was US$19 million. In this Landfill, a cell is approximately 50 feet long by 50 feet wide by 14 feet high (15.25 m × 15.25 m × 4.26 m). The amount of trash within the cell is 2,500 tons and is compressed at 1,500 pounds per cubic yard! This compression is done by heavy equipment (tractors, bulldozers, rollers and graders) that go over the mound of trash several times).

The cross section of the landfill and photographs of key components of the landfill are reproduced here.

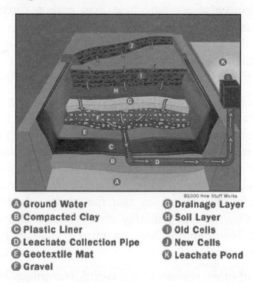

Ⓐ Ground Water   Ⓖ Drainage Layer
Ⓑ Compacted Clay   Ⓗ Soil Layer
Ⓒ Plastic Liner   Ⓘ Old Cells
Ⓓ Leachate Collection Pipe   Ⓙ New Cells
Ⓔ Geotextile Mat   Ⓚ Leachate Pond
Ⓕ Gravel

Fig. 7.4 The cross-section drawing shows the structure of a municipal solid waste landfill. The arrows indicate the flow of leachate

A bulldozer prepares

Collection pond catches storm water

A leachate collection pond catches the contaminants that can get into water that goes through the trash in a landfill

A methane collection pipe helps capture the hazardous gas.

A methane "flare" is used for burning landfill gas.

An experimental tarp provides daily cover of the landfill cells.

Grass and other plants cover the municipal solid waste landfill.

7.10 Landfills in Developing Countries

Three basic types – Open dumps, semi-controlled or operated dumps, and sanitary landfills – are under practice in developing countries. Of the three types, open dumps are the most primitive and crude, while the sanitary landfills are the best. An operated or semi-controlled dump is the next stage of open dump/ landfill. This is a compromise between the open dumping and sanitary landfills. Semi-controlled dumps have some form of inspection, recording and monitoring arrangements. In some of these landfills, compaction of waste at the control points and/or the tipping stations is practiced. In some of these landfills, soil cover is applied but only limited measures are taken to mitigate other environmental impacts from release of leachate and landfill gas. Conversion of open or operated dumps to engineered landfills and sanitary landfills is one of the important steps towards better disposal practices. Operated landfills reduce the impact of landfilling over the environment and public health but cannot eliminate it. A number of characteristics distinguish a sanitary landfill from a semi-controlled dump, although these characteristics vary from region to region, from nation to nation, and even from site to site. Table 7.1 lists few distinct characteristics of the different types of landfills in Asia.

Table 7.1 Characteristics of Different types of landfills

	Engineering measures	Leachate management	Landfill management	Operation measures
Semi-controlled dumps		Unrestricted contaminant release	None	Few, some placement of waste-still scavenging
Controlled dump	None	Unrestricted contaminant release	None	Registration and placement/compaction of waste
Enginered landfill	Infrastructure and liner in place	Containment and some level of leachate treatment	Passive ventilation of flaring	Registration and placement/compaction of waste; uses daily soil cover
Sanitary lanfill	Proper siting, infrastructure ; liner and leachate treatment in place	Containment and leachate treatment (often biological and physiochemical treatment	Flaring	Registration and placement/compaction of waste; uses daily soil cover. Measurement for final top cover

Table Contd...

	Engineering measures	Leachate management	Landfill management	Operation measures
Sanitary Lnadfill with Top Seal	Proper siting infrastructure; liner and leachate treatment in place, Liner as top seal	Entombment	Flaring	Registration and placement compaction of waste; use daily soil cover
Controlled containment release landfill	Proper siting infrastructure, with low-permeability liner in place potentially low-permeability final top cover	Controlled release of leachate into the environment, based on assessment and proper siting	Flaring or passive ventilation through top cover	Registration and placement/compaction of waste; uses daily soil cover. Measures for final top cover

(*Source*: Johannessen and Boyer 1999)

Open dumps have the lowest initial capital investment and operating costs. That is why; these sites are most common in developing countries especially in Africa and in some parts of Asia. Frequently, municipalities dump wastes in low-lying land, rather than at designated dump sites, literally as landfill; for this reason the site in these cases is not permitted to rise above ground, as it is designated for development. Private landowners who wish to have depressions filled accept municipal wastes. Filling of wetlands with wastes has been important, as witnessed in the land development of Kolkata, Mumbai, Chennai, and Colombo. Sometimes, wastes are illegally dumped in water bodies of all kinds, especially by settlements that are denied the facility of municipal waste collection, polluting the water bodies.

The practice of open dumping is a problem for the poorer and smaller cities and towns of developing countries. For example, most of the Central America except for Costa Rica, the Guyanas, and most Caribbean countries, and all non-capital cities in Bolivia, Ecuador, and Peru, and many medium-sized cities with the exception of those in Chile, Cuba, Trinidad and Tobago, and Colombia dispose waste in open-air dumps, posing significant environmental health to waste pickers who enter freely.

Since most large dumps have hundreds of extra workers in the form of waste pickers, and the municipal workers are not being provided with protective stuff, the health risks at dumps are very high. These workers are exposed to risks from human feces, slaughterhouse wastes, toxic dust, infectious biomedical wastes, snakes, scorpions, broken glass, landfill gases, and explosions. In cities where plastic shopping bags are used to put out wastes for collection, waste pickers sometimes set the bags on fire in order to find metal cans. Spontaneous fires also break out on dumps and contribute to a great extent to the air pollution like in Karachi and Tehran. As seen earlier, open dumps attract numerous birds that feed on the wastes, which can make them carry more serious disease vectors than flies or rodents. The groundwater which is contaminated may never be returned to usable condition; other environmental impacts may take many decades to improve. However, for very poor countries where cities are located near deserts, like in North Africa and the Middle East, unimproved open dumps may possibly be considered sound provided (a) the collection service is improved, (b) open dumping practices are reorganized, and (c) gradually upgrade the sites. Generally, many open dumps start off as controlled dumps and degrade due to lack of equipment, management and other resources. Shortage of cover, lack of leachate collection and treatment, inadequate compaction, poor site design, and many pickers working at the site are common problems especially in Asia.

In South and West Asia, there is rarely any controlled disposal of hazardous, biomedical, or slaughterhouse wastes, although certain areas of dumps are usually designated for slaughterhouse and biomedical wastes.

Another major problem is that of development at or on top of landfills; many shantytowns are built from disposed-of waste and in some cases entire neighborhoods are located on top of existing landfills (Zerbock 2003). For example, the Smoky Mountain dump in Manila (Philippines) had as many as 10,000 families living in shacks on or adjacent to the dump site (UNEP 1996). Aside from the obvious health implications, these concentrations of people further complicate transport and unloading procedures and present numerous safety and logistical concerns (Blight and Mbande 1996). UNEP estimates that approximately 100,000 people currently scavenge wastes at dump sites in the Latin American region alone.

Planning for environmentally safe landfills, monitoring their future impacts and site remediation are rarely undertaken in the poorer countries of South and West Asia. Dump sites are almost used immediately after closure, either as building sites or for farming. Lack of planning, use of inappropriate equipment, and involving untrained personnel adversely affect improvement.

There is now considerable experience in a number of countries with low-cost methods of upgrading for healthy operation of landfills. The first step is the construction of boundary drains to catch run-off and leachates; then, the site be graded to minimize leaching through the wastes. Machines can be rented to periodically adjust the grading, construct trenches for the deposit of waste, and dig up cover material. The work of maintaining the grading and applying cover material can then be done manually. In some cases, a provincial ministry acquires the necessary earthmoving equipment and it is rotated among the dumps of the jurisdiction. In cases where equipment is obtained by the authority operating a dump, such equipment should be kept as simple as possible to make operation and maintenance costs feasible. It is important to demonstrate to municipal workers that improvements can be made to open dumps with little capital outlay and few increased costs.

Here are two examples of utilizing sound practices: In the newly industrialized city of Jubail, Saudi Arabia, the landfill is divided into three areas, for hazardous, putrescible, and inert wastes. The site is lined and continually monitored with systematic data collection. The Ministry of Environment in Israel has closed down and remediated a number of improper dumping sites recently. The country is now planning for environmentally sound landfills using state-of-the-art technologies.

Some large cities in Latin America such as Belo Horizonte, Buenos Aires, Guayaquil, Medellin, Mexico City, Santiago, and Sao Paulo do have state-of-the art landfills. Landfill design in these cities typically consists of an initial clay layer, followed by a sand or ground stone layer. Synthetic liners are not usually used except for some new landfills in Argentina, Brazil, and Chile. Leachate collection systems are used, the landfills are subdivided into cells, and they have chimneys for gas ventilation. Wastes are covered daily with topsoil. When full, landfills are closed by covering with a clay layer and topsoil. Then the site is developed with vegetation. In East Asia also, some cities like Bandung, Jakarta, and Manila have well-designed and properly operated sanitary landfills. In well designed and properly sited landfills there is the potential for methane recovery; few landfills in the developing world are designed to capture and make use of methane. In all of Latin America, only three such landfills were in operation, all in Chile (UNEP 1996). The methane produced is supplied by the gas companies in Santiago and Valparaiso to approximately 30% of the population in each of the cities. The landfill gas (methane) management is generally required at sanitary landfills. At controlled dumps, there should at least be monitoring to determine if dangerous amounts of gas are being released. A low-cost design to handle landfill gas may consist of buried vertical perforated pipes, using the natural pressure of the gas to collect and vent or flare it at the surface. This is called a passive collection system. More costly active collection systems utilize a buried network of pipes and pumping to trap the gas (described earlier). Gas capture has been tried on an experimental basis in just a few cases; for instance, in New Delhi, gas is supplied to a nearby hospital. In India, there is some cultural inhibition to using gas from dumps for domestic cooking. Generally the required capital for methane recovery installations is lacking, and the low price of commercially produced gas does not make methane recovery a viable enterprise economically.

The situation in Africa, in general, requires a great deal of improvement. Landfills in Africa are primarily open dumps without leachate or gas recovery systems. Several are located in ecological or hydrological sensitive areas such as Algeria, Libya, Sudan, Cameroon and Zaire. The landfills are generally operated below the standards of sanitary practice. Waste pickers remove materials of economic value for recycling without a fee to the facility owner and operator. Operation and maintenance costs are provided from municipal budget allocations and often do not cover the entire amount needed. The result is substandard and unsafe facilities which pose public health risks and aesthetic burdens to the citizens. Though the standards of modern sanitary landfills with leachate and gas recovery may be too expensive for most African cities, efforts have been made in countries like Egypt and South Africa through policy changes to upgrade landfills; in Tunisia, to develop nationwide sanitary landfill programme; and in Zambia to improve landfilling to upgrade MSW collection services, etc. The ocean dumping is prohibited by law in African countries; however, the practice is still illegally followed in some coastal cities of Africa.

In Latin American cities with less than 50,000 inhabitants, *manual* landfills are being developed. Manual landfills are similar in design to mechanized landfills except for their size and the equipment they require. These landfills have the capacity to receive 10-50 tons per day of wastes. They sometimes require the use of heavy equipment, but only for periodic preparation of the terrain. Otherwise, landfill operation is carried out manually, including cell preparation, compaction, daily cover, and cell closure. The capital and operation and maintenance costs of these landfills are lower than a mechanized landfill. The most successful cases are in Colombia, although Chile, Costa Rica, Honduras, Peru, Ecuador and Panama all have manual landfills. Manual landfills are often the best option for small cities and towns. Those involved in manual landfills in Colombia believe that, in general, the maximum that such facilities can reasonably handle is 20 tons/ day.

Landfilling is one of the most widely used methods of disposal for E-waste and requires special attention. It is highly prone to hazards because of leachate which often contaminates water resources. Uncontrolled dumps and older landfill sites pose a much greater danger of releasing hazardous emissions. Mercury, lead and cadmium are the most toxic leachates. Mercury will leach when certain electronic devices such as circuit breakers are destroyed. Lead leaches from the broken cone glass of cathode ray tubes from TVs and monitors which contain lead. When brominated flame retarded plastics or plastics containing cadmium are landfilled, both PBDE and cadmium may leach into soil and ground water. Landfills are also prone to uncontrolled fires which can release toxins.

Landfills can be a part of an integrated system for the management of MSW in developing countries. If carefully designed and well managed within the framework of the local infrastructure and available resources, landfills can provide safe and cost-effective disposal of a city's MSW. But they are not designed for the routine disposal of industrial or hazardous waste, used oil, or other special wastes. If they are pressed beyond their design limits, the landfill degrades into a potentially toxic open dump and results in adverse consequences for human health and the environment.

An integrated MSWM system may prioritize its waste management options according to waste minimization, materials recovery/ recycling, composting, incineration, and landfilling. Incineration is a sound practice only under particular conditions. But it is not generally used in MSWM systems due to high capital and technical resources required. The other components of the integrated approach can improve landfill operations and extend the life of the facility.

The benefits of Waste minimization or source reduction, materials recovery and recycling, composting process are already explained. It is more cost-effective to perform these operations close to the site of waste generation. This reduces the cost of transporting the materials to the landfill and minimizes the difficulty of separating mixed wastes at the landfill.

It must be recognized that an effective MSWM which avoids pollution and helps produce useful energy, heat or electricity, from waste is essential for a country's sustainable development. Hence, adequate financing and supporting institutional and policy environment must be provided for a successful MSWM.

7.11 Landfills in Developed Countries

Landfilling is still the primary means of managing solid waste in North America, handling about 65 to 70% of MSW. MSW landfills in the US are allowed to accept only non- hazardous solid waste, such as household garbage, except for small quantities of residential and commercial hazardous waste exempted from hazardous waste management laws. A state-of-the-art landfill in North America contains sophisticated engineering features to prevent the release of hazardous substances to the environment, including liners, leachate collection, final covers, and other features. Some landfills in the US now reinsert leachate into the landfill to speed biodegradation. Landfill gas is recovered as a source of energy at landfills that generate sufficient quantities of methane. The technology necessary to recover landfill gas is proven and commercially available. Most landfills located in the US recover methane gas. Many landfills have approached the end of their useful lives and the authorities are facing the problem of siting new landfills. This has already occurred in many areas, particularly in the Northeast and the Midwest. To overcome this situation partly, larger and more environmentally sound regional landfills ('megafills') are built to handle waste disposal needs. These new landfills, which provide considerable low-cost capacity, are designed to comply with stricter federal and state regulations, and are built in part with private sector investment. Siting of regional landfills can be difficult.

Landfills in North America are seen as a necessary component of any integrated MSW management system. Although recycling and composting can divert a significant portion of MSW from landfills, not all MSW is recyclable or compostable. Similarly, although WTE technologies can significantly reduce the MSW volume, all WTE facilities produce residual ash that must be landfilled. In addition, as WTE facilities are shut down for repairs or maintenance, MSW will have to be diverted to landfills.

In European waste systems, landfilling has become an inevitable part. In certain Northern European countries, less than half of the waste may be landfilled; while in southern countries like Greece and Spain, or

Eastern European countries such as Hungary and Poland, virtually all waste finds its way to landfill. The European Union Draft Landfill Directive identifies three kinds of landfills: for hazardous waste, for municipal waste, and for inert materials. Monofills, i.e., landfills for one particular material are also recognized in the directive.

The 'modern' landfills are carefully sited; admission and dumping are controlled and monitored. They require incoming waste to be weighed, and to be paid for on a per-ton basis. Design and construction of modern landfills is more expensive than simple dumping, and these facilities may also be difficult to site. Public resistance is not as much in Europe as in North America, but still plays a significant role in siting. The costs have shot up due to environmental controls and the increased costs force developers to build larger landfills, which serve a region rather than a single municipality. These are typically more cost-effective. In a few cases, gas recovered at landfills in Europe is simply flared, while the energy is recovered in others.

There is also considerable experience in Europe with bio-reacting landfills, in which leachate is recirculated to maintain optimal moisture levels for bio-degradation to occur.

Australia and Japan normally classify landfills into three categories, based on whether they are intended for hazardous wastes, special wastes, or MSW. The design specifications of landfills for hazardous wastes are very stringent. These are constructed like a bathtub with several layers of impermeable liners and with leachate and gas control systems. In these cities, even for MSW, modern landfills are planned and constructed to minimize soil, groundwater, and surface water contamination from landfill leachate and the migration of landfill gas to surrounding areas. Landfill gases are sometimes collected for fuel.

Some Japanese coastal cities (e.g., Kityakushu) use solid wastes for land reclamation, with sophisticated pre-treatment and compaction. In smaller towns in rural areas, MSW contains fewer hazardous substances as compared to MSW in large cities, and regulations for landfill disposal of MSW tend to be less stringent.

CHAPTER 8

MSW Management in India

8.1 Introduction

Traditionally, agriculture and livestock rearing and related activities have been the main livelihood of most of the Indian people. Still India is an agriculture-oriented country – a rural India – despite rapid growth of wide range of industrial activity in the last half-a-century. India has been witnessing two developments in the recent decades: a desirable development such as fast economic growth through rapid industrialization, and an undesirable development such as population explosion. These have led to unplanned and rapid urban growth and extensive slums. Though increasing urbanization in India is a part of the global trend with 27.8 percent of India's population (285 million) of the total 1027 million living in urban areas (as per the 2001 census). The number of towns and cities have increased to 4378 of which 393 are Class-I towns, 401 are Class-II towns, 1,151 are Class-III towns and remaining are classified as small towns with populations ranging between 20,000 to less than 5000. The number of metropolitan cities having million plus population has increased to 35 as per 2001 census. This growth has seen growing public concern with exponential increase in sanitation and environmental concerns (WB- Hanrahan, D 2006). Sanitation and environment issues are clearly the contributors to basic health conditions in urban areas but MSWM has a lower priority than water supply and sanitation.

The changing urban consumption patterns consequent to economic growth and improved incomes, and local production of goods and services have resulted in an increase in per capita waste generation. This increase is city-dependant.This recent development of growth in waste generation is exerting significant additional pressures on already stretched MSWM systems across cities in India (MoEF-GOI 2009, Sharholy et al 2007). The generation of solid waste is projected to increase significantly as the country strives to attain the status of an industrialized nation by the year 2020 (Sharma and Shah 2005; CPCB 2004; Shekdar et al. 1992*).*

Municipal Corporations/Urban local bodies (ULBs) traditionally provide SWM services in India because these bodies generally oversee the issues related to public health and sanitation. In Indian cities, to a great extent, these services are measured substandard as the systems applied are unscientific, outdated, inefficient and do not cover the entire population. The apathy of municipal authorities who do not consider MSWM as a priority is another reason. As a result, the waste is found littered all over creating insanitary conditions.

Overview of Main Components of MSW in India: MSW management covers the full cycle from collection of waste from households and commercial establishments through to acceptable final disposal. In the process, efforts are made to reduce the final volumes, through recycling and materials recovery, as well as processing/treatment. The Fig. 8.1 outlines the typical system of waste management in India. An analysis along these lines should be carried out for any municipality, as a first step to understanding and dealing with the necessary upgrading of the system.

In India, there are many challenges in MSW management: analysis of quality and quantity of wastes, and appropriate institutional mechanisms for collection, storing, transportation, processing/ treatment and related activities.

Rarely there are sincere efforts to adopt recent methods and technologies of waste management except in a few cities. The fundamental underlying problems are in fact, financial and institutional. There are some individual good examples and, not surprisingly, the larger municipalities tend to have better systems in place.   By and large, financial and human resources, and institutional mechanisms are limited.

In the absence of a facility to collect waste from sources (houses or shops or restaurants etc.), people are prone to dump wastes on streets, drains, open spaces, and near-by water bodies creating insanitary conditions and causing an adverse impact on the environment and public health. The outbreak of plague in Surat in 1994 was the best example of how unsanitary conditions in the cities cause environmental and health hazards.

People generally believe that waste thrown onto the streets would be collected by the municipal street sweepers. The municipalities, probably, have to do much more to educate the citizens on the basics of MSW managemant, and proper storing of the waste in their own bins in the households (Asnani 2006; Rathi 2006; Sharholy et al. 2005; Ray et al. 2005; Jha et al. 2003; Kansal 2002; Kansal et al. 1998; Singh and Singh 1998; Gupta et al. 1998).

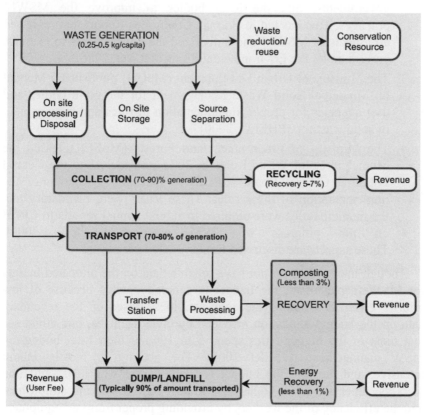

Fig. 8.1 Typical system of Waste management in India (Source: WB 2006)

The subject of public health has assumed prominance in the last one to two decades due to population explosion in municipal areas, growing public awareness towards cleanliness and proper sanitation, and emergence of newer technologies (Kumar *et al.*, 2004). Some of the important developments that took place in India are the following:

1. The Bajaj Committee was appointed in 1994 to draw up a long term policy to be adopted, and made several recommendations in all elements of SWM. The Ministry of Health and Family Welfare and the Central Public Health and Environmental Engineering Organization (CPHEEO) of GOI jointly organized a national workshop in April 1995 which emphasised the necessity to improve the SWM as a priority.

2. In 1998, a Public Interest Litigation (PIL) was filed in the Supreme Court of India seeking a direction to the central and state governments, and the local bodies to improve the MSWM practices. This has led to Barman Committee Report that reviewed all aspects of MSWM and authorized the governments to exercise powers under the Environmental Protection Act, 1986.

3. The Ministry of Urban Development (MoUD) published a Manual on Municipal Solid Waste Management for the civic bodies and user agencies for proper implementation of rules and management of solid wastes (CPHEEO, 2000). .

4. The Ministry of Environment and Forests (MoEF) released the Municipal Solid Waste Rules in 2000, and identified the Central Pollution Control Board (CPCB) as the agency to monitor the implementation of these rules. These Rules were mandatory and the municipalities were required to submit annual reports to CPCB on the progress of SWM practices (CPCB, 2000). These aspects are discussed further in the later pages.

Though many cities do not have reliable data on the allocated budget for MSW management, the budget is generally paltry because of low priority given to it among the civic activities. Based on the secondary data on the budget allocation provided by civic agencies, one could see that most of the bigger cities spend 5 to 10% of their total budget on MSW management (NEERI 2005). The activity is mostly labour intensive, and most of the budget is spent on the wages of sanitation workers, supervisors and higher ups. Improving the working atmosphere and the efficiency of the workers by providing proper training, equipment and gadgets receive little attention.

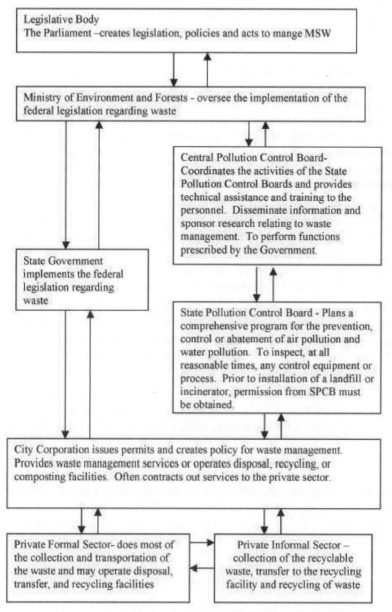

Fig. 8.2 Municipal solid waste management (MSWM) system in India.

There are other reasons for poor management of MSW, like powerful workers'unions, politicization of the unions, and more importantly the indiscipline among the workers arising out of poor working conditions and clumsy handling of labour issues. Most of these problems arise, in general, due to poor socio-economic conditions of the workers, and lack of understanding, on their part, of the serious health problems that would arise out of poor MSW services.

Institutional/Management structure: The first level is Central government under which the Ministry of Environment and Forests, and Central Pollution Control Board are the Policy and regulatory bodies; the second level is State government under which the Ministry of Urban development and District administration are responsible for MSW management with State Pollution Control Board as a Policy and regulatory body; and the third level is Urban local bodies under which the municipal corporations and municipalities are responsible for MSWM.

In many cities, the Health Officers/ Chief Medical Officer/ Additional or Deputy Commissioner/Assistant Health Officer is in charge of the MSWM activities. In a few smaller cities, the activities are administrated by the Chief Officer/Special Officer/CEO/Jt. Secretary/Tax Officer, etc (Kumar et al 2009). Mega cities such as Mumbai, Delhi, and Kolkata have separate SWM departments.

MSW management system in Bangalore (source: BBMP): Bruhath Bengaluru Mahanagara Palike (BBMP) was established in 2007 after the merger of surrounding areas with erstwhile Bangalore Mahanagar Palike (BMP). The City is divided into eight Zones comprising 198 wards. BBMP manages delivery of SWM services in all 198 wards which falls under its limits.

The Solid Waste Management department of BBMP is headed by Deputy Commissioner (Health). Other key officials include Engineer-in-Chief, Executive/Assistant/Environmental Engineer. They are assisted by Environmental Officers and Health Inspectors in discharging solid waste management services effectively in their respective zones. There are 200 Health Inspectors in addition to Environmental officers handling the supervisory functions. Furthermore, BBMP employs around 4300 employees including supervisors and operators referred to as 'poura karmicas'. The actual manpower involved in BBMP's MSWM activities are the following:

Deputy Commissioner (Health)	1
Engineers-in-Chief	4
Assistant and Environmental Engineers	34
Environmental Officers	64
Health Inspectors	200
Drivers	200
Poura Karmicas (Supervisors and Operators)	4300

In addition, there are outsourced contracts (source: BBMP).

Typically in smaller municipalities, the public health department is responsible for collection, street sweeping, transport and disposal of solid wastes generated in the local body's wards.

The head of Health Department reporting to Municipal Commissioner is generally head of the MSW management system but there is often poor coordination between the engineering department (which is responsible for transport and disposal) and health department.

Each city has its own arrangement covering all aspects of MSW services.

In small towns with population below one hundred thousand, the SWM services are unprofessional because they are handled by sanitary inspectors with the help of sanitary workers. In smaller towns, even sanitary inspectors are not employed, and SWM is attended by unqualified supervisors. The services are better in towns with population above 100 thousand because qualified health officers and engineers head the SWM services.

Manpower provisions range between 2–3 workers per thousand in 32 out of 59 cities. Manpower deployment in the range between 1–2 workers per thousand has been reported for cities such as Ludhiana, Thiruvananthapuram, and Surat. Cities with less than 1 worker per thousand are Agra, Dhanbad, Ranchi, Aizawal, Gangtok, Imphal, Kanpur, Silvasa, etc. The largest workforce was observed at Port Blair and the lowest at Gangtok (Kumar et al 2009).

For providing effective solid waste services in a city, NEERI's studies, undertaken in more than 40 Indian cities, have shown that the desired strength is 2–3 workers per thousand. However, this number may change based on local conditions. For MSWM, every municipal agency can decide the strength of workers by considering the productivity of workers, which can be considered to be 200–250 kg/worker/8 h shifts.

8.2 Analysis of MSW

Many categories of MSW such as compostable organic waste, industrial waste, construction and demolition waste, and sanitation waste, so on are generated in India like in other developing countries. MSW also contains recyclables (paper, plastic, glass, metals, etc.), toxic substances (paints, pesticides, used batteries) and medical waste (Jha et al. 2003; Reddy and Galab 1998; Khan 1994). The quantity of MSW generated depends on factors such as food habits, life styles, and the nature and extent of commercial activities. Still, the waste generation rates in India are lower than the low-income countries and much lower compared to developed countries (Asnani 2006). See Tables in the earlier Section. With increasing urbanization and changing life styles, Indian cities now generate eight times more MSW than they did in 1947. The per capita waste generation is increasing by about 1.3% per year. With the urban population growing at 2.7% to 3.5% per year, the annual increase in the total quantity of solid waste in the cities will be more than 5% (Asnani 2006). Presently, about 90 million tons of solid waste is generated annually as byproducts of industrial, mining, municipal, agricultural and other activities (Pappu et al., 2007; Shekdar 1999; Bhide and Shekdar 1998).

Several studies report that the MSW generation rates in small towns are lower than those of metro cities. The per capita generation rate of MSW in India ranges from 0.2 to 0.6 kg/ day amounting to about 115,000 metric tonnes (MT) of waste per day, and 42 million MT annually (Asnani 2006; Siddiqui et al. 2006; Sharholy et al. 2005; CPCB 2004; Kansal 2002; Singh and Singh 1998; Bhide and Shekdar 1998; NEERI 1995). It is also estimated that the total MSW generated by urban people increased from 23.86 million tonnes /year in 1991 to more than 39 million tonnes in 2001.

Despite several of these studies, CPCB has undertaken to assess the status of the MSWM services in 59 identified cities, covering 35 metro cities with population greater than 1 million, as well as 24 state capitals and union territories. The Supreme Court of India has asked CPCB to retain NEERI to complete this work. Under this study (Kumar et al 2009) extensive field investigations have been carried out to determine waste quantification, characterization of waste, financial and institutional aspects, and assessment of MSWM status as per MSW Rules 2000. Further an action plan for better managementof MSW has been suggested. The map shows the cities selected covering all the metropolitan areas and state capitals representing the geography of the country (Fig. 8.3). The methodology adopted and other details are covered in that paper.

The study has estimated *waste generation* rates in kg/capita/day for various population ranges:

(i) Cities with a population < 0.1 million (8 cities): 0.17-0.54 kg/capita/day,

(ii) Cities with a population of 0.1–0.5 million (11 cities): 0.22-0.59 kg/capita/day,

(iii) Cities with a population of 1–2 million (16 cities): 0.19-0.53 kg/capita/day,

(iv) Cities with a population > 2 million (13 cities): 0.22–0.62 kg/capita/day.

Among the 59 cities, despite similar populations in a few cities, variations in waste generation rates have been observed. The reasons could be many: differences in standard of living, food habits, employment status, road conditions, difference in equipment, machinery and implements, climatic conditions, geographical status, etc.

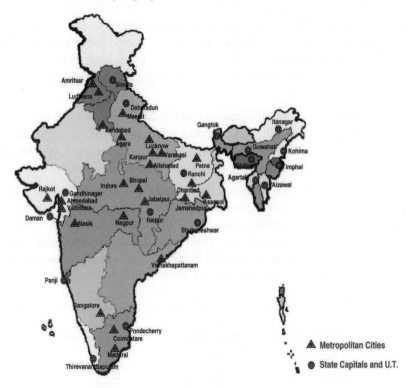

Fig. 8.3 Metropolitan cities and state capitals covered
(*Source*: S Kumar et al, 2009)

Regarding the waste characteristics, the study has found the following:

The compostable organic and recyclable fractions were observed to be higher in some cities probably due to higher standard of living.

(a) For cities having *population < 0.1 million* and between *0.11–0.5 million* (19 cities), the characteristics are C/N ratio = 18 to 37; the compostable fraction: 29 to 63%; total recyclables: 13.68–36.64%. The moisture content was high at 65% in the MSW at Shillong, Kohima, Simla, and Agartala due to heavy rains. High calorific value on a dry weight basis was observed to vary from 591 to 3766 kcal/kg.

(b) For cities with a *population of 0.5–1 million* (16 cities), the constituents were varying; compostable matter: 35– 65%, recyclables: 11–24%, C/N ratio: 17–52, high calorific value on a dry weight basis: 591–2391 kcal/kg, and moisture content: 17–64%.

(c) For cities having a *population of 1–2 million* (11 cities), the ranges for various constituents varied; compostable fraction: 39–54%, recyclables: 9–25%, C/N ratio: 18–52, high calorific value (on dry weight basis): 520–2559 kcal/kg, and moisture content: 25–65%.

(d) For cities with populations *greater than 2 million* (13 cities), the constituents varied; compostable fraction: 40–62%, recyclables: 11–22%, C/N ratio: 21–39, high calorific value (on a dry weight basis): 800–2632 kcal/kg and moisture content: 21–63%.

The study noticed that in some cities located in hilly and coastal areas, and on islands, the implementation of MSWM system is constrained by problems specific to localities (Kumar et al 2009). Cities with 100,000 plus population (423 in number) contribute 72.5% of the total waste generated in the country as compared to other 3955 urban centres produce only 17.5% of the total waste. Table 8.1 shows the details.

Table 8.1 Waste generations in Class 1 Cities with Population above 100,000

Type of cities	Tonnes/day	% of total garbage
7 mega cities	21,100	18.35
28 metro cities	19,643	17.08
388 class 1 towns	42,635	37.07
Total (423) cities	83,378	72.50

Note: Mega cities are above 4 million population and metro cities (also known as million plus cities) are the same as the identified cities under the proposed JNNURM. Class 1 cities with population in the 100,000 to 1 million ranges are 388 in number

Source: Ministry of Urban Development 2005, Asnani 2006.

In India, MSW differs greatly with regard to the composition and hazardous nature, when compared to the developed countries (Gupta et al., 1998; Shannigrahi et al., 1997; Jalan and Srivastava 1995). Analysis of physical composition of waste indicates: compostable matter (40- 60 %), recyclable fraction (10 and 25%), moisture content (30 to 60%), and C/N ratio (20: 40) (CPCB 2004). It has been noticed that the physical and chemical characteristics of MSW change with population density (Garg and Prasad 2003; CPCB 2000; Bhide and Shekdar 1998). It is also observed that the differences in the MSW characteristics indicate the effect of urbanization and development. In *urban areas*, the major fraction of MSW is compostable materials (40–60%) and inert (30–50%). The organic waste component in MSW is generally found to increase with the decreasing socio-economic status; that is, rural households generate more organic waste than urban households. For example, in south India the extensive use of banana leaves and trees in various domestic functions and events results in a large organic content in the MSW. Also it has been noticed that the fraction of recyclables (paper, glass, plastic and metals) is very low due to the picking of these materials at the points of generation and collection, and disposal sites by poor people (waste pickers) to make a living. Tables 8.2 to 8.5 provide details of the physical and chemical characteristics, and composition of waste in Indian Cities.

Table 8.2 Physical Characteristics of Municipal Solid Waste in Indian Cities

Population range (in millions)	No. of cities surveyed	Paper	Rubber, leather and synthesis	Glass	Metal	Total compostable matter	Inert material
0.1 to 0.5	12	2.91	0.78	0.56	0.33	44.57	43.59
0.5 to 1.0	15	2.95	0.73	0.56	0.32	40.04	48.38
1.0 to 2.0	9	4.71	0.71	0.46	0.49	38.95	44.73
2.0 to 5.0	3	3.18	0.48	0.48	0.9	56.67	40.07
> 5.0	5	6.43	0.28	0.94	0.80	30.84	53.90

Note: All values are in per cent calculated on wet weight basis. *Source:* NEERI (1995)

Table 8.3 Chemical Characteristics of Municipal Solid Waste in Indian Cities

Population range (in millions)	Nitrogen as total nitrogen	Phosphorus as P_2O_5	Potassium as K_2O	C/N Ratio	Calorific value kcal/kg
0.1 to 0.5	0.71	0.63	0.83	30.94	1009.89
0.5 to 1.0	0.66	0.56	0.69	21.13	900.61
1.0 to 2.0	0.64	0.82	0.72	23.95	980.05
2.0 to 5.0	0.56	0.69	0.78	22.45	907.18
5.0 and above	0.56	0.52	0.52	30.11	800.70

Source: NEERI (1995)

8.3 Storage and Collection of MSW

Storage of MSW at the source is considerably lacking in most of the urban areas. There is no system, in general, of primary collection from household or shop or office, the sources of generation. The waste is discharged at all places which are collected by municipal workers through street sweeping. The tools used for street sweeping are inefficient and obsolete, and no uniform rules or methods are adopted for street sweeping. Sometimes residents usually store the waste in 16–20 litre-capacity plastic buckets (of different design) and then dispose of into community bins. The type of container generally reflects the economic status of its user (Kumar et al. 2009). Table 8.6 indicates how the different categories of waste generators can manage waste at source for easy collection.

The bins are common for both decomposable and non-decomposable waste since no segregation of waste is performed by the generator. Round cement concrete bins, masonry bins or concrete structures are used. These are either movable or fixed ones. The movable bins are generally not durable, but flexible for transportation, while the fixed bins are more durable whose positions cannot be changed once they have been constructed (Nema, 2004; Malviya et al., 2002).The collection of MSW in most of the cities is through community bins placed at various points along the streets. The observation is that community bins have not been installed at proper locations resulting in poor handling efficiency. Also, the situation becomes worse due to lack of public awareness leading to overflowing or creating open heaps of waste which is unsightly, generating bad odours and health problems.

Table 8.4 Physical Composition of Municipal Solid Waste in 1 million plus Cities and State Capitals in India (average values)

Name of the city	Total compostable	Recyclabes				Inert	Other including inert					Total
		Paper etc.	Plastic	Glass	Metal		Rubber and leather	Rags	Wooden matter	Coconut	Bones	
Indore	48.97	6.10	5.77	0.55	0.15	31.02	2.95	2.41	1.17	0.91	0.00	100
Bhopal	52.44	9.01	12.98	0.55	0.98	18.88	0.09	2.65	1.35	2.25	0.01	100
Dhanbad	46.93	7.20	5.56	1.79	1.62	26.93	2.77	4.14	1.56	1.58	0.00	100
Jabalpur	48.07	7.67	8.30	0.35	0.29	26.60	2.15	4.42	1.49	0.66	0.00	100
Jamshedpur	43.36	10.24	5.27	0.06	0.13	30.93	2.51	2.99	4.29	0.22	0.01	100
Patna	51.96	4.78	4.14	2.00	1.66	25.47	1.17	4.17	1.13	2.34	0.89	100
Ranchi	51.49	3.17	6.48	1.79	1.45	25.92	1.45	4.97	2.74	3.19	0.38	100
Bhubaneshwar	49.81	5.74	5.70	0.46	0.79	27.15	2.10	3.21	2.85	2.20	0.00	100
Ahmedabad	40.81	5.28	5.29	0.79	0.30	39.28	0.92	5.00	1.22	1.02	010	100
Nashik	39.52	9.69	12.58	1.30	1.54	27.12	1.11	2.53	0.34	4.12	0.15	100
Raipur	51.40	8.31	7.07	0.76	0.16	16.97	1.47	3.90	1.43	6.44	0.08	100
Asansol	50.33	10.66	2.78	0.77	0.00	25.49	0.48	3.05	3.00	2.49	0.95	100
Bangalore	51.84	11.58	9.72	0.78	0.35	17.34	1.14	2.29	2.67	2.28	0.01	100
Agartala	58.57	8.11	4.43	0.98	0.16	20.57	0.76	2.17	0.00	2.56	1.69	100
Agra	46.38	6.12	8.72	0.85	0.11	30.07	1.97	3.92	1.68	0.19	0.00	100
Allahabad	35.49	7.27	10.33	1.23	0.40	31.01	1.83	7.34	2.08	2.74	0.30	100
Damam	29.60	10.54	8.92	2.15	0.410	34.80	2.60	4.90	4.60	4.48	-	100

Name of the city	Total compostable	Recyclabes				Other including inert						
		Paper etc.	Plastic	Glass	Metal	Inert	Rubber and leather	Rags	Wooden matter	Coconut	Bones	Total
Meerut	54.54	4.95	54.48	0.30	0.24	27.30	0.49	4.98	0.95	0.66	0.12	100
Nagpur	47.41	6.87	7.45	0.92	0.29	18.01	5.38	9.48	2.10	2.09	0.00	100
Vadodara	47.43	5.98	7.58	0.47	0.47	27.80	1.28	4.86	1.55	2.57	-	100
Gandjinagar	34.30	5.60	6.40	080	0.40	36.50	3.70	5.30	3.70	3.30	-	100
Vishakapatanam	45.96	14.46	9.24	0.35	0.15	20.77	0.47	2.41	0.68	5.51	-	100
Dehradun	51.37	9.56	8.58	1.40	0.03	22.89	0.23	5.60	0.32	-	-	100
Ludhiana	49.80	9.65	8.27	1.03	0.37	17.57	1.01	11.50	0.80	0.00	-	100
Guwahati	53.69	11.60	10.01	1.30	0.31	17.66	0.16	2.18	1.39	1.38	0.26	100
Kohima	57.48	12.28	6.80	2.32	1.26	15.97	0.18	1.86	1.70	0.00	0.35	100

Note: Increasing use of plastics is changing the composition of municipal solid waste and causing harm in the processing of waste. The use of plastics has increased 70 times between 1960 and 1995. *Source:* CPCB (2000)

212

Table 8.5 Chemical Characteristics of Municipal Solid Waste (Average Values) of 1 million plus Cities and State Capitals

Name of the city	Mositure	pH Range	Volatile matter	C per cent	N per cent	P per cent as P_2O_5	K per cent as K_2O	C/N ratio	Hev Kcal/kg
Indore	30.87	6.37–9.73	38.02	21.99	0.82	0.31	0.71	29.30	1436.75
Bhopal	42.66	6.99–9.03	35.78	23.53	0.94	0.66	0.51	21.58	1421.32
Dhanbad	50.28	7.11–8.01	16.52	9.08	0.54	0.55	0.44	18.22	590.56
Jabalpur	34.56	5-84–10.94	46.60	25.17	0.96	0.60	1.04	27.28	2051
Jamshedpur	47.61	6.20 – 8.26	24.23	13.29	0.69	0.54	0.51	19.29	1008.84
Patna	35.95	7.42–8.62	24.72	14.32	0.77	77	0.64	19.39	818.82
Ranchi	48.69	6.96–8.02	29.70	17.20	0.85	0.61	0.79	20.37	1059.59
Bhubaneshwar	59.26	6.41–7.62	25.84	15.02	0.73	0.64	0.67	20.66	741.56
Ahmedabad	32	6.2–8.0	63.80	37.02	1.18	0.67	0.42	34.61	1180
Nashik	74.64	5.2–7.0	59	34.22	0.92	0.49	-	38.17	3086.51
Raipur	29.49	6.65–7.99	32.15	18.64	0.82	0.67	0.72	23.50	1273.17
Asansol	54.48	6.44–8.22	17.73	10.07	0.79	0.76	0.54	14.08	1156.07
Bangalore	54.95	6.0–7.7	48.28	27.98	0.80	0.54	1.00	35.12	2385.96
Agartala	60.06	5.21–7.65	49.52	28.82	9.96	0.53	0.77	30.02	2427
Agra	28.33	6.21–8.1	18.90	10.96	0.52	0.60	0.57	21.56	519.82
Allahabad	18.40	7.13	29.51	17.12	0.88	0.73	0.70	19.00	1180.12

213

Name of the city	Mositure	pH Range	Volatile matter	C per cent	N per cent	P per cent as P_2O_5	K per cent as K_2O	C/N ratio	Hev Kcal/kg
Damam	52.78	5.88–6.61	52.99	30.74	1.38	0.47	0.6	22.34	2588
Faridabad	34.02	6.33–8.25	25.72	14.92	0.80	0.62	0.66	18.58	1319.02
Lucknow	59.87	4.8–9.18	34.04	20.32	0.93	0.65	0.79	21.41	1556.78
Meerut	32.48	6.16–7.95	26.67	15.47	0.79	0.80	1.02	19.24	1088.65
Nagpur	40.55	4.91–7.80	57.10	33.12	1.24	0.71	1.46	26.37	2632.23
Vadodara	24.98	-	34.96	20.28	0.60	0.71	0.38	40.34	1780.51
Gandhinagar	23.69	7.02	44	25.5	0.79	0.62	0.39	36.05	698.02
Vishakapatanam	52.70	7.5–8.7	64.4	37.3	0.97	0.66	1.10	41.70	1602.09
Dehradun	79.36	6.12–7.24	39.81	23.08	1.24	0.91	3.64	25.90	2445.47
Ludhiana	64.59	5.21–7.40	43.66	25.32	0.91	0.56	3.08	52.17	2559.19
Guwahati	70.93	6.41–7.72	34.27	19.88	1.10	0.76	1.06	17.71	1519.49
Kohima	64.93	5.63–7.7	57.20	33.17	1.09	0.73	0.97	30.87	2844

Source: Akolkar (2005)

In the last few years, efforts are made to organize house-to-house collection in many cities – Delhi, Mumbai, Bangalore, Chennai and Hyderabad – with the participation of NGOs. Many municipalities have employed private contractors for transportation of the waste from the community bins or collection points to the disposal sites. Some cities/towns have employed NGOs and citizen's committees to supervise separation of recyclables at the collection points or other locations between sources and dumpsites. In addition, the Resident/Welfare Associations of residential colonies arrange collection in some urban areas on specified monthly payment. To clean the roads, the municipalities appoint sweepers and each one is allotted an area of about 250 m^2. The sweepers use wheelbarrows to transfer the collected road wastes to dustbins or collection points (Colon and Fawcett 2006; Nema 2004; Malviya et al. 2002; Kansal et al. 1998; Bhide and Shekdar 1998). In most cities, a fraction of MSW generated remains uncollected on streets, and the collected is transported to processing or disposal sites. The collection efficiency is generally defined as the quantity of MSW collected and transported from streets to disposal sites divided by the total quantity of MSW generated during the same period. Many studies on urban environment have revealed that MSW collection efficiency is dependent on two major factors: manpower availability and transport capacity. The average collection efficiency for MSW in Indian cities and States is about 72.5% (Rathi 2006; Siddiqui et al. 2006; Nema 2004; Gupta et al. 1998; Maudgal 1995; Khan 1994). Around 70% of the cities lack adequate waste transport capacities (TERI 1998). The MSW collection efficiency is high in the states/cities, where private contractors and NGOs are drawn in for the collection and transportation.

Most of the cities are unable to provide waste collection services to cover all parts of the city. Generally, overcrowded low-income settlements or unorganised slums where the people are unable to pay for the services do not have MSW collection and disposal services, an observation in many of the developing countries. They throw away the waste near or around their houses creating public health problems. The CPCB has collected data for the 299 Class-I cities to determine the mode of collection of MSW. It is found that manual collection comprises 50%, while collection using trucks comprises only 49% (CPCB 2000).

The Door to Door Refuse / Garbage Collection system implemented in Surat Municipal Corporation Area as the 'Best Practice' is detailed in the Annexure 10.

8.4 Transfer Stations and Transportation

Except in cities like Chennai, Mumbai, Delhi, Ahmedabad and Kolkata, transfer stations are not planned, and the vehicle which collects the waste from individual dustbins, takes it to the processing or disposal site (Colon and Fawcett, 2006; Khan, 1994). Various types of vehicles are used for transportation of waste to the disposal site. In smaller (rural) towns, bullock carts, tractors and trailers, three-wheelers etc., are mainly used for the transportation of MSW. Light motor vehicles, ordinary trucks, tippers and compactors are generally used in big towns or cities. General-purpose open body trucks holding 5 to 9 tonnes of waste are common in big cities. They are usually loaded manually. In a few cities, compactor vehicles are also being used. Municipal agencies use their own vehicles for MSW transportation, though in some cities they are hired or leased (Kumar et al 2009 WMP, Goose et al. 2006; Siddiqui et al. 2006; Nema 2004; Bhide and Shekdar 1998). The municipality-owned vehicles are poorly maintained and no schedule is observed for preventive maintenance. Due to shortage of finances, many of the vehicles have outlived their standard life, resulting in high fuel consumption and low efficiency. Since the trucks that transport MSW are usually kept uncovered, the waste tends to spill onto the roads during transportation making them unhygienic. The traditional transportation system does not synchronise with the system of primary collection and secondary waste storage facilities resulting in multiple manual handling of waste. In some cities, modern hydraulic vehicles are gradually introduced (Bhide and Shekdar 1998; Reddy and Galab 1998). The collection and transportation of waste constitute major part of the budget, about 80–95%; hence, it forms a key component in determining the economics of the entire MSWM system.

Table 8.6 Waste Management at Source (source: cpreec)

Source of waste generation	Action to be taken
Household	• Not to throw any solid waste in the neighbourhood, on the streets, open spaces, and vacant lands, into the drains or water bodies • Keep food waste/biodegradable waste in a non corrosive container with a cover (lid) • Keep dry, recyclable waste in a bin or bag or a sack • Keep domestic hazardous waste if and when generated separately for disposal at specially notified locations
Multi-storeyed buildings, commercial complexes, private societies	• Provide separate community bin or bins large enough to hold food/biodegradable waste and recyclable waste generated in the building or society. • Direct the members of the association to deposit their waste in community bin
Slums	• Use community bins provided by local body for deposition of food and biodegradable waste
Shops, offices, institutions, etc	• If situated in a commercial complex, deposit the waste in bins provided by the association
Hotels and restaurants	• The container used should be strong, not more than 100 litre in size, should have a handle on the top or handles on the sides and a rim at the bottom for easy handling
Vegetable and Fruit Markets	• Provide large containers, which match with transportation system of the local body. • Shop keepers not to dispose of the waste in front of their shops or open spaces. Deposit the waste as and when generated into the large container placed in the market.
Meat and fish markets	• Not to throw any waste in front of their shops or open spaces around. Keep non-corrosive container/containers not exceeding 100-litre capacity with lid handle and the rim at the bottom and deposit the waste in the said containers as and when generated. • Transfer the contents of this container into a large container provided by the association.

Table Contd...

Source of waste generation	Action to be taken
Street food vendors	• Not to throw any waste on the street, pavement or open spaces. Keep bin or bag for the storage of waste that generates during street vending activity • Preferably have arrangements to affix the bin or bag with the hand–cart used for vending.
Marriage halls, community halls, kalyanamandapas	• Not to throw any solid waste in their neighbourhood, on the streets, open spaces, and vacant lands, into the drains or water bodies. • Provide a large container with lid which may match with the transportation system of the local body and deposit all the waste generated in the premises in such containers.
Hospitals, Nursing homes, etc	• Not to throw any solid waste in their neighbourhood, on the streets, open spaces, and vacant lands, into the drains or water bodies. • Not to dispose off the biomedical waste in the municipal dust bins or other waste collection or storage site meant for municipal solid waste. • Store the waste as per the directions contained in the government of India, Ministry of Environment Biomedical Waste (Management and Handling) Rules, 1998.
Construction/ demolition waste	• Not to deposit construction waste or debris on the streets, footpaths, pavements, open spaces, water bodies etc. • Store the waste within the premises or with permission of the authorities just outside the premises without obstructing the traffic preferably in a container if available through the local body or private contractors.
Garden waste	• Compost the waste within the garden; if possible trim the garden waste once in a week on the days notified by the local body. • Store the waste into large bags or bins for handing over to the municipal authorities appointed for the purpose on the day of collection notified.

8.5 MSW Treatment/Disposal

In general, no processing of municipal solid waste is done as a rule in India. The following technologies, however, would be relevant, considering the quantity and quality of waste generated, for the various WTE applications in the urban and industrial sectors in India.

Urban Waste:

  (a) Municipal Solid Waste: Biomethanation, Gasification, Composting, Incineration, Landfill with Gas Recovery (LFG), Refuse Derived Fuel (RDF)

  (b) Municipal Liquid Waste: Biomethanation

Industrial Waste:

  (a) Liquid waste: Biomethanation

  (b) Solid waste: Gasification, Incineration

  (c) Semi-solid waste: Biomethanation, Incineration / Gasification.

In a few cities, (i) composting (aerobic composting and vermi-composting) and (ii) incineration (refuse derived fuel), and (iii) biomethanation are utilized for processing. There are also trials of other technologies such as gasification and pyrolysis, and plasma pyrolysis.

WTE projects are relatively recent in India and are taking off the ground in the past few years. The factors that determine the techno-economic viability of WTE projects are quantum of investment, scale of operation, availability of quality waste, statutory requirements and risks involved in the projects. In the developed countries, the plants are economically viable because of the tipping fee charged for the service of waste disposal by the facility, in addition to the revenue generated from power sales. But, in India, earnings from the power sales are the only main source of revenue. Though technologically it is feasible to set up projects with smaller capacity 1 to 5 MW corresponding to around 100 to 500 TPD waste treatment, sustainability of such projects is yet to be firmly established. The economics of scale generally favour centralized large scale projects. The terms for MSW supply, allotment of land, and sale of power directly affect net revenue and in turn, the financial viability of projects as well as private sector participation.

The methods currently in practice for the disposal and treatment of MSW in the country are briefly outlined along with the related issues and opportunities.

8.5.1 Landfilling

In many metropolitan centres, open uncontrolled and poorly managed dumping is commonly practiced, giving rise to serious environmental and health problems. More than 90% of MSW in cities and towns are directly disposed of on open land. Such dumping practices in many coastal towns have led to heavy metals and other contaminants rapidly leaching into the coastal waters. In larger cities like Delhi, the availability of land for waste disposal is highly limited (Mor et al. 2006; Siddiqui et al. 2006; Sharholy et al. 2006; Gupta et al., 1998; Das et al., 1998; Kansal et al., 1998; Chakrabarty et al., 1995; Khan 1994). Hence, MSW is disposed by dumping in low-lying areas outside the city ignoring the principles of sanitary landfilling. The incoming MSW vehicles are not weighed and no specific plan is followed when filling the dumpsites. Compaction and leveling of waste and final covering by earth/inert material are rarely done at most disposal sites. These sites are devoid of a leachate collection system or landfill gas monitoring/collection equipment (Bhide and Shekdar 1998; Gupta et al. 1998). The poorly maintained landfill sites are prone to groundwater contamination because of leachate production posing a serious threat to human health. Most of the disposal sites are unfenced and the waste picking is common, creating problems during operation of the sites. Further, open dumping of garbage facilitates the breeding for disease vectors such as flies, mosquitoes, cockroaches, rats, and other pests (CPCB 2000). The situation is similar to the one in any developing country. Smoke and/or fire nuisance is caused by unauthorized burning of waste by rag pickers. Open firing of MSW at disposal sites is the regular method of reducing the volume of wastes. It is also carried out to make picking of recyclables easier (Kumar et al 2009).

Organic matter content in the deposited MSW at the landfill site tends to decompose anaerobically leading to emission of volatile organic compounds and gaseous by-products. As we know, the landfill gas (LFG) contains methane (50 to 60 per cent) and carbon dioxide as major constituents, and has potential for non-conventional energy. Since methane is more potent than carbon dioxide in contributing to greenhouse gas effect, LFG has to be properly handled without letting into atmosphere. TERI has estimated that the country released about 7 million tonnes of methane into the atmosphere in 1997, and if no efforts are made to reduce the emission through methods like composting, it may increase to 39 million tonnes by 2047 (Asnani 2006).

As no segregation of MSW at the source takes place, all kinds of wastes including contagious waste from hospitals generally end up at the disposal site. The industrial waste is also deposited quite often at the landfill sites meant for domestic waste (Datta 1997). Until recently there was not a single sanitary landfill site in India. Of late, four sites have been constructed at Surat (Gujarat), Pune (Maharashtra), Puttur and Karwar (Karnataka), and a few more sites are coming up. The Municipal Solid waste (Management and Handling) Rules 2000, make it essential for all local bodies in the country to have sanitary landfills. Some states like West Bengal, Rajasthan and Gujarat are considering constructing regional facilities, as the construction of landfills are expensive and require professional management.

Despite several problems, landfilling would continue to be the most widely acceptable and adopted practice in the country for many more years/decades. The municipal authorities need to effect improvements to the existing ones to ensure that the new landfills as well as the existing ones follow the accepted norms of sanitary landfilling (Kansal 2002; Das et al. 1998; Dayal 1994) and MSW 2000 Rules.

8.5.2 Composting

Throughout India, a large number of small-scale decentralized composting schemes are operating with various levels of success. In the composting, the waste volume can be reduced to 50–85%. Manual composting is the practice in smaller urban centers, and mechanical composting in big cities (Bhide et al 1998; Chakrabarty et al.1995). Government of India (GOI) primarily concentrated in 1960s, on promoting composting of urban MSW and offered soft loans to urban local bodies. In the Fourth 5-year plan period (1969–1974), block grants and loans were provided to state governments for setting up MSW composting plants. In 1974, GOI introduced modified scheme to revive MSW composting, particularly in cities with a population over 0.3 million. Many mechanical compost plants with capacities ranging from 150 to 300 tonnes/ day were set up in Bangalore, Baroda, Mumbai, Kolkata, Delhi, Jaipur and Kanpur during 1975–1980. A survey (UNDP/WB RWSG-SA 1991) undertaken in 1991, analysed 11 heavily subsidized mechanical municipal compost plants that were put up during 1975-1985 with input capacity of 150-300 tonnes/day, and found that only three were under operation, operating at much lower capacities than expected. This survey suggested setting up of several small-scale (decentralized) compost plants instead of one large mechanical compost plant. The decentralized compost plants have the merits to: (a) enhance environmental awareness, (b) create employment in the neighbourhood, (c) create more flexibility in operation and maintenance, (d) allow the residents close examination of the services and products, (e) reduce waste management costs for the municipality, and (f) decrease dependence on municipal services (Mansoor Ali 2004). NGOs and Community groups have initiated and established subsequently small-scale compost plants in many cities.

However, the first large-scale aerobic composting plant in the country was set up in Mumbai in 1992 to handle 500 t/day of MSW by Excel Industries Ltd. Currently, the plant is working at 300 t/day capacity, but very successfully and the compost produced is in demand. Another plant with a capacity, 150 tonnes/ day has been in operation in Vijaywada. Over the years a number of plants have been installed in the principal cities of the country – Delhi, Bangalore, Bhopal, Ahmedabad, Hyderabad, Luknow and Gwalior. Many other cities are in the process of establishing composting facilities. It is estimated that about 9% of MSW is treated by composting in the country (Gupta et al. 1998; Gupta et al, 2007; Sharholy et al. 2006; Srivastava et al. 2005; Malviya et al. 2002; CPCB 2000; Reddy and Galab 1998; Dayal 1994; Rao and Shantaram 1993).

Approximately 35 composting projects have been set up in India with private sector participation in the states of Maharashtra, Tamil Nadu, Andhra Pradesh, and Kerala. Typically the arrangement has been on a BOO or BOOT structure. The treatment capacity of these facilities ranges from 80 to 700 TPD and their combined capacity is about three millions tons per year. More projects are being finalized with PSP arrangement. Capital investment required for such facilities (capacity 100 to 700 TPD) typically ranges from Rs. 30 to 75 million. The promoter equity has largely been the project financing. The private partner recovers the investment by selling compost derived from waste processing. For example, Kolhapur Municipal Corporation (KMC), Maharashtra selected Zoom Developers Ltd. to implement the solid waste composting project, in association with Larsen Engineers in 1999. KMC and Zoom signed a 30-year, Build-Own-Operate-Transfer (BOOT) contract in 2000. The facility would handle 160 TPD in the initial year, increasing to 270 TPD in the final year. The KMC would deliver solid waste to the treatment site (a weekly average of 770 tons), for which the concessionaire would compensate it with a fixed annual payment of Rs. 0.48 million (escalated annually at eight percent). The concessionaire would pay the city one rupee per square meter per year for the land lease. The city would receive an estimated Rs. 0.65 million in the first year of the facility's operation.The Composting projects in India are listed (see Annexure 4). But not all of them are working.

Vermi composting is under operation in Hyderabad, Bangalore, Mumbai and Faridabad, the Bangalore plant being the largest with a capacity of 100 MT/day. Small-scale units are operating in many towns and cities. Normally, vermi-composting is preferred to microbial composting in small towns as it requires less mechanization and is easy to operate. Experiments to develop household vermicomposting kits have also been conducted. However, the progress has not been much (Ghosh 2004; Bezboruah and Bhargava 2003; Jha et al. 2003; Sannigrahi and Chakrabortty 2002; Gupta et al. 1998; Reddy and Galab 1998; Jalan 1997; Khan 1994).

8.5.3 Anaerobic Digestion (Biomethanation)

It is comparatively well-established technology for disinfections, deodorization and stabilization of sewage sludge, animal slurries, farmyard manures, and industrial sludge. Its application to the organic fraction of MSW is more recent. Anaerobic digestion leads to energy recovery through biogas generation, in addition to residual sludge. The method offers advantages over composting in terms of energy production/ consumption, compost quality and net environmental gains. The method is suitable for kitchen waste and other putrescible wastes which are too wet for aerobic composting. It is a net energy-producing process, around 100-150 kWh per tonne of waste input. The biogas, which has 55–60% methane, can be used directly as a fuel or for power generation. It is estimated that in controlled anaerobic digestion, 1 tonne of MSW produces 2 to 4 times more methane in 3 weeks in comparison to what 1 tonne of waste in landfill will produce in 6–7 years (Ahsan, 1999; Khan, 1994).

In India, Western Paques have tested the anaerobic digestion process to produce methane gas. The results of the pilot plant show that 150 t/day of MSW produce 14,000 m^3 of biogas with a methane content of 55–65%, which in turn can generate 1.2 MW of electric power. The government is eager to promote biomethanation technology as a secondary source of energy by utilizing industrial, agricultural and municipal wastes. A great deal of experience with biomethanation systems exists in Delhi, Bangalore, Lucknow and many other cities. But, there is little experience in the treatment of solid organic waste, except with sewage sludge and animal manure (e.g., cow dung). Several schemes for biomethanation of MSW, vegetable and yard wastes, are currently in operation and are also planned for some cities (Ambulkar and Shekdar 2004; Chakrabarty et al. 1995).

The biogas technology developed at BARC (Bhabha Atomic Research Centre) in India for treating all biodegradable waste and commercialized Nisarga-Runa technology is an improvement on this technology (see Annexure 7). BARC in collaboration with Bangalore Corporation (BBMP) will set up 16 biomethanation plants covering City area that convert garbage into gaseous fuel which can be converted into electricity. It will resolve the problem of solid waste management disposal as well as resolve the power woes of the city (Times of India, Apr 12, 2011). Unlike conventional biogas plants that can process only human waste and cow dung, Nisargruna plants can process all biodegradable waste. The Bangalore city generates about 3,200 tonnes of garbage every day. The first project will be started at Mathikere in April 2011. One tonne of waste can generate 60 metric cubes of gas and 50 kg of manure. Further, the gaseous fuel can be converted into electricity with the help of generator and can be used to light 250 street lights for 10 hours. Methane gas can be used as kitchen fuel to run a canteen. The project can be successful only if garbage segregation begins at household level.

The Ministry of Non-conventional Energy Sources has been promoting setting up of Waste-to-Energy projects in the country through two schemes: (i) National Programme on Energy Recovery from Urban and Industrial Wastes, and (ii) UNDP/ GEF assisted Project on Development of High Rate Biomethanation Processes as a means of Reducing Green House Gases Emission.

The first scheme is applicable to private and public sector entrepreneurs and organisations as well as NGOs for setting up of waste-to-energy projects on the basis of Build, Own and Operate (BOO), Build, Own, Operate and Transfer (BOOT), Build, Operate and Transfer (BOT) and Build Operate Lease and Transfer (BOLT). It is being implemented through State Nodal Agencies who certify the financial, managerial and technical capabilities of the promoters and on assured availability of waste materials on a long term basis (over 10 years) for operating the project.Three projects with capacity of 5 MW, based on palm oil industry waste, cattle dung and poultry waste were completed and commissioned during 2004-05. Two projects were under the National Programme and the third project was completed under the UNDP/GEF project. Some details about these and other projects are given.

3.0 MW power project based on Palm Oil Industry Waste by M/s Sai Renewable Power Pvt. Ltd. Hyderabad, at Eluru, Andhra Pradesh: The plant has been set up based on combustion of empty bunches from palm trees and residues of palm fruits. These remains are rich in volatile substances and will be used as a fuel for the boiler in the plant. About 100 tonnes per day of palm oil industry waste is being used to produce 3 MW of power. MNES finds the project working satisfactory to the capacity.

1.5 MW power project based on poultry droppings at Namakkal, Tamil Nadu: The plant has been installed by M/s G. K. Bio-Energy Pvt. Ltd., Namakkal, based on biomethanation technology to generate power using poultry droppings from one million birds of nearby poultry farms. It is based on BIMA (Biogas Induced Mixing Arrangement) technology developed and commercialised by M/s Entec, Austria. The engines used for generating power are 100% biogas engines imported from Austria. The total cost of the project is Rs.180.4 million.

5 MW municipal solid wastes (MSW) based project at Lucknow: The project has been executed by M/S Asia Bio-energy Pvt. Ltd (ABIL), Chennai on Build, Own, Operate and Maintenance basis in association with Lucknow Nagar Nigam (LNN) who are responsible for supply of required quality and quantity of MSW at the plant site. The plant based on Biomethanation technology started its commercial operation in August 2003 but could reach to a maximum generation capacity of 1.5 MW only by March 2004. The plant is presently facing problem in its operation mainly due to non-availability of the required quality of MSW free from debris, sand and silt.

0.5 MW power project based on slaughterhouse solid waste at M/s Hind Agro Ltd., Aligarh, U.P: M/s Hind Agro has a 100% export oriented modern integrated abattoir cum meat processing plant at Aligarh. The biogas plant being installed at their place is designed to treat solid waste generated from slaughtering of 1600 buffaloes everyday. The project for biomethanation of slaughterhouse solid wastes to produce about 4000 cum. biogas per day for generation of 0.5 MW power from about 50 tonnes per day solid wastes was installed. The plant was installed by M/s RSB Japan on turnkey basis under the technical supervision of Central Leather Research Institute, Chennai.

A typical *Farmyard Biogas plant* generally applicable to rural locations is described below (Ludwig Sasse's article on Biogas plants): A biogas plant operates on the principle of anaerobic digestion and supplies energy (in the form of biogas) and residue as fertilizer. It improves hygiene and protects the environment. A biogas plant is a modern energy source, and improves working conditions especially for the rural people in the country.

The methane content in the biogas depends on the digestion temperature. Low digestion temperatures give high methane content, but less gas is produced. The methane content also depends on the feed material. Some typical values are as follows: Cattle manure 65%, Poultry manure 60%, Pig manure 67%, Farmyard manure 55%, Straw 59%, Grass 70%, Kitchen waste 50%, Algae 63%, Water hyacinths 52% and Leaves 58%.

A plant with long retention times is beneficial to a farmer with few animals and to the national economy. The personal benefit of a biogas plant to the owner-farmer depends on how his energy and fertilizer requirements are met earlier: the benefit is greater the more energy had to be bought in (diesel oil, coal, wood) and the higher the cost of that energy. However, there is always a close relationship between energy costs and those of construction of the plant.

A floating-drum plant with internal gas outlet is shown in Fig.8.4. The gas pipe is securely mounted on the wall and leads directly to the kitchen. Ideally, as in this example, the digester should be located directly beside the animal shelter, which should have a paved floor. Urine and dung can be swept into the inlet pipe with little effort. The plant has a sunny location, and the vegetable garden is situated directly adjacent to the digested slurry store. The well is an adequate distance away from the biogas plant.

Fig. 8.4 A farmyard Biogas plant (Courtesy: Ludwig Sasse – Biogas Plants).

The benefit of the fertilizer depends primarily on how it is used by the farmer. If the digested slurry is immediately utilized and properly applied - as fertilizer, each kg of slurry can be expected to yield roughly 0.5 kg extra nitrogen, as compared with fresh manure. If the slurry is first left to dry and/or improperly applied, the nitrogen yield will be considerably lower.

If parasitic diseases had previously been common, the improvement in hygiene also has economic benefits (reduced working time). If the sludge is completely digested, the more pathogens are killed. High temperatures and long retention times are more hygienic.

The principal organisms that are killed in biogas plants are: Typhoid, paratyphoid, cholera and dysentery bacteria (in one or two weeks), hookworm and bilharzia (in three weeks). But the tapeworm and roundworm totally die only when the fermented slurry is dried in the sun.

Different types of Biogas plants and their relative merits and demerits, as discudded by Ludwig Saaae are given in Annexure.

Biomethanation as a WTE technology option has several positive favorable attributes for its adoption by the industrial sectors in India. Biomethanation has a long track record of over a decade in India in sectors like distillery and paper. Other sectors with full-scale biomethanation plants operating in the country include dairy, starch, yeast (based on sugarcane molasses), pharmaceutical (antibiotic, vitamin plants), poultry, tannery (wastewater and fleshings) and cattle farm manure. The full scale operating units in the country have provided a wealth of cumulative knowledge/information.

Several proprietary designs have evolved to offer compact commercial systems with low hydraulic retention time (HRT) and high solids retention time (SRT). These vary from simple contact reactors to advanced versions of fixed film, UASB, fluidised bed and other hybrid types of bio-reactors. Organic loading rate (kg $COD/m^3/d$) is the main process design parameter for liquid substrates and HRT is used as a convenient parameter for solid / semi-solid / slurry substrates. These parameters can be established from laboratory feasibility studies or pilot plant trials.

8.5.4 Incineration

The most attractive feature of the incineration process is very substantial volume reduction of combustible solid waste. In some newer incinerators designed to operate at temperatures high enough to produce a molten material, it may be possible to reduce the volume to about 5% or even less (Jha et al. 2003; Ahsan1999; Peavey et al.1985). In India, incineration is not utilized on a large scale. This may be due to the high organic material (40–60%), high moisture content (40–60%), high inert content (30–50%) and low calorific value or energy content (800–1100 kcal/kg) in the average municipal solid waste (Kansal 2002; Joardar 2000; Bhide and Shekdar 1998; Sudhire et al. 1996; Jalan and Srivastava 1995; Chakrabarty et al. 1995).

The first MSW incineration plant capable of generating 3.75 MW power from 300 t/day waste was constructed at Timarpur, New Delhi in 1987 at a cost of Rs. 250 million (US$5.7 million) by Miljotecknik volunteer, Denmark. The plant was out of operation after 6 months due to its poor performance. Another incineration plant was constructed at BARC, Trombay, Mumbai, for burning only the institutional waste, mostly paper. In many cities, small incinerators are used for burning hospital waste (Sharholy et al.2005; Lal 1996; Chakrabarty et al. 1995; Dayal 1994).

8.5.5 Gasification

In India, there are a few gasifiers in operation, but they are mostly for burning of biomass such as agro-residues, sawmill dust, and forest wastes. Gasification can also be used for MSW treatment after drying, removing the inerts and shredding for size reduction. Two different designs of gasifiers exist in India. The first one (NERIFIER gasification unit) is installed at Nohar, Hanungarh, Rajasthan by Narvreet Energy Research and Information (NERI) for the burning of agro-wastes, sawmill dust, and forest wastes. The waste-feeding rate is about 50–150 kg/h and its efficiency is around 70–80%. About 25% of the fuel gas produced may be recycled back into the system to support the gasification process, and the remaining is recovered and used for power generation. The second unit is the TERI gasification unit installed at Gaul Pahari campus, New Delhi by TERI (CPCB 2004; Ahsan 1999). No commercial plant has come up for the disposal of MSW, as it is still an emerging technology for MSW.

8.5.6 RDF Plants

The main purpose of the refuse derived fuel (RDF) method is to produce an improved solid fuel or pellets from mixed MSW as a fuel feed for thermal processes. The RDF fluff or pellets can be stored and conveniently transported over long distances, and can be used as a substitute for coal at a lower cost. The Technology Information,

Forecasting and Assessment Council (TIFAC) of Department of Science and Technology (DST), GOI has developed the technology to separate combustible fraction from MSW and carry out densification of the rest into pellets, and set up a demonstration plant at the Deonar, Mumbai in the early 1990. These fuel pellets have consistently recorded a calorific value of over 3000 k cal/kg. However, the plant has not been in operation for the last few years. The DST has transferred the technology to M/s. Selco International Ltd., and to M/s Sriram Energy Systems Ltd to set up RDF plants at Hyderabad and Vijayawada respectively. The Hyderabad plant, commissioned near the Golconda dumping ground with a 1000 t/day capacity, is currently working with a lower capacity of 700 t/ day. Selco is using the RDF produced (fluff and pellets) for generating electric power in a plant of capacity 6.6 MW setup by them. The fraction of agro-waste used along with MSW claimed by these two operators of these facilities is being challenged and the matter is under judicial scrutiny (Asnani 2006).

A similar project has been established in Bangalore and has regular production of fuel pellets, since October, 1989. The plant utilises 50 t/day of garbage, and converts into 5 t of fuel pellets which are designed both for industrial and domestic uses (Yelda and Kansal 2003; Reddy and Galab 1998; Khan 1994).

In summary, different technologies to recover useful energy from Municipal Solid Wastes are extensively available and are utilised mostly in developed countries. For India, biomethanation, incineration, gasification and RDF are promising technologies, and the last three can be used for power production. Particularly, the RDF plants reduce pressure on landfills. Combustion of the RDF from MSW is technologically sound and is capable of generating power. The technology developed by Department of Science and Technology, GOI for RDF production is very sound and reliable. RDF may be fired along with the conventional fuels like coal without any adverse effects for generating heat. In the operation of the thermal treatment systems, however, higher costs and relatively a higher degree of expertise are involved.

However, an Integrated Waste Management system supported by legislative and control measures is necessary in India, for the success of these technologies. A detailed feasibility study needs to be conducted in each case, duly taking into account the available waste quantities and characteristics and the local conditions as well as relative assessment of the different waste disposal options. Suitable safeguards and pollution control measures need to be incorporated in the design of each facility to fully comply with the environmental regulations and safeguard public heath. If the Waste-to-Energy facilities are set up with such consideration, they can effectively bridge the gap between waste recycling, composting and landfilling, for tackling the increasing problems of waste disposal in the urban areas in an environment-friendly manner. Further, they can help augmenting power generation in the country.

Relative Costs of Variuos WTE Technologies: The cost of waste treatment will vary for various treatment options. The application of a particular technology may be limited by requirement of land, operational costs and management complexities of various available treatment techniques. The background study undertaken by Environment Unit, South Asia Region, World Bank (WB 2006) suggested the approximate cost comparison for various technologies in India as shown in Table 8.6(a)

Table 8.6(a) Relative capital costs of various WTE technologies in India.

Technology	Assumed MSW quantity (Metric ton)	Land required (acre)	Cost (Rs. in million)
Biomethanation	150	6-7	60-90
Pelletisation	125	3-4	40-50
Incineration	100	2-3	60-70
Composting	150	7-8	15-20

8.6 Recovery of Recyclable Materials

Recycling of waste in India can be considered as a highly organised and profit making venture, though it is largely informal in nature. Community based organisations and NGOs play an important role especially in collection and segregation of wastes, as well as their recycling. A number of recyclable materials, for example, paper, glass, plastic and rubber, ferrous and non-ferrous metals present in the MSW are recovered and reused. It has been estimated that in Indian cities, the recyclable content in the waste varies from 13% to 20% (in Mumbai 17% and in Delhi 15%). A survey conducted by CPCB during 1996 in some Indian cities has revealed that rag pickers play a key role in collecting the recyclable materials from the streets, bins and disposal sites for their livelihood; only a small quantity of recyclable materials is left out. In India, about 40–80% of plastic waste is recycled compared to 10–15% in the developed countries. However, the recovery rate of paper was 14% of the total paper consumption in 1991, while the global recovery rate was higher at 37% (Pappu et al. 2007; CPCB 2004; Yelda and Kansal 2003; Shekdar 1999; Ahsan 1999; Dayal 1994; Khan, 1994).

The role of the informal sector is very significant in recovering materials compared to municipal authorities (Fig. 8.5). In Delhi, there are more than 100,000 rag pickers and the average quantity of waste materials collected by one rag picker is 10–15 kg/day. According to another source, in Delhi, there are about 200,000 self-employed waste pickers comprising of men, women, and children collecting about 2,000 tons of rubbish daily (CSE-publications). The rag pickers handle nearly 17% of solid waste in Delhi who collect, sort and transport waste free of cost, as part of the informal trade in scrap. This exercise saves the government Rupees 600,000 (US$13,700) daily. In Bangalore, the informal sector involved in waste handling prevents about 15% of the MSW going to the dumpsites. The waste pickers in Pune save around Rupees 9 million/ year (US$200,000) for the municipal corporation by their involvement.

In Hyderabad, the cost of MSWM per ton is less in the areas where the private sector participates compared to the areas serviced by municipality. In Mumbai, the observation is that the cost per ton of MSWM is US$35 with community participation, US$41 with public private partnership (PPP) and US$44 when only Municipal Corporation of Greater Mumbai (MCGM) handles the MSW. Hence, community participation in MSWM is the low-cost option. Several studies undertaken by different institutes and authorities also revealed that the role of the informal sector in MSWM is very important because it provides a livelihood to many poor, unskilled and marginalized people. The informal collection avoids environmental costs and reduces capacity problems at dumpsites; and the rag pickers can provide excellent segregation of MSW (Sharholy et al. 2005, 2006, 2007; Rathi 2006; Joseph 2006; Agarwal et al. 2005; Srivastava et al. 2005; CPCB 2004; Kansal 2002; Reddy and Khan 1994).

Fig. 8.5 Women scavenging a dump, a health risk (source: ARRPET 2004)

8.7 Healthcare Waste Management

Healthcare centres in India generate substantial quantities of waste, and it is their responsibility to manage that waste. Rural hospitals generate much less waste. The healthcare waste is generally estimated based on the number of beds in the centres. The health information sources reveal that 20% of total beds are in rural hospitals and 80% are in urban hospitals. Extrapolating from previous statistics on the number of beds and an average quantity of waste generation at the rate of 1 kg/bed/day, it has been estimated that about 0.33 million tonnes of hospital waste is generated annually (Patil and Shekdar 2001).

The quantities generated vary from hospital to hospital and depend on the nature of healthcare facility and local economic levels; the hospitals generally do not have a system of maintaining data on waste quantities. In a study carried out in Indore City (Patil and Shekdar 2001), quantities of waste were weighed in different hospitals serving with specialized units. Waste quantities are invariably estimated assuming 100% bed occupancy because many hospitals in India are over-occupied. In large hospitals, infectious wastes are usually disinfected and disposed of along with general waste. Waste generated in out-patient departments is treated in the same way. Wastes from operating theatres, wards and pathology laboratories are disposed of without any disinfection/sterilization.

Amputated body parts, anatomical wastes and other highly infectious wastes are incinerated wherever incinerators are available; the rest is burnt in a corner of the hospital grounds, mostly in open pits.

In small towns, healthcare waste is often buried in pits in the space available on-site or sent to MSW disposal sites. Smaller private nursing homes and clinics do not take any precautions and often dispose of their waste in the community bins intended for storage of MSW.

It is common practice to dispose of healthcare waste along with MSW. Open burning is also used for its disposal. Large numbers of waste pickers collect recyclable items such as plastics, needles and glass materials from the waste to sell. Cleaners/sweepers in healthcare units often pick out recyclables such as plastics, glass and metals from the waste and sell them. This is an unsafe practice, as it is associated with high risk of infection and serious disease.

8.8 Hazardous Waste Management

There is a steady increase in the generation of hazardous waste in India. Most of the industry and companies are unaware that the waste they are generating is hazardous and are not keen to make the information available. Hence, comprehensive and reliable data on the generation of hazardous waste in India is unavailable and the data obtained is far from authentic. In such a situation, it is necessary to rely on estimated and projected data on hazardous waste generation for planning and development purposes.

For hazardous waste management in India, the key issues are the environmental implications of: (a) uncontrolled waste generation, (b) improper waste separation and storage prior to collection, (c) multiple waste handling, (d) low standard of disposal practices, and (e) non-availability of treatment/ disposal facilities.

The scarcity of funding and sufficient skilled human power further aggravates the situation. Due to economic considerations, the companies which generate hazardous waste in rich countries look for the cheapest and easiest dumping grounds that exist in developing countries (including India) and ship their waste to those countries. Thus the trans-boundary movement of hazardous wastes (for example, E-waste) from developed countries to India where the environmental laws and their enforcement are not stringent has become another serious issue (Kumar et al 2008).

Some hazardous wastes are disposed of at a treatment, storage and disposal facility (TSDF), a centralized location catering for the hazardous waste generated from nearby sites. The TSDF provides small- and medium-scale industries generating hazardous waste with an efficient disposal channel. However, for small-scale waste producers who do not have their own treatment facilities, services for hazardous waste collection, treatment and disposal are highly inadequate. Hazardous wastes from these sources are collected and disposed of with MSW. In addition, following legislation designed to contain hazardous waste in secure sites, large stocks of hazardous or partially treated hazardous waste are stockpiled in the vicinity of industrial sites (Kumar et al 2009)

The site selection criteria for a TSDF depend upon the receptors and pathways of likely waste movement, waste characteristics and waste management practices. The planning for hazardous waste management involves a number of aspects ranging from identification and quantification of hazardous waste to development and monitoring of the TSDF. Recycling of hazardous waste is also practiced but its potential is low.

The issue of hazardous waste treatment is complex and requires a broad approach. The selection of an affordable and environmentally sound treatment technology necessitates a great deal of waste characterization. Though there are established guidelines for separation, storage, collection and transportation for hazardous waste, the wastes from industrial and non-industrial sectors still find their way to public landfills, nearby dump sites or waterways resulting in serious environmental concerns such as groundwater pollution. The people who depend on such ground water sources are affected. The disposal of hazardous wastes is a very critical issue and need to be addressed aptly.

8.9 E-Waste Management

The collection of E-waste from corporates and households was done earlier by scrap dealers and rag pickers, and the collected waste used to get into an informal e-waste recycling system. The informal recycling system not only includes processes such as dismantling and sorting but also very harmful processes such as burning and leaching in order to extract metals from electronic equipment. The formal e-waste recyclers are just emerging; corporate companies are actively started working on the Collection and Recycling and the demand seems to be increasing.

More than 332 000 tonnes of e-waste was generated in India in 2007; the same is expected to touch 467 000 tonnes by 2011 and is likely to increase to 800 000 tonnes by 2012. The top states in order of highest generation of E-waste are Maharashtra, Andhra Pradesh, Tamilnadu, Uttar Pradesh, West Bengal, Delhi, Karnataka, and so on. E-waste is mostly generated in large cities like Delhi, Mumbai, Bangalore, Chennai, Kolkata, Ahmedabad, Hyderabad, Pune, Surat and Nagpur. Mumbai ranks first among the top 10 cities that generate e-waste in the country, with the generation of 23 000 tonnes of e-waste every year. Bangalore's innumerable IT (information technology) and related companies produce 11000 tonnes of e-waste every year.

Most of the E-waste collectors and recyclers, only do size reduction (shredding), segregation and export them. Companies like 'Eco Recyclying Ltd' and 'Trishyiraya Recycling'are involved in exporting the waste mainly to Belgium. Three categories of WEEE account for almost 90% of the generation:

Large household appliances: 42.1%, Information and Communications Technology   equipment: 33.9%, Consumer electronics: 13.7%. An estimated 30000 computers become outdated every year from the IT industry in Bangalore alone (Source: WEEE, Waste Guide, MAIT, Eco Recycling Ltd, Trishyiraya).

Like in developing countries, E-waste is most sought after item in India for scavengers and recyclers. In Delhi alone, there are about 25,000 workers employed at scrap-yards, where 10,000 to 20,000 tons of E-wastes are handled every year, with computers accounting for 25 percent (Indian Express, 2005).

The *challenges* of managing E-waste in India are very different from those in other countries because of the complexity of the e-waste issue. Some of the challenges are: rapidly increasing e-waste volumes, both domestically generated as well as through imports, no accurate estimates of the quantity of e-waste generated and recycled, poor awareness amongst manufacturers and consumers of the hazards of faulty e-waste disposal, widespread e-waste recycling in the informal sector using undeveloped techniques resulting in severe environmental damage, workers have no knowledge of toxins in e-waste, inefficient recycling processes result in substantial losses of material value so on.

Table 8.7 Informal recycling of E-waste in Chennai.

Computer component	Recovered component	Mechanism employed
Monitor	Cathode ray tube, circuit board, copper, plastics	Dismantling using screw drivers (the broken CRTs are dumped)
Hard disk	China steel, aluminum, actuator (magnet), platter, circuit board	Broken using hammer
Circuit board	Capacitor, condenser, copper, gold, chipped board	Gold recovery - acid treatment, Copper recovery - heating, Crushing of boards by custom-made crushers
Printer	Motor, plastics	Dismantling using screw drivers
Cables and wires	Copper, aluminum	Burning or stripping

8.10 Rules, Legislation and Legal Provisions

Many laws concerning solid waste handling are framed to improve the quality and efficiency of solid waste management and to regulate the disposal activity which affects adversely public health, the environment and economics. Many laws are intended to deal with the issues/problems related to solid waste management.

The Acts, Rules and Notification regarding Solid Waste Management in Inida are the following:

Law of Torts, Indian Penal Code 1860, Code of Civil Procedure 1908,

Constitution of India 1950, Code of Criminal Procedure 1973,

The Water (Prevention and Control of Pollution) Act 1974,

The Air (Prevention and Control of Pollution) Act 1981,

The Environment (Protection Act) 1986,

The Hazardous waste (Management and Handling) Rules 1989, 2003, 2008

The Coastal Regulation Zone Notification 1991,

The Bio-medical wastes (Management and Handling) Rules 1998, 2003

The Recycled plastics (Manufacture and Usage) Rules 1999

The Municipal Waste (Management and Handling) Rules 2000

Environmental Impact Assessment Notification, 2006 as amended 2009

National Environment Policy 2006

To live in a clean and healthy environment is not only a fundamental right guaranteed under Article 21 of our Constitution but also a right recognized and enforced by the Courts of Law under different laws, like Law of Torts, Indian Penal Code 1860, Civil Procedure Code 1908, and Criminal Procedure Code 1973. The Constitution of India 1950, the earliest legislation and the supreme law of the land has *imposed a fundamental duty on every citizen of India* under Article 51-A (g) *to protect and improve the environment.* The obligation on the State to protect the environment is expressed under Article 48 A. The right to live in a healthy environment is also a basic human right. The Universal Declaration of Human Rights 1948 gaurantees everyone the right to life under Article 3, and the right to a standard of living adequate for health and well being of himself and of his family under Article 25 (CPREEC).

The National policy of 'waste management' was legally enacted by the Government of India (Ministry of Environment and Forests, MoEF) as the *'Municipal Solid Waste (Management and Handling) Rules 2000'* mentioned already, in exercise of the power conferred under Sections 3, 6 and 25 of the Environment Protection Act, 1986. These rules shall apply to every municipal authority responsible for collection, segregation, storage, transportation, processing and disposal of municipal solid wastes. These Rules are finalized after inviting suggestions and objections to the draft document circulated among **all** the municipalities in the country. The key elements in The Municipal Solid Waste (Management and Handling) Rules 2000 are:

(i) *Collection of municipal solid wastes*: Organising doorstep collection of biodegradable and non-degradable municipal solid waste from houses, hotels, restaurants, office complexes and commercial areas, slums and squatter areas,

(ii) *Segregation of municipal solid wastes*: Municipal authority shall organize awareness programmes for segregating the waste at source in two bins: one for biodegradable and another for recyclable material, and promote recycling or reuse of segregated materials,

(iii) *Street sweeping*: Municipal authorities organize street sweeping covering all the residential and commercial areas on all the days of the year including Sundays and public holidays,

(iv) *Storage of municipal solid waste*: Municipal authorities shall establish and maintain storage facilities such that wastes stored are *not exposed to open atmosphere* and shall be aesthetically acceptable and user-friendly and it should be easy to operate, design for handling, transfer and transportation of waste.

(v) *Transportation of municipal solid wastes*: Vehicles used for transportation of waste shall *be covered and waste should not be visible* to public, or exposed to open environment and shall be so designed that multiple handling of wastes prior to final disposal is avoided.

(vi) *Processing of MSW*: Municipal authorities shall adopt suitable technology or combination of such technologies to treat wastes so as to minimize burden on landfill.

(vii) *Disposal of municipal solid waste*: Landfilling shall be restricted to non-biodegradable, inert waste and other waste that are not suitable either for recycling or for biological processing. Landfilling of mixed waste shall be avoided unless the same is found unsuitable for waste processing.

The guidelines also cover upgrading the existing facilities to arrest contamination of soil and ground water. The rules are to be implemented and monitored within a timeframe as given in the Table 8.8.

Table 8.8 Timeframe for the Implementation of the MSW Rules 2000 (source: Asnani 2006)

S.No.	Compliance criteria	Schedule
A	Setting up of waste processing and disposal facilities	By 31 December 2003 or earlier
B	Monitoring the performance of waste processing and Disposal facilities	Once in six months
C	Improvement of existing Landfill Sites as per provisions of these rules	By 31 December 2001 or earlier
D	Identification of landfill sites for future use and making site(s) ready for operation.	By 31 December 2002 or earlier

Recycled Plastics Manufacture and Usage Rules 1999 were amended in 2003 and the Rules are applicable in all the States/Union Territories. The rules lay much stress on the manufacturing of plastics using virgin materials and recycled plastics. The Rule also details the standard size and thickness of the plastics to-be-manufactured. The rules clearly state that the existing plastics manufacturing and recycling units register with the State Pollution Control Board/ Pollution Control Committee by fulfilling the conditions (CPCB, India).

Major developments on 3Rs (Reduce, Reuse, Recycle) in India:

(a) *Non-biodegradable Garbage (Control) Ordinance 2006, Maharashtra, India.* This Ordinance has come into force with Immediate effect. The ordinance controls ways in which non-biodegradable materials are to be disposed. It also bans the manufacture, transport and use of polythene bags. Maharashtra is the third state to enact such legislation. The state has set 50 microns as the least permissible thickness for polythene bags (Goa and Himachal Pradesh have specified a thickness limit of 40 microns and 70 microns respectively). However, polythene bags used for food items, medicines and milk and oil packets are omitted from the ambit of this ban, with a specification that such bags are to be manufactured *using virgin plastic raw material in its original*. The ordinance makes it mandatory for polythene bags to mention the details of the manufacturers, including the registration numbers issued by the Maharashtra Pollution Control Board (MPCB). It also enjoins manufacturers to provide information on the size and quality (virgin or recycled) of polythene. Moreover, no unit is allowed to manufacture polythene bags in the state without the consent of the Directorate of Industries and Commerce and MPCB (Source: MPCB –India*)*.

(b) *Recycling Schemes*: The MoEF, Government of India has launched a 'Registration Scheme' to channelize indigenously generated and imported recyclable waste to only those units with necessary facilities/technology to reprocess such waste in an environmentally sound manner. The Ministry reported a total of 476 registered Plastic Reprocessing, and other 252 registered units for Used Oil Reprocessing, Lead waste Reprocessing and Non-Ferrous Reprocessing.

(c) *Charter on Corporate Responsibility for Environmental Protection (CREP)*: After a series of industry specific interaction meetings, the Charter on Corporate Responsibility for Environmental

Protection (CREP) was adopted in March, 2003 for 17 categories of polluting industries. It is a road map for progressive step up in environmental management. Eight task forces comprising of experts and members from institutions and industry associations have been constituted for effective implementation of the Charter. These task forces meet regularly to monitor and to provide guidance to the industries for adopting necessary pollution abatement measures (Source: SOM 2006).

There are other Municipal Corporation Acts by different States such as the Delhi Municipal Corporation Act 1959, Uttar Pradesh Municipal Corporation Act 1959 and Karnataka Municipal Corporation Act 1976. These Acts also deal with environmental pollution caused by improper disposal of MSW. For example, The Delhi Plastic Bag (Manufacture, Sales and Usage) and non-biodegradable garbage (control) Act 2000 was enacted to prevent contamination of foodstuff carried in recycled plastic bags, to reduce the use of plastic bags, and to ban throwing or depositing non-biodegradable garbage in public drains, roads and places open to public view.

Local authorities often see MSWM services as poor compared to other basic services because MSWM can barely recover operating costs. Most of the municipalities are unable to provide adequate level of conservancy and SWM services for a variety of reasons (Siddiqui et al. 2006; Kansal 2002; MoEF 2000; Gupta et al. 1998).

Responsibility for implementation of MSW Rules 2000: These rules are applicable and mandatory to all Municipal authorities responsible for MSWM in the country. They are also accountable for implementation of these Rules and development of infrastructure. These authorities have to obtain authorization from State Pollution Control Boards for setting up waste processing and disposal facilities, and furnish annual compliance reports. The Urban Development Department of the State Government is responsible for the enforcement of the provisions in metropolitan cities. The State Boards and the other committees are required to monitor compliance of the standards regarding not only groundwater and ambient air but also leachate and compost quality including incineration standards. In addition, they are required to examine proposals for waste processing facilities giving due consideration to the views of other agencies.

The CPCB is required to coordinate with the State Boards (SPCB) for proper implementation of the rules. The central and the state governments, CPCB, State Pollution Control Boards and several national and international institutions have conducted training programmes to the personnel involved to help the cities and towns for prompt implementation of the rules.

Apart from implementing the MSW 2000 rules, other desirable actions required are: waste processing and disposal facilities to be monitored every six months, identification of landfill sites for future use, MSW not to be mixed with biomedical (healthcare) and industrial wastes, citizens to be encouraged to segregate the wastes, and processing of biodegradable waste by composting and anaerobic digestion (Kumar et al 2009).

An expert panel set up by the Ministry of Urban Development, Government of India has prepared and published a Manual on SWM in May 2000. All the states are provided with this Manual so that the Municipalities may adopt appropriate systems of solid waste management.

Compliance/Non-compliance of MSW Rules 2000: Complete compliance by 31st December 2003 did not happen. Many cities did not even initiated measures while a few cities started. A study was conducted to ascertain the status of compliance of the Rules by class I cities. 128 class I cities have responded. The collected responses on progress on each of the main components of MSW covered in the Rules as on 1st April 2004 is shown in Fig.8.6 (Asnani 2004).

According to municipal authorities, the compliance in waste collection is constrained by (Asnani 2006): lack of public awareness, motivation and education; lack of wide publicity through electronic and print media; lack of finances to create awareness; resistance to change; difficulty in educating slum dwellers; lack of sufficient knowledge on benefits of segregation; non cooperation from households, trade and commerce; unwillingness on part of citizens to spend on separate bin for recyclables; lack of sufficient litter bins in the city; non availability of primary collection vehicles and equipment; lack of authority to levy spot fines; lack of financial resources for procurement of tools and modern vehicles.

The creation of sufficient treatment and disposal facilities are limited by:

- dearth of financial resources as well as lack of support from state government
- non-availability of appropriate land;

- exorbitant time and cost considerations in land acquisition and setting up of treatment and landfill technologies;
- lack of adequate technical know-how and skilled manpower for treatment and disposal of waste;
- low quality of municipal solid waste;
- delay in clearance of disposal sites.

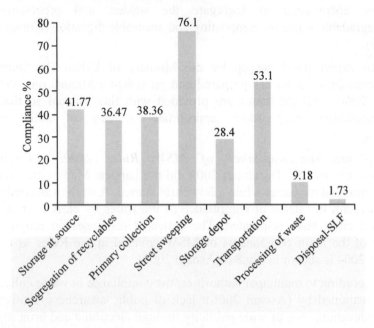

Fig. 8.6 Status of Compliance of MSW Rules 2000 as on 1st April 2004

Though the progress has not been at a desirable level, there has been a perceptible awareness among local bodies and policy makers to promote solid waste management systems. Recent years have witnessed the conditions improving in the country due to (a) regular monitoring by the Supreme Court, (b) offer of incentives by state governments, (c) large financial support from Central government on the advice of 12th Finance Commission, (d) provision of urban renewal funds to the states (JNNURM), and (e) technical and financial support from various Ministries of Government of India and national and international organizations.

To implement the rules expeditiously, one of the strategies could be to contract out most of the SWM services to Private sector, NGOs, RWAs and CBOs. The overall merits in inviting Private sector participation in MSWM have been discussed in later pages.

Rules for Healthcare waste management: The Biomedical Waste (Management and Handling) Rules 1998 (amended in 2000) set out requirements for treatment and disposal of healthcare waste. The rules include the following provisions:

(a) biomedical waste must not be mixed with other wastes,

(b) all operators of biomedical facilities must set up on-site waste treatment facilities (e.g. incinerator, autoclave, microwave system) for the treatment of the waste generated, or ensure proper treatment of the waste at a common or any other waste treatment facility,

(c) biomedical waste must be segregated into containers/bags at the point of generation (as described by the Rules) prior to its storage, transportation, treatment and disposal,

(d) containers of biomedical waste must be labelled as specified by the Rules. If a container is taken to an external waste treatment facility, the container must also carry the information prescribed by the Rules,

(e) despite any provisions of the Motor Vehicle Act 1988 (or rules there under), the biomedical waste must be transported only in vehicles authorized for the purpose by the government/ authority,

(f) untreated biomedical waste must not be kept or stored for more than 48 hours. In case of necessity to store the waste for longer, measures must be taken to ensure that human health and the environment are not adversely affected,

(g) all operators submit a report to the specified authority specifying the nature and quantities of biomedical wastes handled during the preceding year in a specified format by 31 January, to be sent to the CPCB by 31 March every year,

(h) every authorized person for Biomedical waste handling must maintain records relating to the generation, collection, reception, storage, transportation, treatment, disposal and/or any form of handling of biomedical waste in accordance with these rules and guidelines issued.

(i) accidents that occur at any institution or facility or any other site where biomedical waste is handled or during transportation of such waste must be reported immediately by an authorized person on the designated form to the authority.

Rapid urbanization in the country has led to improved medical facilities in urban centres. But separate systems for the disposal of healthcare waste are available in only a few cases. In all other systems, the infectious waste from healthcare units is mixed with MSW aggravating the problem of waste management, already overburdened in the urban centres. The existing system of management for healthcare waste suffer from certain defects that need attention: mixed collection of wastes increases the quantity of waste classified as infectious, absence of colour-coded storage containers for different categories of waste, non-availability of treatment and processing devices compatible with waste generation, lack of common treatment and processing facilities, unplanned waste management systems, inadequate budget allocation, lack of awareness of better waste management practices, and lack of waste management training for people working in the healthcare centres.

Rules for Hazardous waste management: The Hazardous Wastes (Management and Handling)1989 (amended in 2003) stipulate that every operator or a recycler generating/handling hazardous waste must obtain permission to dispose of hazardous waste by applying on a prescribed Form (together with an application processing fee) as laid down by the State Pollution Control Board or other appropriate authority. Operators or recyclers not having their own TSDF (Transport, Storage, and Disposal Facilities) but operating in an area under the jurisdiction of a State Pollution Control Board/ Committee for a TSDF must (a) become a member of this facility, (b) pay charges as may be required, and (c) take all other necessary steps to ensure proper treatment and disposal of any hazardous waste generated. The rules also contain the following provisions:

(a) Every State Pollution Control Board/ Committee must maintain a register containing details of the conditions imposed for the disposal of hazardous wastes from any land or premises. This register must be open for public inspection during office hours. Entries in the register will provide proof of the grant of authorization for the disposal of hazardous wastes from such land or premises and the conditions against which it was granted;

(b) All operators seeking to import hazardous wastes must apply to the State Pollution Control Board/ Committee using the prescribed form at least 120 days before the intended date of start of the shipment for permission to import hazardous wastes (together with an application fee if required);

(c)   The owner/ operator of a facility is liable for the entire cost of any remediation or restoration. An amount must be paid in advance as estimated by the State Pollution Control Board/Committee, which will plan and arrange for the implementation of the programme for remediation/ restoration. The advance paid will be adjusted once the actual cost is known and any further expenditure will be recovered from the owner/ operator of the facility.

In addition to notified MSW Management Rules 2000, Biomedical Waste Handling Rules 1998, and Hazardous Wastes Management Rule 1989, the Indian Government has taken further initiatives such as Reforms Agenda (fiscal, institutional, legal), Technical Manual on MSW Management, and Publications by the Technology Advisory Group on MSWM.

The most significant initiative was to set up the Jawaharlal Nehru National Urban Renewal Mission (JNNURM) by the Ministry of Urban Development with an assured grant of Rs 12,500 million ($291 million) to 63 chosen cities. Since most of the cities are unable to raise funds, JNNURM provides funds to meet their infrastructure needs. These beneficiaries have obligation to embark on reforms in governance to help improve efficiency and become self-sustaining in the future. These include: income tax relief for waste management agencies, public - private partnerships in solid waste management, capacity building, and creating Urban Reforms Incentive Fund.

A developing country like India that is shifting from the traditional economy to a technology-based industrialized economy faces problems and challenges in the management of wastes. The solid waste management systems are yet to emerge as a well-organized practice despite their existance in most of our urban centres for the last few decades. For instance, significant variations in the MSW characteristics among urban centres exist; but little effort is made to shape the relationship of the waste management system to the waste characteristic

8.11 Financial Resources

A committee appointed by the Supreme Court in 1999, had estimated that Rupees 15 million per 100,000 populations would be required to improve SWM services – collection, transportation, processing and disposal of waste – using scientific methods. On this basis, a total funding of Rupees

42750 million would be necessary. Out of this budget, Rupees 17100 millions would go towards the cost of the tools, equipment and vehicles, and Rupees 25650 for the treatment and disposal of waste. A Manual on SWM prepared by the expert committee set up by Ministry of Urban Development (MOUD), Government of India has given standard estimates for modernizing the SWM practices in different categories of cities and towns (Asnani 2006). See Table 8.9.

Table 8.9 Estimated Cost for Vehicle, Tools, Equipments and Composting (Source: Asnani 2006)

City population (in million)	Cost of vehicles, tools and equipment (in Rs lakh)	Cost of composting (Rs lakh)
< 0.1	50.97	20
0.1 – < 0.5	295.00	150
0.5 – < 1.0	511.00	500
> 2.0	948.00	1000

The municipalities do not have the capital to introduce these practices in SWM; and they have to look to central and state governments for finances. The Ministry of Urban Development has formulated a SWM Scheme for 423 Class I cities/towns that costs Rupees 25000 million. The Ministry has got this amount included in the 12th Finance Commission allocation of funds to ULBs. To draw more funding and to improve efficiency in the implementation, private sector partnership has been suggested as part of this Scheme.

The 12th Finance Commission has allotted Rupees 50,000 million to the ULBs and Panchayats in the country, a substantial allotment, for improving urban infrastructure as seen in Table 8.10. From this allocation, ULBs are allowed to utilize and spend 50 per cent for improving the SWM services during 2005-2010. The urban renewal fund of the central government also earmarks certain portion to SWM services. The central government, therefore, has come up with a significant allocation of funds for improving SWM services. If the state governments and ULBs could match this funding, it should be possible to offer effective management of MSW to the public and keep the cities/towns clean.

Many of the state governments, Andhra Pradesh, Karnataka, Haryana, Gujarat, Tamilnadu, Maharashtra, Madhya Pradesh, Rajasthan, and Uttar Pradesh have announced policy framework to encourage setting up of

WTE projects, in terms of allotment of land at nominal lease rent, free supply of garbage, evacuation facilities, and sale and purchase of power. However, there are issues at the state level, especially with regard to power tariff, though the tariff for power purchase is agreed upon as per the guidelines issued by the Ministry of New and Renewable Energy.

Other Sources of Funds:

(i)   It is a common practice in almost all ULBs, to use certain percentage of property tax for performing SWM services. It is easy to administer this tax, called conservancy tax, because a separate collection system is not required. The main drawback is that in many cities/towns, the assessment and collection of property tax is poor, and hence the income generated is small.

(ii) Private Sector investments in resource recovery facilities such as composting and waste-to-energy plants.

(iii) Revenues from waste recycling, composting and waste-to-energy programmes.

(iv) Loans from financial institutions such as HUDCO and other Banks for financing vehicle and equipment purchase.

Table 8.10 12th Finance Commission Allocation (2005-10),
Source: Asnani 2006

No. State	Panchayats		Municipalities	
	per cent	(Rs crore)	per cent	(Rs crore)
1. Andhra Pradesh	7.935	1587	7.480	374
2. Arunachal Pradesh	0.340	68	0.060	3
3. Assam	2.630	526	1.100	55
4. Bihar	8.120	1624	2.840	142
5. Chhattisgarh	3.075	615	1.760	88
6. Goa	0.090	18	0.240	12
7. Gujarat	4.655	931	8.280	414
8. Haryana	1.940	388	1.820	91
9. Himachal Pradesh	0.735	147	0.160	8
10. Jammu and Kashmir	1.405	281	0.760	38
11. Jharkhand	2.410	482	1.960	98
12. Karnataka	4.440	888	6.460	323

Table Contd...

No. State	Panchayats per cent	(Rs crore)	Municipalities per cent	(Rs crore)
13. Kerala	4.925	985	2.980	149
14. Madhya Pradesh	8.315	1663	7.220	361
15.Maharashtra	9.915	1983	15.82	791
16. Manipur	0.230	46	0.180	9
17. Meghalaya	0.250	50	0.160	8
18. Mizoram	0.100	20	0.200	10
19. Nagaland	0.200	40	0.120	6
20. Orissa	4.015	803	2.080	104
21. Punjab	1.620	324	3.420	171
22. Rajasthan	6.150	1230	4.400	220
23. Sikkim	0.065	13	0.020	1
24. Tamil Nadu	4.350	870	11.440	572
25. Tripura	0.285	57	0.160	8
26. Uttar Pradesh	14.640	2928	10.340	517
27. Uttaranchal	0.810	162	0.680	34
28. West Bengal	6.355	1271	7.80	393
Total	100.000	20000	100	5000

Subsidy for compost and Waste-to-Energy Plants: The Ministry of Agriculture (MOA) and Ministry of Environment and Forests (MoEF) have separate Schemes to promote Composting. The MOA has proposed 'Balanced and Integrated Use of Fertilisers' in 8th Five year Plan (1992-97), under which financial support would be given to local bodies and private sector for building compost plants (includes machinery) using MSW. Each project, with a capacity of 50 to100 tonnes/day, is eligible for a maximum grant of Rupees 50 lakhs (5 million). The total assistance proposed during Ninth Plan was Rupees 180 million, and the budget provided for 2002-03 was Rupees 50 million. Thirty eight projects have been taken up under this scheme, and most of the grant remains unutilized.

The MoEF provides financial subsidy up to 50 per cent of the capital costs to set up demonstration plants on MSW composting. Limited financial assistance is also available for waste chracterisation and feasibility studies. Three pilot projects have been granted for qualitative and quantitative assessment of solid waste in Hyderabad, Simla and Ghaziabad. A few more demonstration projects in North Dumdum, New

Barrackpore, Chandigarh, Kozikode, Udumalpet (Tamilnadu) would be implemented.

Both the Schemes of MoA and MoEF do not include mechanisms to follow up on implementation and performance monitoring. Hence, the impact of these schemes is not available even with the Ministries.

There are other schemes such as JNNURM (Jawaharlal Nehru National Urban Renewal Mission) and UIDSSMT (Urban Infrastructure Development Scheme for Small and Medium Towns) which support projects under MSWM.

Table 8.11 Government of India Subsidy on SWM Plants

A).	Project for power generation from MSW involving refuse derived fuel (RDF)	Rs 1.5 crore/ MW
B).	Power project based on high rate bio-methanation technology	Rs 2 crore/ MW
C).	Demonstration project for power generation from MSW based on gasification/Pyrolysis and plasma arc technology	Rs 3 crore/ MW
D).	Biomethanation technology for power generation from cattle dung, vegetable market and slaughterhouse waste above 250 KW capacity	50 per cent of project cost up to a maximum of Rs 3 crore/ MW
E).	Bio-gas generation for thermal Application	Up to Rs 1 crore/ MW-eq.
F).	Project development assistance	Up to Rs.10 lakh/project
G).	Training course/seminar/workshop,etc.	Rs 3 lakh/ event

Note: The financial assistance for any single project will be limited to Rs 8 crore. *Source:* Government of India, Ministry of Non-Conventional Energy Source Scheme, 25 July 2005.

The MNRE, Government of India has launched a National programme as early as 1995 on *energy recovery* from urban and industrial waste. MNRE has estimated that the current potential of generation of power from urban and industrial waste is around 2600 MW, and actual achievement was only 55 MW as of December 2007. By 2012, the potential may rise to 3650 MW (FICCI survey 2007). An accelerated programme has been notified by this Ministry on energy recovery from urban waste during 2005-06. The incentives offered for different schemes are given in Table 8.11. If these projects are set up in North Eastern region and in Himachal Pradesh, Jammu and Kashmir, Sikkim, and Uttaranchal, the financial assistance will be 20 per cent higher than those specified. But these projects are on hold by an order from Supreme Court not to grant any subsidy because the apex court is examining the allegation that the provisions are being misused.

Carbon Finance/Sale of Carbon credits: Kyoto Protocol is an international initiative created to commit industrialized countries that are responsible for increasing greenhouse gas emissions in the atmosphere to reduce their emissions. The protocol has proposed three mechanisms to help industrialized countries achieve their objective. The industrialized countries could implement projects which result in emission reductions anywhere in the world and earn carbon credits to count towards their effort to reduce greenhouse gas emissions globally. The three mechanisms to achieve measurable and cost effective emission reductions are: Clean Development Mechanism (CDM), International Emission Trading, and Joint Implementation (UNFCCC 1997, Jayarama Reddy 2011).

The waste treatment and disposal projects can be implemented by cities/towns under this provision, especially Clean Development Mechanism, and avail financial benefits through the sale of certified emission reduction credits to the industrialized (developed) countries. CDM helps in overcoming technological and financial barriers associated with MSW management projects.

In India, only 11 projects have been registered till 2008 in MSWM, of which almost 50 % are on waste water and only 2 projects are of waste-to-energy.

Landfills generate a gas consisting of 50 per cent methane which is a potent greenhouse gas. Construction of landfills and compost plants, and setting up WTE projects can earn substantial carbon credits in large cities because of heavy generation of MSW. By selling these credits, municipalities can generate funds which can help to recover the cost of installations and plants, and to improve SWM services. The MoEF has a Cell which serves as a nodal agency for the CDM projects.

There are no examples of Carbon finance (CF) revenues generated by Indian municipalities or private operators from the MSWM business. However, Table 8.12 provides a rough estimate of comparative and potential CF revenues for various treatment technologies. Such finance could be very important in covering the costs of activities which are otherwise non-revenue generating. Although simple in principle, the approach is new and the details are not well established. So, there are a number of technical issues which have to be resolved. Also, there is lack of experience with the procedural requirements.

Table 8.12 *Indicative Carbon revenues Potential for different Technologies*

MSW treatment & disposal options	CO₂ Emissions (t CO₂E/tMSW)	Potential Emission Reductions (tCO₂E/tMSW)	Carbon finance for treatment of MSW Rs/tMSW
Assuming Landfill without LFG recovery as baseline			
Landfill with LFG recovery & flare	0.20-0.25	0.95-1.20	175-200
Landfill with LFG recovery and energy generation	0.21 (may be less if energy component is considered)	More than 0.95	More than 175 Rs/ton
Composting	0 (may be less if replacement of chemical fertilizer is considered)	More than 1.16	More than 200 Rs//ton
Biomethanation	0 (may be less if energy and fertilizer components are considered)	More than 1.16	More than 225 Rs/ton
			More than 225 Rs/ton

International Funding: Funds are also available for MSWM from UNEP, GTZ and other International agencies.

8.12 Future Scenario

In India, the amount of waste generated per capita is projected to increase at an annual rate of 1 to 1.33% (Shekdar 1999). TERI (Singhal and Pandey 2001) outlines various future projections for estimating the growth of MSW and the impacts of such growth, and discusses possible interventions to mitigate such adverse impacts. The projections have been made under the BAU (business-as-usual) scenario for the year 2047 taking 1997 as the base year (Fig. 8.7). The observations of the study are the following: Assuming the daily per capita waste generation in 1995 as 0.456 kg (EPTRI 1995) and the per capita increase in waste generation as 1.33%, the total projected waste quantity in 2047 would exceed 260 million tonnes, more than five times the present level of 55 million tonnes. This huge increase in solid waste generation will have significant implications such as the land required for disposing this waste, methane emission etc. To dispose this quantiy of waste, the requirement of land (base year 1997) would be around 1400 km^2 by 2047 which is equivalent to the size of the city of Delhi. The estimates under the BAU scenario are made considering the average collection efficiency as 72.5%, average depth of landfill site as 4 metres, and average waste density as 0.9 tonne/m^3 (NIUA 1989). Locating land of that magnitude for waste disposal would be physically impossible since areas with the largest generation of solid waste would generally suffer from scarcity of vacant land. It means if the current methods of solid waste disposal continue, the waste would have to be carried over longer distances, requiring the creation of massive transport facilities and infrastructure and enormous extra finances.

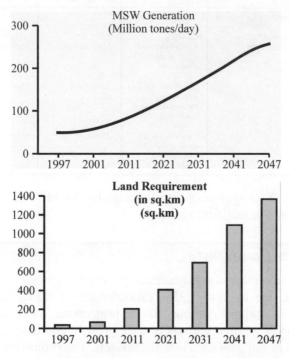

Fig. 8.7 Waste generation trend and implications for land requirement (source: TERI)

Further, indiscriminate landfilling would deteriorate the quality of groundwater in areas of landfill sites due to contamination by leachates from the landfills causing adverse health impacts on people living nearby. Landfill gas comprising 50%–60% methane, contributes significantly to global warming. It is estimated that in 1997, the landfills released about 7 million tonnes of methane into the atmosphere, which would increase to 39 million tonnes by 2047 under BAU scenario. Emissions have been calculated using Bingemer and Crutzen's (1987) approach, which assumes that 50% of the carbon emissions in the landfills are transformed into methane.

Only an efficient and sustainable system of solid waste management would facilitate to meet the situation. To achieve this, three aspects need consideration:

(a)   waste reduction at source

(b)   technological interventions, and

(c)   efforts towards institutional and regulatory reforms.

To achieve waste reduction at source, initiation of the following actions may be advantageous (Marcin et al 1994): (i) Promotion of market-mechanisms by offering tax incentives, or (ii) setting compulsory standards and regulations, or (iii) education and voluntary observance of policies by businesses and consumers. These initiatives are elaborated further:

(i)   *Market actions for waste reduction:* The efficiency of waste management can be improved by charging for the environmental and economic costs of production and disposal of waste. By incorporating disposal costs in the production expenditure, tendency to use less packaging or adoption of the recyclable/ reusable packaging material, and the tendency to reuse the material by the consumer would be promoted.

(ii)   *Mandatory standards for waste reduction:* Setting mandatory standards could make industry/ business responsible for the waste it generates. For instance, Germany has implemented a mandatory recycling programme in which, theoretically, the seller of consumer goods must take back all the package waste that is produced or used. In India the regulatory agencies should take the lead in setting up rules prescribing targets for waste reduction in various manufacturing sectors.

(iii)   *Education and voluntary compliance:* A voluntary programme of consumer education and business initiatives is beneficial, and can be achieved by the adoption of Environmental Management System (EMS). This is essentially a voluntary initiative. The industries adopting EMS have achieved economic benefits as well as better environmental performance in many countries.

India has lagged behind in terms of adopting suitable technologies for solid waste management. Waste collection, treatment and disposal require urgent and extra consideration.

Collection of waste: The existing collection service structure must be systematically changed. The community must be provided with waste bins conveniently placed, for the people to deposit domestic waste, and for undertaking door- to-door collection of waste. Further, separation of waste at source, into biodegradable and non-biodegradable components needs to be done. The MSW Rules 2000 demand these steps. These actions would not only reduce the cost of transportation for final disposal but also provide segregated organic waste stock for WTE activities.

Treatment and disposal: Proper segregation of waste would help to choose better options and provide opportunities for its scientific disposal. Recyclables for example, could be sorted out and transported directly to recycling units which would pay the municipal corporations towards the cost and services, increasing their income. The inert material that is required to be sent to landfill would then get reduced compared to un-segregated waste, which helps to increase the life of the existing disposal facilities.

The choice of technology needs to be based on its techno-economic viability, sustainability and the environmental concerns. The local conditions and the available physical and financial resources are also to be considered. The key issues which require serious deliberation are: the origin and quality of waste; presence of hazardous or toxic waste; the market for the energy generated; market for the compost or the anaerobic digestion sludge; energy prices for energy purchase; cost of land price and capital and labour costs and alternatives; and capabilities and experience of the technology provider. These aspects are explained in Chapter 6 in detail.

In general, the financial constraints, institutional problems within the departments, fragile links with other concerned agencies, lack of appropriate staff, and so on prevent the urban local bodies from delivering and maintaining an efficient waste management system. In this context, it is essential to bring in three other stakeholders - the private sector, NGO's, and rag pickers (informal sector) - into the overall institutional framework and integrate their individual roles. The private sector has now become a key player in delivering effective MSWM services in most of the industrialized countries.

The experience so far has shown that Private sector participation can help improve technical and managerial expertise, increase efficiency in operation and maintenance, and improve customer services, apart from bringing in the capital to support the government/ municipalities in their efforts at waste management. There is a strong case for a large-scale involvement of the private sector and encouraging it to invest in waste management.

Non-governmental organizations also can play an important role in effectively projecting the community's problems and highlighting its basic requirements for urban services. They could help in organizing the rag pickers into waste-management associations/groups under the supervision of the urban local body and the relevant Residents' associations.

Public Sector Participation and role of NGOs are discussed in the next Chapter.

CHAPTER 9

Private Sector Participation in India

9.1 Introduction

There are several issues which hold back an effective delivery of MSWM services by the Municipalities/ULBs. They include the following: (a) low priority given to SWM services among other activities, (b) poor community participation, (c) low quality services despite substantial costs, (d) engaging huge labour force having low productivity, (f) women and members of weaker sections mostly constituting the workforce, (g) lack of technically competent staff, (h) lack of technological know-how and (e) low cost recovery or no cost recovery. These issues are not simple and have socio-economic implications.

Looking at the disappointing performance of MSWM services by the municipalities in the country, it is better to change their role of being a 'service provider' to that of a 'service facilitator'.

Involving Private Sector Participation (PSP) as 'service provider' by the municipalities may probably improve the situation. The global experience has shown that PSP also called Public Private Partnership (PPP) involvement in MSWM has several advantages:

- bring finances for modernisation of SWM services and new investments,

- help in providing efficient MSW management services,
- help in cost savings as it raises productivity of manpower and machinery,
- deliver quality and speedy service as the private sector can be more insulated from bureaucratic and political interference,
- more flexibility in management, for example, in hiring qualified staff and paying according to their performance, in terminating the services of workers with poor performance, and in adjusting working hours according to service demand,
- access to technology and expertise,
- insistance on accountability, and
- focus on customer satisfaction.

There are a few disadvantages also, such as, insufficient competition, risk of commercial failure resulting in breakdown of essential public services, chances of paying very low salaries to the workers causing social suffering and labour problem, and lack of transparency. PSP requires close collaboration between public and private stakeholders and management, and monitoring skills of the public authority. Many of these disadvantages can be avoided with appropriate contracting mechanisms and improved tendering process. Further, PSP should create a congenial setting for an improved performance to the benefit of citizens. In developed countries the private sector plays a key role in managing most of the SWM services. However, the success of a private service provider strongly depends on the political will for a change, the skills of the provider and mutual trust between all partners.

In India, due to resistance from labour unions and misreading of labour laws, the municipalities, by and large, provide the SWM services departmentally and are not enthusiastic to engage private sector on a large-scale. Moreover, absence of a comprehensive state policy and legal frame work do not offer comfort to many municipalities to facilitate PSP in providing complete SWM services.

Implication of the Labour laws on SWM: Under the Contract Labour (Regulation and Abolition) Act 1970, there is a provision to prohibit contracting out any service if the same service is already being provided departmentally by any municipal authority, which limits their authority to act. But the Supreme Court of India has cleared the legal implications by its decision in Special C.A. No.6009-6010 of 2001 in Steel Authority of India Ltd and others Vs National Union Water Front Workers and others in August 2001, which has paved the way for municipal authorities to contract out certain SWM services.

The State Government of Tamil Nadu has exempted the Chennai Municipal Corporation from the purview of contract labour (Regulation & Abolition) Act 1970; vide its order No. 40 MS No. 99 dated 8th July 1999, allowing the municipal corporation to engage contract labour for sweeping and scavenging activities. Karnataka has declared state policy on Solid Waste Management; and States of Gujarat, West Bengal, Kerala and a few others have created high-power state missions to facilitate expeditious implementation of MSW rules in their respective states. Initiatives taken by the States are briefly outlined in the later pages.

9.2 Options in PSP Arrangement

A few Municipalities *have tried Private sector* in the following areas complying with the existing legal provision and labour laws:

(a) door-to-door collection of waste as this service is not currently provided by municipal authorities,

(b) street sweeping in areas which are not served, while retaining the existing street sweepers,

(c) providing large containers for secondary waste storage in various locations of the city,

(d) construction, operation and maintenance of transfer stations,

(e) expansion of transportation of waste without replacing existing work force, and providing fleet of vehicles and equipment for transportation of waste,

(f) construction, operation and maintenance of waste treatment facilities such as composting, waste to energy etc.,

(g) construction of engineered landfills, and

(h) operation and maintenance of landfills.

Currently, the capacity of municipalities in the country to manage the privatization process is very restricted. It is essential to develop the in-house financial and managerial capability to award contracts to private sector and to monitor the services provided because the ultimate responsibility of delivering quality services rests with municipalities. There are several options available that can be utilized from a PSP arrangement (Asnani 2006):

(i) *Service Contract:* The private partner has to provide a clearly defined service to the public partner. This can be used for waste collection, transport, transfer and operation of treatment and disposal facilities. The payment for such services can be based on quantity of waste, number of beneficiaries or total sum; and the contract can be awarded following competitive procurement procedure. The facilities are owned by the municipal authority but mobile equipment in most cases is owned by private contractors. The contract period is 1-2 years to 5-8 years depending on local conditions. The contracting authority is responsible for fee collection and cost recovery. The private sector absorbs the risk of operation.

Several types of 'service contract' are adopted in Indian cities: Bangalore, Nagpur, and Jaipur have involved the private sector in door-to-door collection and transportation. The contractor appoints own manpower, uses own tools, vehicles, and equipment and is paid by the municipal authority for the services rendered.

In some places such as Gandhinagar, North Dumdum, New Barrackpore, the contract is given for the door-to-door collection of the waste, and the contractor is expected to collect the fee from the citizens directly.

In Ahmedabad, the door-to-door collection of waste is entrusted to Resident welfare associations and backward class associations. They are provided with monthly grants to maintain the workers, and annual grants for purchasing tools and equipment.

Street sweeping contracts involving private sector are very few, because municipalities face stiff resistance from the existing staff employed under public regulations. However contracts for street sweeping are currently executed successfully in selected areas of Hyderabad, Chennai, Surat, and Rajkot. It is essential, in such contracts, to prescribe norms for the work and the minimum wages to prevent exploitation of labour by the private operators. Nearly 75 per cent of streets are awarded to 161 small contractors in Hyderabad, applying a unique unit area method of 8 km road length per 18 sanitation workers. Street brushing in Surat in the nights has made the city one of the cleanest in the country.

(ii) *Management Contracts*: The private company takes over the management and operation of a selected department or unit with its staff, facilities and machineries. The company brings in its own staff for key management pitions, but is unlikely to meet major capital costs. Management contracts are more often used in the 'drinking water delivery' and 'collection of waste water' sectors, but comparatively less in SWM due to the complex asset structure. The public authority makes the payment to the contractor which is typically a fixed fee and a performance fee.

In these contracts, the contractor must be given enough freedom to implement commercial reforms. There should be provision for penalties for failing to meet the agreed performance levels as well as incentives for good performance.

(iii) *DBO, BOT, BOOT*: Many cities and towns are not adequately equipped to handle *treatment and disposal of waste* which require a high technical expertise. Private participation is preferred in these areas.

- DBO (Design, Build, Operate) is one such most suitable approach. Here, the financing is met by the contracting authority, and the systems are designed, constructed and operated by professional agencies. DBO contracts minimize critical interface and makes the contractor responsible for successful implementation of the facility. Cities such as Mysore, Kochi, Calicut, Puri, and Shillong have adopted this approach to set up compost plants.

- BOT (Build, Operate and Transfer) is another arrangement. The private sector is responsible for construction, financing and operation of the facility during the contracting period. After the completion of the contracting period, the facility and assets are transferred to the public municipality. This option is more suitable for waste processing like composting and waste-to-energy facilities, and for waste disposal like sanitary landfills.

- BOOT (Build, Own, Operate and Transfer) is another arrangement. The contractor is expected to build the facility at his own cost, and operate for a long term as per the agreement. The contractor recovers his cost and transfers the assets to the Municipal authorities in working condition at the end of the agreement with or without any compensation as agreed between the parties. The municipal authorities prefer this type of contract to get the assets free at the end of the agreement.

Both BOT and BOOT are most popular models, and are implemented in Kolkata, Hyderabad, Vijayawada, Ahmedabad,

Thiruvananthapuram etc for setting up compost plants and waste-to-energy plants. In most cases, the municipalities provide free garbage and land on a token lease rent.

Build Operate and Own (BOO) arrangement is very much similar to the BOT arrangement except for the fact that the private party/contractor is not obliged to transfer the facilities or assets to the public entity.

(iv) *Concession Agreements*: Under a concession arrangement, the private operator manages the infrastructure facility, operates it at commercial risk and invests in creating new facility or to regenerate the existing facility. A typical contract has a fixed term, and at the end of the term the assets have to be returned back to the municipality. Concession contracts are generally awarded for 25 to 30 years facilitating the private party to recover the invested capital costs.

(v) *Lease*: In this contract, a private agency operates a service with assets owned by the public sector. The private agency collects certain fee to cover the operational and maintenance costs. The profits are shared with the public sector as the contracting party is responsible for new investments. Such a contract might be suitable for the operation of waste treatment facilities such as composting plants or biogas plants. Leasing is awarded generally for 25-30 years.

9.3 Examples of PSP in MSW Services

(A) W*aste collection service:* Contracting to private sector is very significant due to its transparency and cost effectiveness. These contracts can be handled with small investment. They help to save in operational cost without compromising on efficiency. They can be packaged in small units, and can be terminated easily if the performance is unsatisfactory.

Before bidding the contract, the contractor needs to clearly recognize the scope of the work and the level of service to provide so as to avoid conflicts in future. Standard tender documents could be prepared for such contracts.

These contracts should be of reasonable size. If they are too large, they would be monopolized by one or two parties, and if too small the contracts might suffer from economic viability.

Bigger cities can split service areas into zones and award contracts for each zone to promote healthy competition among the contractors.The contractors may involve Resident welfare associations and NGOs also.

(B) *Transportation of waste* is another important sector of SWM where the efficiency of operations is generally very low. PSP can be utilized for the purpose. Municipal authorities generally do not have modern covered vehicles for transportation of waste. Private sector participation would help to (a) bring new vehicles, (b) make large number trips in each shift, (c) maintain the fleet of vehicles in working condition to avoid dislocation to work, and (d) realise substantial cost saving.

The private sector participants range from individuals and proprietorship/ partnership firms to Companies and Consortia.

(C) *Waste Treatment facilities* have also been set up in a few cities in India through PSP arrangement. The most common PSP arrangements are for composting. Pelletisation of waste is another treatment facility considered by some ULBs in the recent past; examples are Hyderabad and Vijayawada.

Small biomethanation plants have been set up at Vijaywada and eight other towns in Maharashtra State recently which are functioning well.

The Municipal Corporation of Bangalore is using an integrated and disposal facility for the treatment and disposal of 1000 tonnes of waste /day with private participation. The contractor is paid a tipping fee of Rupees 195/tonne only for the disposal of rejects not exceeding 30 percent of the total quantity of waste delivered.

Various profiles of contracting are emerging based on the technology and investment requirements. Mega cities such as Delhi, Mumbai, Bangalore, Chennai, Kolkata, Ahmedabad, and Hyderabad have gone in for national and international contractors (Asnani 2006). Waste-to-Energy plants operating through PSP in important cities are given in the Annexure. Different models of Private Sector Participation in SWM explained above, are listed in the Table 9.1.

Table 9.1 Models of Private Sector Participation in MSW

1.	Collection & Transportation & Cleaning	Service contract /BOOT/ Management Contract
2.	Development of Transfer Station/ MRTS & Transportation	BOOT/DBFOT
3.	Waste Processing facility	BOOT/DBFOT/BOO
4.	Development of Sanitary Landfill & Post Closure Maintenance	Management contract/ DBFOT
5.	Integrated MSWM systems	Mostly on BOOT

(Source: Position paper on PPP in SWM, Ministry of Economic Affairs, GOI, 2009)

However, new Public Private Participation Models are emerging in Dehradun, i.e. collection, transportation and treatment of waste on lease arrangement of PPP wherein capital finance would be borne by the Government, and working capital finance, O&M of the plant would be taken care of by the private operator. However, the authority to earn and generate revenue shall vest with the private operator.

Special attention is, however, required while preparing and entering into contract for treatment of waste. The following aspects are very important:

(i)   Appropriate site selection,

(ii)   Environmental clearances from the State Pollution Control Board,

(iii)   Provision for the adjustment of cost during execution, as the period of contract would be generally long,

(iv)   Clear estimation and definition of finances to ensure the availability of required funds,

(v)   Sharing of finances as an option,

(vi)   Municipal authority facilitating loan to the private operator, and

(vii)   Careful handling of long term contracts to avoid termination and other problems.

The size of the facility should be large enough to attract a private operator; and for the economic viability, it should be designed to last at least for a period of 15 to 20 years, so that it allows depreciation of the works, buildings and equipments.

Disposal of Waste (Sanitary landfills): Some landfills have been constructed in Surat and Ahmedabad in Gujarat, Karwar and Puttur in coastal Karnataka, and Navi Mumbai and Pune in Maharashtra. But these plants are yet to be fully operationalized. PSP has not been introduced so far but the Municipalities of these cities are actively considering private sector participation.

Contract has been awarded in Bangalore for construction of compost plant and landfills on the basis of Build-Own-Operate (BOO) contract.

9.4 Important Contractual Issues

1. To attract private sector participation and cutting costs of service, it is important to have appropriate contract periods. For effective and bankable private sector contract, their duration and compensation should be sufficient.

   They should be for a period long enough to enable the contractor to repay the loans taken to purchase the equipment or refinance the facilities for the work.

2. The start of operation must be clearly defined. Adequate preparation time should be provided to the contractor to start the operation. Special consideration should be given when certain tools and equipments are to be manufactured or imported which need several clearances. In case of large contracts for collection of waste, the insistence should be to start in phases and scale it up over a reasonable period of time for the smooth operations of the contract.

3. Terms of Payment.

   The most commonly used payment methods are lumpsum and unit price: In the lump sum contract, the contractor has no risk, gets a fare deal and the risk is reduced. In unit price method it is necessary to verify the measurement procedure from time to time and adequately supervise the same.

   The concept of "tipping fee" payable by the contracting agencies, of late, is being accepted.

   Collection & transportation services are usually paid either based on quantum of MSW handled or on the number of vehicle trips for transportation.

Street sweeping contracts are either lump sum based or manpower based.

Urban local bodies are inclined to treat MSWM as a profitable operation. Hence, they have been demanding sharing of revenues/ royalty payments for providing waste; for example, compost plants at Thiruvananthapuram and Mysore, and Biomethanisation plant at Lucknow.

4. Delay in Payments.

In waste collection contracts generally a large work force is engaged and they are to be paid on a monthly basis. Therefore, the contractor needs to be paid on a regular basis in order to ensure the cash flow of the company.

Delay in payment can cause labour unrest and adversely affect the services. A provision should be made to pay penal interest by the contracting authority if payments are delayed beyond the reasonable time prescribed.

5. Risks involved and potential influence of partners: There are risks within a public-private partnership such as:

   (a) Country risk/political risk: No refund of profits, Cancellation of credits, and Expropriation/ breach of contract/war,

   (b) Financial risk: Delayed or cancelled payment, Fluctuation of interest rates, and Fluctuation of foreign exchange rates,

   (c) Demand risk: Change in demand, Cost increase for resources, and Change of tariffs ,

   (d) Operational risk: Delay in construction, Cost overrun, Increasing operation costs, and Quality and performance failures.

   The figure 9.1 shows the potential influence of the partners in the PPP.

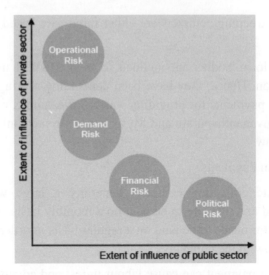

Fig. 9.1 Potential influence of partners.

Preparation of project for PPP by contracting:

1. *Data gathering and preliminary assessment:* Municipalities have to make an assessment of the present infrastructure delivery service to determine how best to improve service provision.

2. *Creation of an independent review team:* It is desirable to have a financial expert, a management expert, a legal expert, a technical expert and an accounting person in the review team. Consideration could also be given to an economist who understands tariff issues.

3. *Evaluation of current system:* If the municipality decides to move forward with a public-private partnership, the evaluation should include financial, technical, legal and administrative assessment.

4. *Further,* during the project development phase, an understanding on the following aspects needs to be attained.

    (a) Current practices of collection and transportation system:

    Service coverage, Current segregation practices, Pre-collection & collection practices, Availability of resources, their adequacy and cost of delivery of the services, Zoning practices, Design and capacity of transportation resources, and Mapping of current transportation practices.

    (b) Assessment of economics of service delivery:

Feasibility report for outsourcing the collection and transportation activities, if applicable

(c) Critical steps that need to be assessed for Transportation of the waste:

Design, capacity and adequacy of the transportation resources,

Mapping of the system to ascertain if the same adheres to the characteristics of the local collection system,

Need for transfer stations, and Resource requirements.

(d) Critical steps that need to be assessed for Treatment of the waste:

Adequacy of available capacities, if any,

Status of existing resource and waste recovery systems (rag pickers, recycling of paper, plastic, glass, metal etc.), and the consequent Characterization of waste received at the treatment plant, Options for treatment facilities and their suitability to ULB, and Cost- benefit analysis.

(e) Important factors impacting Sanitary Landfill development:

Selection of site, Selection of technology, Project implementation structure, and Operations, maintenance and monitoring protocols

5. *Investigation of alternative delivery mechanisms:*

The choice of the type of public-private partnership arrangement depends on three factors, namely, the municipality, review team's evaluation of the current system, and the municipality's needs.

6. *Involvement of stakeholders:*

Stakeholder- analysis is a vital tool for strategic managers as it gives an indication of whose interests should be taken into account and why they should be included in the decision making process.

Bidding Process:

Proper Bidding process is essential for the success of PPP. The main elements are Prequalification, Preparation of Bid document, Invitation of bids or expression of interest, and Invitation of technical Proposal.

It is also necessary to decide on the experience for 'collection and transport of waste projects' as well as for 'treatment and disposal projects.'

(a) The following are the Indicative experience criteria for collection & transportation projects:

1. The price (stated as price or tariff /ton, price/year or total price of the work) is the most important criteria and often the one that is used for final ranking of bidders and selection of the contractor.

2. In standard collection and transportation contracts, the price will usually remain the only variable selection criteria.

3. Quality and technical criteria become more important as the services and works become more complex and challenging.

(b) The following are the Indicative experience criteria for treatment & disposal projects:

1. Experience in operating and maintaining a waste processing facility of requisite capacity.

2. Experience in operating and maintaining a landfill of requisite capacity.

3. Experience in construction and/or operating and maintaining a power plant of requisite capacity.

4. Experience in development of core sector projects developed for government agency.

5. Experience in mining operations.

9.5 Survey on Privatization of SWM

In February 2007, Federation of Indian Chambers of Commerce and Industry (FICCI) conducted a survey on Solid Waste Management titled "Survey on Scope of Privatization of Solid Waste Management in India". Thirty five Municipal Corporations with a population more than 1 million (2001 census) are approached; only the following 25 cities have responded to the survey: Agra, Ahmedabad, Asansol, Bangalore, Chennai, Cochin, Coimbatore, Delhi, Hyderabad, Indore, Jabalpur, Jaipur, Jamshedpur, Kolkatta, Ludhiana, Madurai, Meerut, Mumbai, Nagpur, Nasik, Patna, Pune, Surat, Rajkot,Varanasi,

The survey presents the current status of MSW Management in major cities, efficiency of operations, and extent of privatization of operations in these cities with a view to define future scope of privatization in MSW management.

Despite the ULBs facing a range of problems like resistance from community and lack of funds and technology etc., majority of the surveyed ULBs who have *privatized* some of their SWM activities have indicated satisfaction in SWM efficiency. The observations from the survey are the following:

(a) Out of 35 cities, 10 have fully privatized and 23 have partially privatized few of their activities related to MSW management. 5 cities namely Surat, Nagpur, Nasik, Patna and Ahmedabad have fully privatized their door to door collection of the waste whereas 12 cities which include Asansol, Bangalore, Chennai, Coimbatore, Indore, Jabalpur, Jaipur, Kolkatta, Madurai, Mumbai, Rajkot and Hyderabad have partially privatized *their door to door waste collection.*

(b) About 6 cities, Coimbatore, Jabalpur, Kolkatta, Ludhiana, Nagpur and Surat have fully privatized their *secondary storage activities,* and 8 cities, Ahmedabad, Bangalore, Chennai, Delhi, Jaipur, Madurai, Mumbai and Varanasi have partially privatized their secondary storage activities.

(c) *Street sweeping* has been fully privatized in Jabalpur, Jamshedpur and Patna and partially privatized in Agra, Bangalore, Chennai, Coimbatore, Jaipur, Kolkatta, Ludhiana, Madurai, Meerut, Mumbai, Nagpur, Rajkot, Varanasi and Hyderabad.

(d) Municipal Corporations of Jabalpur, Jamshedpur, Ludhiana, Nagpur and Surat have *fully privatized the transportation of waste,* whereas cities of Ahmedabad, Bangalore, Chennai, Coimbatore, Delhi, Indore, Jaipur, Kolkatta, Madurai, Meerut, Mumbai, Patna and Varanasi have *partially privatized.*

(e) *Disposal of waste* has fully been privatized in Kolkatta, Ludhiana and Nagpur and partially privatized in Bangalore, Delhi, Jaipur, Jamshedpur, Meerut, Patna and Surat.

(f) *Treatment and processing* of waste is fully privatized in Ludhiana, Nagpur and Nasik and is partially privatised in Ahmedabad, Bangalore, Delhi, Jaipur, Jamshedpur, Kolkatta, Mumbai, Rajkot and Hyderabad.

(g) Biomedical waste has been fully privatized in Nagpur and partially privatized in Mumbai.

(h) *Segregation at source* has not been fully privatized in any city but it has been partially privatized in Ahmedabad, Bangalore, Jaipur, Jamshedpur, Mumbai, Nagpur, Nasik and Hyderabad.

The ULBs have shown confidence in adopting the concept of privatization of SWM and have expressed interest in furthering the scope of privatization in their cities.

The survey clearly indicates the positive impact and interest in privatization of SWM. It is necessary that public and private bodies should coordinate to create a suitable environment for facilitating privatization of SWM in the country. The survey also points out that SWM offers an opportunity for private sector participation, which can be given a momentum by establishing a suitable framework for ULBs to ensure sustainability of PPP models.

Absence of user charges in most cities is another obstacle. The door-step collection of waste adds to the cost of SWM service and affects the financial aspect of the system, unless the beneficiary pays. This is lacking in many cities and the contractor is paid from the general revenue of the municipality. The privatization effort followed in Gandhinagar (Gujarat) and elsewhere is a good example of user charges levied to sustain door-to-door collection. Absence of a labour rationalization policy also affects in some cities, because the staff employed is more than adequate and yet underutilized

9.6 Role of NGOs and CBOs

In the recent years, NGOs have taken up initiatives to work with local residents to improve general sanitation. They have been playing an active role in organizing surveys and studies in specified aspects of waste management. Such studies are useful in identifying potential areas of commercial value to attract private entrepreneurs. They can play an important role in segregation of waste, its collection and handing over to local authorities for taking further steps in SWM.

A large number of NGOs are working in the field of solid waste management. For example, Clean Ahmadabad Abhiyan, Ahmadabad; Waste-Wise, Bangalore; Exnora, Chennai; Mumbai Environmental Action Group, Mumbai; Swabhiman, Bangalore, Vatavaran and Srishti in Delhi; CDC, Nagpur and so on. These organizations are successfully creating awareness among the citizens about their rights and responsibilities towards solid waste and the cleanliness of their city. These organizations promote environmental education and awareness in schools and involve communities in the management of solid waste. By and large, the major programmes that NGOs undertake are:

1. Creating mass awareness, ensuring public participation in segregation of recyclable material and storage of waste at source;

2. Providing employment through organizing door-to-door collection of waste;

3. Organizing rag pickers into a professional sanitation workforce;

4. Ensuring public participation in community based primary collection system;

5. Encouraging minimization of waste through in-house backyard composting, vermicomposting and biogas generation;

6. Creating awareness among citizens their right to live in a clean environment, and responsibility to keep the environment clean.

Urban poverty is inextricably linked with solid waste. In our country, over a million people (informal sector) find livelihood in the area of waste, engaging in door-to-door waste collection, sorting and recycling through well-organized systems and composting. Substantial populations of urban poor in other developing countries also earn their livelihood through waste. It is important to understand issues of waste in this context.

9.7 Initiatives by some State Governments

Certain States have taken initiatives towards long-term solutions to solid waste management. Examples of Karnataka, Gujarat, Rajasthan, Maharashtra, and West Bengal are briefly outlined (Asnani 2006).

Karnataka:

The Karnataka government has formulated a state policy for an Integrated Solid waste management (ISWM) based on MSW Rules 2000. It lays down guidelines for all the activities under MSWM and specifies roles and responsibilities for all the parties involved - waste generators, NGOs, CBOs, SHGs, elected representatives etc. SWM Action Plan and Management Plans for 56 cities are prepared based on data related to those ULBs. Technical manuals are prepared on design and specifications of the tools and equipment, and treatment and landfill operations. To build technical capability, posts of environmental engineers are created in 123 local bodies. A series of workshops was organized for local body officials, elected representatives, NGOs etc to prepare action plan, to adopt state policy, to recognize and promote best practices, to identify suitable landfill sites for treatment and disposal of waste, and for carrying out Information, Education and Communication (IEC) activities. Booklets and short films on MSWM are prepared for educating the stakeholders.

Government has allocated a budget of Rupees 16.1 million; government land is given to 226 local bodies free of cost to build sanitary landfills, and entire financial assistance is sanctioned to acquire private land if needed. Further, the government has initiated action to develop scientific landfill sites in eight Class-I cities on BOT basis.

Gujarat:

The state of Gujarat has initiated extensive plans for MSWM. A state level committee headed by Principal Secretary, Department of Urban Development and Urban Housing, and a sub-committee headed by an expert have identified systems for SWM for all the cities and towns, and advised them to implement the suggested systems. Workshops are conducted to train all concerned personnel of ULBs at regional and state level. Action plans have been prepared for all the cities through the state nodal agency - Gujarat Municipal finance Board - and the City Manager's Association of Gujarat.

The year 2005 has been declared the Year for Urban Development. A core committee consisting of administrative and technical experts has been formed, and Gujarat Urban Development Corporation has been identified as nodal agency to facilitate the creation of treatment and disposal sites engaging expert agencies and qualified contractors in all the 141 municipalities. The required land is allotted to municipal corporations at 25% of the market value, and to small local bodies on a token lease rent for a period of 30 years. These sites have been cleared by the State Pollution control Board to establish waste treatment and disposal sites. The funding sources for these projects are the grants from The 12th Finance Commission and Urban Renewal Mission (JNNURM). The state has also proposed financial support, ranging from 50 to 90 per cent, to purchase tools and equipment for waste collection, secondary storage, and transportation. The entire project is likely to cost Rupees 3460 millions.

The cost estimates for the construction of landfill sites and standard compost plants of standard designs for different levels (based on population) of cities and towns are given in the Tables 9.2 and 9.3.

Table 9.2 Cost Estimates for Standard Landfill Sites in Gujarat

Population*	No. of landfills	Capacity (CMT)	Optimal population covered	Design Capacity (MT/day)	Cost of cell (5 yrs)	Cost of office, weighbridge etc.	Cost of handling tractors, JCB etc.	Total cost (Rs lakh)	Total cost per category
120 to 193	10	38,500	200,000	15	48.00	12.00	18.50	78.50	785.00
75 to 120	16	30,800	150,000	12	41.00	12.00	5.00	59.40	950.40
60 to 75	12	20,900	100,000	8	31.40	12.00	6.00	49.40	592.80
15 to 60	103	15,400	75,000	6	28.00	3.00	6.00	35.00	3605.00
Grand Total	141								5933.20

Note: *in thousands. The cost of approach road will be Rs 600 per sq m, which will have to be added to this cost depending on the road length required. *Source*: Asnani 2006.

Table 9.3 Estimates for Standard Compost Plants in Gujarat

Population*	No. of cities	No. of Compost plants to be constructed	Capacity (in MT)	Optimal population expected to be covered	Landfill design (MT/day)	Cell capacity (CMT) to last for 5yrs	Add 10 per cent for inert material
120 to 193	10	10	40.0	2,00,000	15	35,000	38,500
75 to 120	16	16	30.0	1,50,000	12	28,000	30,800
60 to 75	18	18	20.0	1,00,000	8	19,000	20,900
40 to 60	19	19	15.0	75,000	6	14,000	15,400
25 to 40	43	43	10.0	50,000	4	9,500	10,450
15 to 25	35	35	7.5	37,500	3	7,000	7,700
Total		141					

*Population in thousands, *Source*: Asnani (2005)

Rajasthan:

In 2001, the state government has announced a policy for solid waste management which outlines rules/guidelines for setting up of waste-to-energy or waste-to-compost facilities with private entrepreneurs. Details regarding the selection of entrepreneurs, the type of facility that would be extended and the responsibilities of the selected entrepreneurs are specified in the policy document. Further, a state level committee headed by the Secretary, Local Self Government is empowered to receive and recommend the proposals for the conversion of waste.

Land has been either allotted or identified for 152 out of 183 urban local bodies for the construction of landfill sites. The District Collectors have been asked to make the land available; and the development of landfill sites in terms of laying approach road, fencing etc has started under most ULBs. The Rajasthan Urban Infrastructure Development Project (RUIDP) would develop landfill sites in six divisional cities, namely, Jaipur, Jodhpur, Ajmer, Kota, Bikaner, and Udaipur, and also provides equipment and tools for SWM.

Door to door collection of waste has been launched by ULBs as per the guidelines in Jaipur, Ajmer, Jodhpur, Kota, Bhilwara, Pali, Bewar, Jaisalmer, Bharatpur, Alwar, and Ramkanj Mandi.

During the financial year 2005-06, the Chief Minister has announced assistance of Rupees 100 million to smaller local bodies to purchase tools, equipment, and vehicles in order to improve the sanitation facilities.

West Bengal:

The state government launched 'West Bengal Solid Waste Management Mission' in May 2005, which would function under the chairmanship of Chief Secretary to Government. The objective of the mission was to promote modernization of collection and transportation of MSW, and to help development of cost-effective technology for treatment and disposal of waste. Provision to provide the technical and financial support to municipal bodies, PRIs, and authorities for setting up regional/common SWM systems was also proposed.

A technical committee headed by Secretary, Department of Environment prepared an action plan for implementing MSW Rules 2000 in the state. It was planned to construct 25 to 30 regional facilities to cover 126 ULBs that include six Corporations. Each regional facility would serve about five ULBs and each city would share the operation and maintenance costs in proportion to the waste delivered for treatment and disposal.

The improvement of SWM services would mean enhancing public awareness, capacity building of municipal authorities, procurement of necessary equipment and vehicles for primary collection, secondary storage and transportation of waste, setting up of transfer stations, purchasing large hauling vehicles for transportation, construction of individual and regional compost plants all over the state etc. The state Technical committee worked out the cost estimates for improving SWM services; and the state government would support municipalities, if they agree to share the cost. The cost estimates for different aspects are shown in Table 9.4.

Table 9.4 Cost Estimates for Improving Solid Waste
Management Services in West Bengal

Item	Quantity	Cost (Rs crore)	Cost sharing by ULBs (Rs crore)	Cost to be borne by State (Rs crore)
Public awareness	–	1.50	–	1.50
Capacity building	–	1.50	–	1.50
Containarized tricycles	25,000	20.00	5.0	15.00
Secondary storage containers	4,000	10.00	2.0	8.00
Transport vehicles	500	33.75	8.45	25.30
Construction of transfer stations	180	21.60	-	21.60
Large containers for transfer stations	500	7.50	-	7.50
Large hauling vehicles	250	50.00	-	50.00
Construction of compost plants	46	125.00	-	125.00
Engineered landfills	25	125.00	-	125.00
Total		395.85	15.45	380.40

Rs. 1 crore = rupees 10 million   *Source:* Asnani (2005)

Out of Rupees 3930 million allocated to the State's municipalities from the 12th Finance Commission allocation and Urban Renewal fund, the Central government has earmarked 50% to solid waste management in urban areas. Besides an additional Rupees 12710 million have been allotted to panchayats out of which the state expects to spend at least 10% on SWM services. These two sources would make up the total estimates given above to undertake improved SWM services.

Additional funding may be available under Central government's Urban Renewal fund as well as internal resources.

Maharashtra:

The state government took the help of All India Institute of Local Self-Government (AIILSG), Mumbai in establishing a Cell to enhance the institutional capacity of ULBs towards understanding the MSW Rules 2000, and selection of technologies for waste management. AIILSG organized a State level consultation on SWM in February 2001 which led to creation of a Cell in AIIILSG to focus on the MSWM issues which became operational in May 2002.

The cell organized several workshops, and study visits to USA for the City managers to understand the latest waste treatment technologies. The Cell referred several policy issues to the state government for decisions based on the inputs from ULBs. The Cell also released status reports of all the cities along with an action plan in early 2005. The Cell prepared and distributed material on the comprehensive criteria of the MSW rules and sustainable waste management.

The Cell has undertaken a study on the marketability of MSW-derived manure, covering all regions as well as all major crops of the state. It has helped to estimate the market potential in terms of the quantity and the price of MSW-derived manure.

The state government, like in other states, has given land free for setting up landfills; except very few cities, all others have acquired the land.

District-level committees have been set up to coordinate the implementation of MSW Rules 2000. The status of implementation of the Rules is given in Table 9.5.

Table 9.5 Status of the Implementation of the Rules in Maharashtra (source: Asnani 2006)

		Compliance by no. of cities/towns out of 247 cities/ towns in the state
1.	Notification on prohibition of littering and storage at source	214
2.	Doorstep collection of waste	95
3.	Identifying land and agency for waste processing	65
4.	Identifying land for landfill for 25 years	202
5.	MPCB authorization for sanitary landfill granted	242

The implementation is lacking in door-to-door collection of waste and waste processing.

The Cell estimated that Rupees 7760 millions would be required to fund the entire capital costs for implementing all the components of MSW Rules. The state government would consider toprovide a capital grant to all the cities for developing the infrastructure for processing and disposal of MSW.

9.8 Case Studies

(A) Chennai, Tamil Nadu

Chennai is the fourth largest metropolitan city in India. It is the capital of the State of Tamil Nadu spreading over an area of 174 sq km. The city has also attained the status of Mega city (NPC 2005). MSW generation in Chennai has increased from 600 to 3500 tonnes per day within 20 years, and doubled during 1996-2006. The per capita generation rate is 0.6 kg/day. MSWM which include street sweeping, collection, transportation and disposal of MSW from the city limit is the primary function of Corporation of Chennai (CoC). The city is divided into ten zones; and three organizations - the CoC, ONYX (a Singapore- based company), and Community Based Organisations (CBOs) such as Civic Exnora - are involved in the solid waste management services.

The responsibility of Solid Waste Management in the city is entrusted to the Mechanical and SWM Departments of CoC, along with the Assistant commissioners of all the 10 zones. MSWM has been privatized in three zones which cover one third of the total area of the city. CoC looks after the other seven zones. The solid waste generated are collected, treated and disposed into the open dumpsites at Perungudi and

Kodungaiyur which are located at 15 kms on south and north sides of the city. The waste is tipped at the site and levelled by bulldozers. MSW generated includes 68 % of residential waste, 16 % commercial waste, 14 % institutional waste and 2 % industrial waste. The properties of the MSW generated showed that the majority of the waste is composed of green waste (32.3%) and inert materials (34.7%) such as stones and glass (CPCB 2000 and Damodaran et.al. 2003).

In compliance with the MSW Rules 2000, several attempts have been made to improve the MSWM in Chennai. The initiatives include: source segregation, door to door collection, abolition of open storage, daily sweeping of streets and transportation in covered vehicles, wastes processing by Energy recovery or Compsting, and Sanitary Landfilling. A public awareness campaign on source segregation of MSW was initiated during 2003. Corporation workers, zonal officers, revenue officers, technical staff, teachers and school children were drawn in this programme, comprised of public rallies, meetings, distribution of pamphlets, street plays and advertisements. NGOs and members of local welfare associations (like Civic Exnora) have started their own campaign and helped in distributing the pamphlets prepared by the CoC. Door-to-door collection scheme was introduced in June 2003 and expanded throughout the city during January 2004 using Tricycles. Abolition of open storage is moderately achieved by the removal of community bins from the streets. Due to inadequate financial resources and indifference of the population, and inaccessible narrow lanes it is difficult to achieve total abolition of open storage. Daily street sweeping is done by the Corporation workers. But implementation of daily street sweepings is constrained by shortage of sanitary workers, lack of financial support and public holidays. Ward level composting units were introduced in 106 places to reduce the transportation cost of MSW and the amount of waste reaching dumpsite. The segregated waste is collected; the organic fraction is sent to composting at ward level composting units, and the non recyclable fraction is transported to the dumpsites for disposal. There were proposals to recover energy from the waste, and composting of organic fractions in centralized mechanical composting units. At the suggestion of Environmental Resource Management (ERM 1996), the open dump sites have been in operation for the past 20 years and their lifetime is expected to last up to 2011; but they can be extended to a further period by upgrading the sites. Based on the recent studies, CoC has initiated the up-gradation process of Kodungaiyur dumpsite. The approach is a phase wise conversion of open dumpsite into a sanitary landfill as per the recommendation of the National Productivity Council (NPC 2005).

Chennai is the first city in India to contract out MSWM services to a private (foreign) agency, ONYX through a transparent competitive bidding process. The scope of the project includes the activities such as sweeping, collection, storing, transporting of MSW and creating public awareness on MSWM in three zones. ONYX has its own manpower (2,000 employees), tools and equipment for its operations. Their compactor can handle garbage of 7 to 8 tonnes. Movable bins are emptied once in a day and are cleaned every 15 days by the sanitation department. More often, depending on the amount of garbage to be collected, ONYX staff work on holidays and collects 1100 tonnes of waste per day and transports to Perungudi dumping ground (www.chennaibest.com). The characteristic features of ONYX services are: imported technologies for MSWM, containerization of household waste before collection, mechanization of handling tasks through lifting, compacting and tipping devices, day and night services of collection, professional equipment for collectors, better machinery, relatively young work force, training programme for workers, transfer system and haulage, and transfer stations. CoC is managing 2000 TPD of MSW with the manpower of 10,000 including administrative staff and workers, while ONYX is managing 1100 TPD using 2,000 persons. Total cost for street sweeping, collection and transportation of one metric ton waste by CoC and Onyx is approximately US$ 33 and 25 respectively. The privatization has proved to be cheaper for MSWM in terms of the waste collection cost reduction to the tune of 8 US$/t.

Waste Processing: The CoC had entered into an agreement with an Australian company to develop a waste-to-power plant through gasification technology in 2001. The plant was proposed at Perungudi dumpsite on a 15-acre plot of land for 15 years at an estimated project cost of Rs. 180 cores. The project was proposed to generate 14.85 MW of electricity using 600 metric tonnes of MSW per day. But the project failed due to the disagreements in the power purchase rate and protests by the environmentalists (www.Toxicslinks.org 2001 and Srinivasan 2005).

Chennai Metropolitan Development Authority (CMDA) with the support of Ministry of New and Renewable Energy (MNRE), Govt. of India has established a biomethanation plant of 30 tpd capacity for power generation from vegetable wastes at Koyembedu market in 2004. It is the first eco-friendly power plant in India. The investment for this project is about Rs.50 million. The special feature of the process is the digestion of vegetable waste in Biogas Induced Mixed Arrangement (BIMA) digester. It will generate around 5MW electricity and ten tonnes/day of bio fertilizer.

A few community based organizations (CBOs) are involved in the MSWM of the city in addition to CoC and ONYX. Among them, Exnora International is a broad based voluntary NGO established in Adyar, Chennai in 1988. Over the past decade and a half, Exnora has been able to motivate and form thousands of CBOs, each comprising 70-75 families. They take "Civic Pride" in their locality, manage their waste in an environment friendly way and are able to participate in the governance of their locality. Figs.9.2a, b, c, d show the private sector participation in SWM in Chennai (source: S. Esakku et al 2007):

Fig. 9.2(a) Manual sweeping   **Fig. 9.2(b)** Mechanical sweeping

Fig. 9.2(c) Mechanized collection vehicle   **Fig. 9.2(d)** Transfer station

The practices implemented by "Civic Exnora" are: (i) Community motivation and encouragement of high level self-involvement, (ii) Income generation through recycling and reusing, and (iii) spreading the message and helping the communities to Zero Waste Management.

Improvements to open dumps: CoC has initiated the improvement process in the open dumping ground following the recommendations of National Productivity Council (NPC 2005) and Centre for Environmental Studies (CES) of Anna University. CES has taken up a research project on 'Sustainable Solid Waste Landfill Management in Asia' under the Asian Regional Research Programme on Environmental Technology (ARRPET), wherein assessment of reclamation and hazard potential of the sites have been carried out. Detailed investigations on solid waste characteristics, leachate quality and methane emission potential of the dumpsites are used to assess the reclamation potential. Landfill mining studies have shown that the soil fraction of the mined waste from the dumpsites is 40 – 60%, which can be reclaimed as compost or cover material. The recovered space can be reused for future dumping.

An integrated risk based approach was also developed for the rapid assessment of the hazard potential of the dumpsite. Validation of the approach indicates that both sites have moderate hazard potential and require rehabilitation (Esakku 2006).

The infrastructure created at the dumpsites include: construction of compound wall at Kodungaiyur landfill site, completion of WBM roads at both the sites, improvement of the Transfer station, and construction of compost yards. The infrastructure needs further improvement, and the proposed infrastructure is: phase-wise improvement of the dumpsites to sanitary landfills, construction of soil bund around the dumpsites to prevent sliding, construction of mechanical compost plant and leachate evaporation ponds, covering of sanitary landfill with top cover and gas venting system, and mechanical composting and RDF plant for waste treatment.

(B) Delhi

The highest percentage of urban population of India lives in Delhi (93.01%) as per Census 2001. There has been a decennial population growth of 46.31% between 1991 and 2001 as against the all-India growth level, 21.34%. The rural to urban mass migration exercise additional population stress on the city. Change in lifestyles of the people has resulted in increased wasteful consumption, leading to a change in the composition and increase in the quantum of solid waste generated (PJ Sarkar). Solid waste management in Delhi has been a poorly managed affair with obligation on simply transporting the mixed waste by trucks and disposing it in sanitary landfills (SLF).

Legal Framework: The Delhi Municipal Corporation Act 1957 has section 42 C, 355-5.8 stating the functions and role of MCD and citizens in disposal of the waste. The violation of the sections 353, 354, 355(2), 356 and 357 are subject to penalties. Section 357 (1) "Keeping rubbish and filth for more than 24 hours", carries an additional daily fine. The responsibility of MCD is to provide receptacles, depots and places for waste disposal; and not necessarily house to house collection. It is the obligation of residents to use them for disposal of their waste.

Institutional framework: Three municipal bodies - the Municipal Corporation of Delhi (MCD), the New Delhi Municipal Council (NDMC) and the Delhi Cantonment Board (DCB) are responsible for solid waste management in Delhi. MCD alone manages almost 95 % of the total area of the city. These authorities are supported by a number of other agencies. The Delhi Development Authority (DDA) is responsible for siting and allotment of land to MCD for sanitary land filling. Delhi Energy Development Agency (DEDA) under Delhi Administration (DA) is accountable for solid waste utilization projects such as bio-gas or power generation in consultation with MNRE and MoEF of GOI. The Department of Flood Control of Delhi Administration looks after the supply of soil to be used as cover for sanitary landfills by the MCD.

The three municipal bodies, MCD, NDMC, and DCB have obligation to carry on MSWM services in their respective areas, shown in the Tables (source: CDP-Delhi, Ch.12):

1. Municipal Corporation of Delhi (MCD): The MCD area includes urban areas, rural and urban villages, slum clusters and regularized unauthorized colonies.

S.No.	Item	Area (sq. km)	Number	Responsible Department
1	Total area	1397.30		Conservancy and Sanitary Engineering Department presentlychanged to the Department of Environmental Management Services (DEMS)
2.	Urban area (approx.)	595.00		
3.	Rural area (approx.)	795.00		
4.	Administrative zones of MCD		125	
5.	Total number of Employees		More than 52000	
6.	Number of workers (*Safai Karmachari*)		About 50000	

The services of Conservancy and Sanitary Engineering department (CSE) include collection, transportation and disposal of municipal solid waste; road sweeping; cleaning of surface drains and construction and maintenance of public conveniences.

2. New Delhi Municipal Council (NDMC): The activities of SWM include daily street sweeping; removal of the garbage deposited in 'dhalao' (masonry dustbins) and metallic bins; and transporting the waste to MCD landfill sites at Ghazipur. The green (mainly horticulture) waste is transported to the NDMC compost plant at Okhla

S.No.	Item	Area (sq. km)	Number	Responsible Department
1	Area	42.74	------	
2.	Sanitation Circles (approx.)	------	13	Health Dept
3.	Number of employees involved with sanitation	------	1800	

3. Delhi Cantonment Board (DCB): In the contonement area, sweeping of the roads and markets and lifting of garbage are the services rendered. The area covered and the number of sanitary employees is given in the Table.

S.No.	Item	Area (sq. km)	Number	Responsible
1	Area	42.97		
2.	Number of employees Involved with sanitation		450	Health Department

Methodology: MCD planned public private partnership project in six zones: City, South, West, Central, Karol Bagh and Sadar Paharganj to improve efficiency and effectiveness of waste management services.

The civic body signed the "concession agreement" for the project with three agencies: (i) Delhi Waste management Private Ltd for south, central, and city zones; (ii) Noida based Ag Enviro infra Projects (P) Ltd for Karol Bagh and Sadar Paharganj zones; and (iii) Metro Waste Handling (P) Ltd (MWH) for west zone. The project went fully operational in June 2005.

The private companies were allotted a concession period of nine years inclusive of the implementation period of 12 months from the date of signing the agreement. The agreement also contained a performance evaluation and monitoring mechanism where the monitoring of the project was to be carried out by an independent engineer, MSV Pvt. Ltd., appointed by mutual consent of the Corporation and the Companies (Ankur Garg 2007).

These contractors have to place sets of two bins (blue and green coloured) for collection of non-biodegradable/recyclable and bio-degradable waste respectively. These bins are emptied into separate vehicles of similar colour daily. The contractors also undertake segregation of biodegradable and non-biodegradable components of waste before the waste is collected into separate vehicles (CDP-Delhi, chap12).

Waste segregation could be achieved to a great extent by proper involvement of the people. This involvement could be elicited by improving awareness among the people towards waste segregation and minimization. In the West zone, the Metro waste Handling Private Ltd has been conducting, slum programmes, street plays, school assembly sessions and student rallies, posters & banner displays, awareness programmes for households, training of staff etc. The response from the people has been good in terms of improvement in the status of waste segregation according to the contractor. NDMC has 900 community bins (masonry built) and 1000 metallic skips (open container of about one m³ capacity).

Waste Generation: In the absence of a streamlined system of solid waste management, the available data is based on per capita generation from some studies (e.g., NEERI, 1991, Delhi Master Plan 2001, State of Environment Report for Delhi, 2001), vehicle trips and fragmented data from landfill records. Table 9.6 gives some idea of the waste generation estimated from such sources.

Table 9.6 Waste Generation

S.No.	Local Body	Existing generation for 2001in TPD	Projected *generation for 2021 in TPD
1	Municipal Corporation of Delhi	6300	15100
2.	New Delhi Municipal Council	400	550
3.	Delhi Cantonment Board	100	100

Source: Public Health Department of MCD, NDMC and DCB

* 700 gm per capita per day for calculation of projected generation in 2021 as per CPEHHO Manual on solid waste management.

1. Quoted in Delhi Urban Environment and Infrastructure Improvement Project (Status Report for Delhi) prepared in 2001: 6000-6300 TPD for MCD, 350-400 TPD for NDMC and about 100 TPD for DCB, the total generation in the National Capital Territory of Delhi shown as around 6500-7000 TPD

2. As per Delhi Master Plan, 2021 (the generation in 2001 being shown as 5250, 245 and 48 TPD for MCD, NDMC and DCB respectively)

3. State of Environment Report for Delhi, 2001, prepared by TERI quote the total figure at 6000-7000 TPD from the NCT (6300, 400 and 100 TPD for MCD, NDMC and DCB respectively)

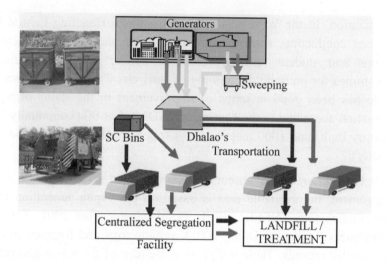

Fig.9.3(a) Waste management system in West zone (source: Ankur Garg)

Fig. 9.3(b) Awareness programme for workers and Households
(Source: Ankur Garg)

A news paper report (Times of India, May 14, 2006, New Delhi) estimates the solid waste generation at 8000 TPD and garbage dumped at the three landfill sites at 7435 TPD. According to a study carried out by IL&FS Ecosmart in 2005, the total generation is around 7700 TPD. The website of NDMC notes lifting of 200-210 TPD garbage from its area. Based on these estimates, it is proposed that the present generation of municipal solid waste may be taken as 6500 TPD for MCD, 400 TPD for NDMC and 100 TPD for DCB (total for NCT 7000 TPD). The figure is corroborated by the figure of waste collection of 6500-7000 TPD presented in a paper by MCD.

There are other stakeholders who participate in the overall scheme of solid waste management in the city are private sweepers and garbage collectors employed by the people for cleaning privately owned premises, waste pickers (rag-pickers), waste dealers and recycling industries, which consume recyclable waste to produce recycled products (PJ Sarkar). A study by an NGO indicates that the number of rag-pickers in Delhi is in the range of 80,000 to 100,000 (Srishti). It is estimated that about 1200-1500 TPD is removed from the municipal collection and disposal chain by these activities. These people carry out in unhygienic conditions and are subjected to unfavourable environmental, occupational health and community health implications (chap12).

Starting from collection of recyclable materials to the final disposal and recycling of waste, many private groups (in addition to the municipal authorities) contribute significantly in Delhi. These stakeholder groups controls the informal sector recycling trade activities, namely, segregation, collection, sale and purchase of recyclable materials, and the process of recycling at recycling units.

Residents and shopkeepers sell recyclable items - newspaper, glass containers, tin cans etc. - to *kabariwallas* or itinerant waste collectors. The waste pickers retrieve recyclable materials from the waste discarded by households, commercial establishments and industries. Larger commercial establishments and industries sell the recyclable waste (in segregated form or otherwise) to waste dealers in bulk, who then sell it to recyclers. Waste pickers pass on the retrieved materials to waste dealers. Then there are agents who facilitate transactions between medium/ large scale waste dealers and recycling unit owners. In the Table 9.7, a list of different recyclable waste materials collected by waste pickers, their colloquial names and prices are given.

Table 9.7 List of recyclable waste materials collected by waste pickers.

Waste material	Colloquial name	Price at which sold to Waste dealer (Rs)/Kg
PLASTIC:		
PET bottles (coke, mineral water bottles etc.)	Raincoat	2
Plastic thread, fibres, rope, chair cane	Cane	6-7
Milk packets	Dudh Mom	6
Hard plastic like shampoo bottles, caps, plastic box, etc.	Guddi	7

Waste material	Colloquial name	Price at which sold to Waste dealer (Rs)/Kg
Plastic cups and glasses, LDPE, PP	Fresh PP	7-8
PAPER:		
White paper used in offices/press cutting	Saphed (White)	3
Mixed shredded paper	Raddi	2
Mixed paper	2 No Raddi	0.5 – 0.75
Cartons and brown packing papers	Gatta	2.50
Fresh News Paper	Gaddi	4.5 – 5.0
Carton sheets	Raddi	4.5 – 5.0
Tetrapack	Gutta Sheet	2
ALUMINIUM:		
Beer and cold drink cans	-	50
Deodrant, perfume bottles	-	50
Electrical wires	-	40
Aluminium foil	Foil	20
Other Metals		
Steel utensils	Steel Bartan	20
Copper wires	Tamba	80
GLASS:		
Broken glass	Shisha	0.50
Bottles (Beer)	Bottle	2

Note: *Selling prices of all items as on January 2002, PET: Polyethylene Terepthalate, LDPE: Low density Polyethylene, HDPE: High density Polyethylene, PP: Polypropylene

(Source: Recycling Responsibility, Traditional systems and new challenges of solid waste in India, Srishti, 2002.)

Processing and Disposal of waste:

Currently, Delhi has 4 compost plants, shown in the Table 9.8 (source: CDP-Delhi, chap. 12).

Table 9.8 Compost Plants in Delhi

S. No	Facility	Capacity (TPD)	Area (ha)	Starting Year	Technology	Remarks
1.	Okhla (MCD) (closed at present)	150	3.2	1981	Aerobic windrow composting	Proposed to be upgraded to 200 TPD
2.	Okhla (NDMC)	200	3.4	1985	-do-	Operated below capacity
3.	Bhalswa (Private sector)	500	4.9	1999	-do-	Operational at 50% capacity
4.	Tikri Khurd (APMC and Private sector)	125	2.6	2011	-do-	Dedicated waste stream from APMC
	Total	975	14.1			

The three compost plants can process at present about 400 TPD out of 7000 TPD, and the balance is assumed to be dumped at the three dumpsites (landfills) listed in theTable 9.9.

Table 9.9 Landfill sites serving Delhi area

No	Name of site	Location	Area (Ha)	Year started	Waste received (TPD)	Zones supplying waste
1	Bhalsawa	North Delhi	21.06	1993	2200	Civil Lines, Karol Bagh, Rohini, Narela, Najafgarh and West
2	Ghazipur	East Delhi	29.16	1984	2000	Shahdara (south and north), Sadar Paharganj and NDMC
3	Okhla	South Delhi	16.20	1994	1200	Central, Najafgarh, South and Cantonment Board

(**Source:** Information provided by MCD, and CDP-Delhi, chap12).

The current expenditure for municipal solid waste management for MCD is Rupees 5030 million (Rs. 3860 million non-plan and Rs.1170 million plan).

MCD does not collect and dispose construction and demolition waste (C & D waste). The regulations place the responsibility of disposal of construction waste with the generator and MCD levies a fee of Rs. 250/ ton of waste disposed at landfill. However, significant quantities of waste are disposed at unauthorized/designated public locations. From these places, MCD is forced to evacuate C&D to the landfills.

Fig. 9.4 Okhla landfill site (source: CDP-Delhi, chap 12)

Typically, demolition activity is undertaken by specialized demolition contractors who bring their own equipment and personnel and transport the residual waste. The property owners pay fee to the demolition contractors, which is decided based on the recoverable value of recycled materials, steel, wood, glass, pipes etc. by demolition contractors. Currently the C&D waste is disposed without any kind of processing at Ghazipur and Bhalswa MSW disposal sites. Considerable quantities are disposed off at unauthorized locations or MCD designated dumping sites in Delhi.

The *main constraints* to the MSWM in Delhi relate to technical, financial and institutional aspects. The technical problems relate to municipal storage, collection and transportation. But the major problem is in the area of processing and disposal. The existing landfills (dump-sites) are almost full. They need to be closed immediately in a scientific manner to the extent possible and new sanitary landfill (SLF) sites need to be developed and commissioned at the earliest possible. The major obstacles for this to happen are:

(i) Non-availability of adequate land for building new sanitary landfills, and

(ii) Arranging for disposal of the waste in the intervening period before the new SLFs are ready to receive waste.

MCD is trying to set up 3 new facilities at Jaitpur, Narela-Bawana, and Bhatti-mines.

The key issues to be addressed in Delhi with respect to solid waste management are (*Source:* CDP-Delhi, chap 12): Efficient service delivery (collection and removal of garbage, construction and demolition debris and other types of waste, street sweeping etc. leading to clean surroundings and a sense of well being among the citizens, Appropriate disposal of waste in conformity with the MSW Rules 2000, and Strategy for reducing land requirement

Of the three issues, the most important one is reduction in land requirement for disposal. In the absence of adequate land, the three landfill sites are being over-used; Delhi is actually in a very serious situation to procure land required for processing and disposal of solid waste.

(C) Greater Mumbai, Maharashtra.

Municipal Corporation of Greater Mumbai (MCGB/ BMC) extends over 437.71sq.km and constitutes two major geographic divisions, 'island city' and 'suburban areas'. Greater Mumbai has been divided into six zones and 24 wards for the administration of the municipal corporation. The SWM is a ward-level activity in MCGB.

Greater Mumbai is a densely populated city in the country - its density is as high as 46,000 persons per km^2 in Mumbai and 20,000 persons per km^2 in suburban Mumbai (SWM Dept. 2004). Another feature is that more than half of the Mumbai's population lives in slums. The population census of 2001 shows that in MCGB area 48.5% of population lives in slums. If other industrial workers' housing (called *chawls,* one room housing units) is included, then close to 70 per cent of Mumbai's population lives in either slums or *chawls* (Mukhija 2000). Their density in some areas can reach as high as 400,000/ km^2 (Mumbai pages, 1997). The slums are considered as vulnerable settlements due to their location on the hilltops, slopes, *nallahs,* low-lying areas (with tendency to flood during high tides), coastal locations, under high tension wires, along highways, along railway lines, within industrial zones, pavements, along water mains, and along open drainage. The garbage clearance, therefore, becomes a major problem in slum areas.

Status of SWM in Mumbai: Under the Mumbai Municipal Corporation Act of 1988, it is mandatory for the Corporation to maintain the area clean to ensure a good and healthy environment. Municipal Solid Waste Rules 2000 make it obligatory for the storage of garbage at the source and its synchronized collection at the doorstep. The MCGM has already declared the segregation and storage of garbage at source mandatory.

The per capita generation of wastes in Mumbai is about 630 gm. per person/day (MCGM 2004). The quantity of municipal solid waste generated within Greater Mumbai is 7,800 MT per day. The solid waste is in the form of regular garbage from households, debris, silt removed from the drains, *nallas*, cow dung and waste matter removed from gullies between the houses. Around 4,500 MT (57.68%) of waste in the whole of the city is biodegradable in nature; 500 MT (6.41%) is the dry waste consisting of paper and cardboards, plastics, metals, glass, etc., 2,500 MT (32%) is the debris and silt, and 25 MT is biomedical waste (MCGM 2004). The data shows that about three-fifths of the waste generated in Mumbai is biodegradable in nature, putting heavy burden on the municipal corporation to organise the disposal of the same.

Part of the recyclable waste generated is sold by the households themselves which do not reflect in the data given; but the part picked up by the rag-pickers to earn their own living do account in the figures of total waste generated. Hence, before the waste reaches the dumping (disposal site), part of it is already recycled by the recyclers. There is therefore a gap between the percentage of recyclable waste generated and percentage of recyclable wastes that reach the disposal site. This is happening in all the three regions of Mumbai, the island city, the eastern suburbs and the western suburbs.

Manual sweeping of all the public roads and streets (total length is 1800 km) is done during night hours. To successfully cover the entire length, the area is divided into 'beats', each beat area is about 4,000-5,000 sq. m. for the city area and 8,000-10,000 sq. m. for the suburban areas. A pair of sweepers is assigned a single beat who uses one handcart and 2 containers and brooms. About 8,400 staff carries out this activity for the entire Greater Mumbai. Wastes thus collected are deposited in nearby community dustbin containers, which are provided by the MCGM.

The MCGM carries out campaigns through newspapers, instructing the citizens/ institutions to collect their own garbage and store the same in bins to be kept at the gates from where the municipal vehicles would pick them up mechanically at specified time.The citizens are further notified that the wet waste would be collected daily and the dry waste would be collected once or twice a week, depending on the amount generated. 83% of total population of Greater Mumbai is served by the community bin collection system and 15% by door-to-door collection. Garbage collectors employed by various housing societies collect the waste manually at the household level and dump it in the garbage bin at specified street corners. There are about 6,300 community dustbins of different designs and construction provided throughout the 5,500 waste collection points in Greater Mumbai for collection and temporary storage of the all waste other than the debris, silt etc.

The Corporation utilizes both manual (22%) and mechanical (78%) means for the removal and transportation of wastes (Jain 2004). Manual handling is carried out at the collection points, where waste is collected by the municipal workers and dumped into transportation vehicles. Both Municipal and contractors' vehicles are used for removal and transportation of garbage, but only municipal labour is used in this work. For debris, silt etc. however only the contractors' vehicles and their labourers are used. 45% of the transportation is through municipal transport and 55% is contracted out. Transportation of waste is carried out by using different types of vehicles depending on the distances to be covered by them. 60% of waste is transported through stationary compactors, mobile compactors and closed tempos; 10% is through partially open dumpers whereas 20% is through tarpaulin-covered vehicles, which includes silt and debris.

There are at present 2 transfer stations (TSs), situated at Mahalaxmi and Kurla. Both the TSs together handle about 600 MT of garbage everyday and the remaining is transported directly to the dumping grounds situated in the northern part of Mumbai. Separate transport transfers the garbage from Mahalaxmi and other parts of the city which is nearly 95 per cent of the waste generated, to the dumping grounds (Coad 1997). The transportation of garbage from the transfer stations is done using 15-20 cubic metre Trailers and Bulk Refuse Carriers. Actually, the TS at Mahalaxmi have the capacity to handle at least twice the present load; but currently it is under-utilized. To remove and transport the garbage, six municipal workers and one *Mukadam* (labour contractor) are deployed with each refuse vehicle. The worker uses two baskets and two iron rakes per vehicle.

For primary collection, transportation and disposal, MCGM deploys 141 refuse vehicles for the city region and 120 for the suburbs. 13 service garages are organised within Greater Mumbai area for maintenance of these refuse-vehicles (Coad 1997).

Disposal of waste: The Corporation disposes waste through landfill or land dumping. Waste is brought from various locations throughout the city as well as from the TSs at Mahalaxmi and Kurla. Refuse and debris are levelled by means of bulldozers and landfill compactors. The land filling carried out here is open dump tipping. At present there are 3 landfill sites in Mumbai area located at Deonar, Mulund and Gorai (Fig.9.5).

Table: Amount of Waste Disposed at Dumping Sites

Location	Area (hectares)	Quantity of MSW received (Maximum) (TPD)
Deonar	111.00	6,826
Mulund	25.30	598
Gorai	14.50	2,200
Total	**150.80**	**9,624**

Source: MCGM, Dec. 2004

1 Deonar Disposal Site ①
2 Mulund Disposal Site
3 Malad Disposal Site
4 Gorai Disposal Site
5 Mahalaxmi Refuse Transfer Station ⑤
6 Kurla Refuse Transfer Station ⑥

Fig. 9.5 Transfer and Disposal sites in Greater Mumbai
(Courtesy: Mahadevia, D)

Two more landfill sites have been proposed: at Kanjurmarg of 82 ha and at Mulund of 40 ha (SWM Cell, AIILSG, 2003).

Transportation costs of waste are quite high and approximate to about Rupees16 lakhs per day. Costs for maintenance of dumping ground, waste transportation and hire charges come to Rupees126 crores per annum and constitute nearly 28 per cent of the total budget allocated for SWM (Davis n.d.). These sites need to be upgraded and the waste appropriately treated as it has been estimated that they will last for another 5 years only (SWM Cell, AIILSG 2004).

Land being scarce in Mumbai, the Corporation has planned for disposal of garbage through manufacture of organic manure and generation of electric power. For the disposal of municipal solid waste, composting, biomethanation of wet garbage, vermicomposting and recycling of dry waste are adopted. Several organizations working in coordination with MCGM have started a number of decentralized units based on these technologies, thus forming an effective public private partnership. The list of decentralized facilities existing now in Mumbai is given in Table 9.10.

Table 9.10 Decentralised Waste Disposal Centres, Mumbai

No	Organisation	Method of Disposal	Quantity of waste disposed (TPD)
1.	M/S Excel Industries Ltd at Chincholi Dumping Ground*	Converting to organic manure	240.0
2.	200 active ALMs through out the city#	Vermi-composting (Individual or Community based)	50.0
3.	5 T plants at Dadar (market waste), Versova and Colaba#	Vermin-culture	15.0
4.	Stree Mukti Sanghatan (SMS) composting units +	Composting	21.0
5.	Waste collected by parisar Bhaginis of SMS+	Recyclable dry waste	
6.	Approx. 30-40 Municipal Gardens#	Composting	1.5
7.	Hotel Orchid#	Vermi-composting	0.15
8.	Composting units under Force Foundation#	Vermi-composting	20.0
9.	Churchgate Plaza#	composting	0.6
10.	Units belonging to Force Foundation#	Composting/Vermi-culture	20.0
Total			**369.25**

Source:* MCGM, 2004, # Discussion with personals from Dept., + Data from Stree Mukti Sangathan

Issus in SWM

The issues in MSWM in Greater Mumbai relate to primary collection and disposal of waste because of peculiar characteristics of the city.

Large floating population and daily commuters, with almost 6500 thousand people travelling daily is a cause for road littering (Jain 2004). In many areas of the city, streets are poorly maintained due to lack of timely street sweeping. There is regular clogging of surface water drains due to dumping of solid waste into it. Only 77 per cent of roads are cleaned 6 days a week, and the major roads only are cleaned 7 days a week (SWM Cell 2003). At present, the door-to-door collection of waste is limited to just 15 per cent of the waste generated; the remaining waste including recyclables is sent for disposal. This has led to increase in the disposal waste quantity.

The problem is further aggravated due to a high density and large proportion of slum population. The slum and pavement dwellers do not have access to proper services and hence they dispose their waste in the public spaces, roads, drains or railway tracks. Added to it, the hawkers contribute significantly to littering of roads.

With increasing urbanization, land for dumping and creation of landfill sites for disposal of waste is becoming unavailable. There are only 4 landfill sites in the MCGM area, whose expected lifespan remains only 5 years. It may be difficult for MCGM to find new waste disposal sites in the near future to take care of not only the present level of waste generation but the increase in generation due to the projected new population.

Non-compliance of MSW Rules 2000 and absence of proper and regular communication between citizens and local body authorities have also added to poor MSWM services in Greater Mumbai. In order to set right the lapses and to face future challenges in MSWM, local NGOs along with the MCGM have taken up certain initiatives to improve the waste management activities.

New Initiatives

Three new initiatives, namely, (i) The Advance Locality Management (ALM), (ii) the Slum Adoption Programme and (iii) the Parisar Vikas programme by the Stree Mukti Sanghatana (SMS) are planned.The features, implementation, and the successes/failures of the schemes are briefly discussed.

(i) *Advance Locality Management:* The scheme has been initiated by MCGM with the main objective of mobilizing citizens' participation in solid waste management in an environment-friendly manner. The focus of the initiative has been 'waste minimization' and 'segregation of waste at source'. It was first started in Joshi lane, Ghatkopar.

A housing society or a group of housing societies in a locality, depending on the size, and the resident and nonresident population would form an ALM with the MCGM. The ALM Society is registered with the local Municipal Ward Office and appoints a Nodal Officer. The Nodal Officer collaborates with the ALM, attends to citizens' complaints, and follows up all the actions required at the MCGM level and co-ordinates the actions among different departments of the MCGM at the ward level. This is the essential aspect of the partnership that ensures success of the ALM.

A common fund is set up by collecting Rupee 1 per apartment per day quarterly. It is estimated by the residents of Joshi Lane that the cost of Integrated Solid Waste Management by the residents is Rupees 8/ capita/ month or Rupees 96 annually (Jain 2004). Contributions are received from the residents (maintenance fund) and are utilized towards the 'maintenance'. Started in July 1997 with only one locality as its participant, the number of societies registered in ALM scheme crossed 1000 (Modi et al 2002).

Various NGOs who have been associated with the work relating to local governance, and groups of senior citizens involved in civic issues have become partners of the ALM process. With the involvement of NGOs, corporates also have joined the process. The rag pickers have been involved in the collection of dry recyclable waste directly from individual houses. NGOs like Stree Mukti Sangathana (SMS), Force and Akkar Mumbai have taken the task of training the rag pickers, supported by MCGM. As a result, rag pickers have become more organized, received fairer prices to their collection, and better health and insurance services, along with work provided by the NGOs.

The private contractors are also involved in the collection, segregation and disposal of solid waste. The role of the beneficiaries is to segregate the waste at source and maintain vigilance on the spot to prevent littering. They are also involved in creating awareness among the community about the need for source segregation along with importance of disposal of waste in the bins to avoid littering on the roads and public places.

An innovative aspect in the ALM approach is the residents' initiative to RRR, i.e. Reduce-Reuse-Recycle. The waste is segregated at the source and recyclables are removed; the rag-pickers take away the recyclables which gives them some income. The wet waste is taken directly for composting at individual level or to community vermi-composting units. This led to 'Zero Garbage' situation. This has eliminated the need for community dust bins. This scheme has considerably reduced the burden of primary collection, transportation and disposal of waste which helped MCGM to save expenditure on the waste disposal process amounting to Rs. 1.5 per kg of waste (Jain 2000). Thus, while the doorstep collection has added to the collection cost, it has been counterbalanced by reduction in waste quantity.

Other benefits:

With the success of the ALMs in SWM, the Municipal authorities have delegated additional functions which included beautification of the localities and maintenance of gardens, parks and roads (Figs.9.6). The ALM movement has been so successful that the citizen groups have, in addition to the responsibility of their immediate neighbourhoods, organized maintenance of open spaces like the Juhu Beach etc., at the ward level (Kundu 2005). The successful ALM societies have also taken up other activities such as tree plantation, prevention of encroachment on pavement and beautification of streets.

A total of 100 MT of flowers and other bio-degradable material offered during worship during Ganesh festival from 500 *mandals* were processed in this park. The corporate houses have undertaken the responsibility of managing sanitation and solid waste and their roads. There are a total of ten Corporates currently that are part of the ALM movement; one of them has even encourages vermi-composting through a 'Trust'. 261 vermi-composting units spread over six zones reduce approximately 20-25 MT of garbage per day from reaching the disposal site. It is estimated that about 25 per cent of the ALM are managing solid waste at the local level through vermi-composting and recycling of dry waste (Redkar 2005).

Fig. 9.6 (left) shows the Waste collection bin at Nani Nani- Munna Munni Park maintained by Pestom Sagar ALM. Figure (right): The Kangra Garden leased to the ALM for maintenance has been turned into one of the finest gardens in the city. A pit of 440 cu.ft volume surrounding the garden is used for vermin composting (Courtesy: Mahadevia Darshini).

(ii) *Slum Adoption Scheme:* MCGM has found that the residents in the slums have no sense of participation to keep the area clean, which results in piling up of garbage and deteriorating health conditions of the residents. This has been the conclusion in a survey conducted in 100 communities of a slum enclave by YUVA, an NGO. In order to motivate and involve the slum population in keeping the slums clean, a scheme called 'Slum Adoption Scheme (SAS)' has been started by the MCGM through community-based organizations and public participation.

A Community Based Organisation (CBO) involved SWM work in the Prem Nagar Slum Community is provided with necessary equipments for the purpose by MCGM who has also taken care of the salaries of the slum cleaners. The project is functioning successfully. The MCGM provides finances to the CBO for the first three years, but the amount reduces gradually over the period. Then, the CBO would raise Rs. 10/ household for collection of segregated waste from house-to-house and for the maintenance of toilet blocks. The scheme has been so designed that by the end of the third year, the CBO would be self- sufficient in managing services related to waste management and sanitation at the primary level. But in reality, the beneficiary's contribution being as low as

8 to 10 per cent of the total budget, the CBOs have found it hard to sustain financially.

SAS has been designed to be a sustainable programme to implement SWM scheme in the slums, but, in reality, it has not been fully successful due to several reasons that included economic unsustainability, excluding some slums by MCGM, interference by local councillors etc.

(iii) *Parisar Vikas Programme:*

The Parisar Vikas Programme has been initiated by the Stree Mukti Sanghatana (SMS) which is an NGO, a Woman's Liberation Organisation established in 1975. The SWM project of the SMS is being funded by 'War on Want', a London based NGO, and the Central Government's Suvarna Jayanti Shahari Rojgar Yojna (SJSRY). The duration of the project has been from 2002 to the end of 2005.

The main strategies of the programme are: Organisation and training of the women ragpickers, improving the living standard of women ragpickers, developing new techniques for treatment of waste, and creating zero waste situations in cities by appropriate waste recycling techniques.

SMS has been recognized as a training institute. The rag pickers are given identity cards and are trained in waste handling, waste collection, transportation of waste to pits and pit management, in addition to health care and hygine. These trained women are addressed as 'Trained Parisar Bhaginis' (TPB). The two training centres established in M-ward (Chembur) have trained Parisar Bhaginis in bio-composting, vermicomposting and gardening. 300 women have been trained in manure and gardening techniques through which 250 women have gained meaningful employment. Simultaneously SMS has developed 5 to 6 composting models in available space within localities.

200 groups, with 10 Parisar Bhaginis each, have been established. A group leader heads each group. Awareness and leadership development camps are organised for the group leaders with the special material developed by SMS over the years for the training.

The 'Parisar Bhaginis' go from house to house and collect garbage already segregated into 'wet' and 'dry' waste. They compost the wet waste, and the product is sold in the market for use in the plant nurseries and gardens in housing societies; the dry waste is sent for recycling. Self Help Groups (SHGs) or micro credit societies of these women have also been established, and six service cooperatives have been registered with 50 women in each Cooperative which function as business enterprises. With the formation of waste cooperatives, they have been able to get the right price for the sale of the dry (recyclable) waste.

The SMS was engaged for Solid Waste collection and treatment in major public and private sector housing colonies and office premises such as Tata Power, Tata Consultancy Services, RBI, Navy, BEST, Pfizer, CIDCO, MCGM, BARC etc. About 250 trained Parisar Bhaginis have brought "Zero Garbage" status in these offices and colonies and also small housing complexes spread over 13 wards in Mumbai. In 2004, their work has spread to the suburban areas of Navi Mumbai, Kalyan and Dombivli.

In Parisar Vikas Programme, 2000 women rag pickers are currently working in Mumbai. The waste management schemes are implemented successfully at 40 places throughout the city. Parisar Vikas has constructed and operating two *Nisargruna* plants (bio gas plants) with the technology developed by Bhabha Atomic Research Centre (BARC) for processing 5 MT of wet waste per day for MCGM.

Fig. 9.7(a) Biogas Unit at BARC (Courtesy: Mahadevia, D)

Fig. 9.7(b) Compost units along the roadside for Garden waste (Courtesy: Mahadevia, D)

As incentives, pre-primary education is made available to the children of the Parisar Bhaginis by starting *Balwadis* (kindergartens) in the communities with the help of Pratham, an organization working for universalisation of primary education. A crèche for the children of Parisar Bhaginis has been started in the Community Centre next to the Deonar dumping ground. Health camps are held for women and children with the help of Family Planning Association of India.

These women are included under the Swarna Jayanti Shehari Rojgar Yojana (SJSRY), and a grant of Rupees10 000 is given to each self help group (SHG) as a running capital to start their micro enterprises. Of the 200 groups, 63 groups with 678 women received the revolving fund of Rupees 678, 000 disbursed under the 'Thrift and Credit Societies' component of SJSRY.

A tempo is provided for the collection of dry waste. In 5 wards, even Parisar Bhaginis have got such tempos from the MCGM.

Under the infrastructure development of SJSRY, grant has been provided to construct sheds in seven wards for the storage of dry waste. These sheds would be operated on a cooperative basis under the aegis of PBVS and 5 such informal sheds are already in use by the Parisar Bhaginis. The success of this scheme facilitated the formation of neighborhood committees (NHC).

The Department of Urban Development, Government of Maharashtra recognizing the importance of this partnership, has recommended to all other Municipalities of the state to adopt this approach.

SMS's approach in Parisar Vikas has helped beneficiaries to gain knowledge and skills, to advocate their rights, and to get organized. Through this empowerment, the Parisar Bhaginis could improve their bargaining power, better their social organization, and increase their income and self-sufficiency.

For long-term sustainability of the project, SMS has strengthened the self-help groups of rag pickers by involving more number of housing colonies in the waste management scheme as well as in gardening activities.

The involvement of MCGM in undertaking these experiments with citizens, NGOs, and CBOs has resulted in new innovations and methodologies in waste management systems. The projects have also made an effort in an indirect manner to link the civil society groups to the governance system, bringing in a sense of ownership and increasing the transparency within its structure. Such kind of third party monitoring has also helped to overcome the problems of the system and of mal-practices, if any. The negotiating position of the ALM societies as well as the CBOs has strengthened and they are able to use this in bargaining for better civic service deliveries from the MCGM.

The rag-pickers have been brought into the formal schemes of SWM which helped them to get organized. As a result, they receive better prices for the collected materials, have better health care, are covered under insurance services and are getting more work than before. All these have legitimized the rag-pickers in the society and have increased their income and well being.

After nearly five years of the implementation of these programmes, the benefits to various stakeholders are evident.

The participatory mechanism can reduce the cost of implementation of any scheme, as seen from MCGM partnering with the CBOs for solid waste management either through ALM or Slum Adoption Schemes.

Despite the new initiatives taken, the experience from Greater Mumbai shows that the city has still a long way to go in SWM. The scale of these new initiatives is inadequate to cover the whole city because of its huge spread. The floods of July 26, 2005 has revealed that the garbage collection has not been enough as not to clog the city drains in times of heavy rains and cause severe inundation of the city areas. Nonetheless, these new approaches have the potential to address the problem of waste management in the most populated and complex Mumbai City.

CHAPTER 10

MSW Management and Planning – Global Examples

MSWM systems outside India are briefly discussed. The only criterion for choosing the countries/cities from Africa, Latin America, East Asia, Europe and North America is the availability of published/ documented information.

10.1 Asia

In countries with high levels of education, approach to integrated solid waste management is normally dictated by the public awareness of resources, economics, and the quality of the environment. In Japan, for example, cities have implemented laws and regulations governing *disposal bans* on substances such as batteries, waste oil, tires, CFC gases, PCBs, etc., and also a mandatory deposit/take-back requirement for articles such as mercuric oxide batteries, aluminum and plastic containers, tires, and non degradable plastic bags. The households in Japan are also required to use transparent plastic bags for waste disposal so that collection people can see the contents. In 1992 South Korea passed a law promoting recycling.

The most common MSWM problems in developing countries of East Asia/Pacific are institutional deficiencies, inadequate legal provisions, and resources constraints. There is considerable overlap of executive and enforcement authorities at the national, regional, and local levels as far as environmental control are concerned. There is a lack of long- and short-term planning due to resources constraints and the shortage of experienced specialists. Many of the laws and regulations are inadequate to effectively deal with the complexities of MSW in large cities. In many cases, the regulations are directly copied from industrialized countries unmindful of the local socio-economic conditions, the expertise availability and administrative structure. As a result, they prove to be unenforceable. While the old regulations often are in existence, lack of authority to effectively enforce existing environmental regulations adds to the problem.

In most Asian developing countries recycling laws are not enforced except in China. Although there are community initiatives to separate and collect recyclables for sale and reuse, these activities are 'informal', not supported by the municipal authorities, except in China, North Korea, and Vietnam. Monitoring of programmes in developing countries is in general not satisfactory. The decision-making process is slow and complicated due to unnecessary paper work and bureaucracy; for example, the illegal dumping of hazardous substances on lands and into the waterways is a result of this situation.

In developing countries, given more recognition, NGOs could play a more effective role in the improvement of solid waste management. Traditionally, there is no input from the local communities in the decision making. In places where the municipal authority does not do primary collection, people have created community organizations to collect wastes. These work well in parts of Jakarta and Hanoi, and are extensive in South Korean cities.

In South and West Asia, however, the planning, management, and decision making depend on a country's administrative structure, bureaucratic style, and political values. Most municipalities do not have any legislation related to MSW, and operate on 'old' regulations. Many of them have no integrated approaches to waste management including citizen participation in decision making.

Municipalities in most South Asian countries operate under the environment, health, or local government ministries of the central or regional governments. In the central part of the region and in some countries in the north, Health Ministries are expanding to directly oversee municipal corporations. In the Indian subcontinent, there is a trend towards decentralization, with municipalities being expected to raise their own funds and take on more responsibilities. MSWM here is characterized by bureaucratic fragmentation, with interlinked aspects being dealt under different departments or ministries.

Although many countries now have environmental protection agencies which are directing their attention to the waste management, legislation relevant to modern waste management is deficient. Decisions on legislation, major capital spending, or administrative changes need approval from the super ordinate ministries or departments. With regard to routine management and planning, procedural variations exist throughout the region. For example, in Oman, the Ministry of Regional Municipalities is responsible for providing municipal services in all the cities except Muscat, where the municipal council makes decision. In the subcontinent, municipalities have responsibility for routine management. In large cities, in democratic countries, major decisions are made by city municipal corporations consisting of elected representatives. Though these urban areas have the advantage of access to citizen opinions through the representatives, the corrupt motives that generally exist may seriously distort financial and technical decisions.

Master plans have been prepared for some of the large cities at considerable expense but very few of their proposals have been implemented. Smaller towns do not attempt long-term planning. Planning for MSWM at the regional level has not yet responded to important worldwide trends. Hence, waste minimization, recycling, helpful procurement policies, etc., for the most part find no place in MSW regulations.

In the subcontinent, several countries suffer from management difficulties in workforce relations. Strikes affecting solid waste collection quickly jeopardize public health in hot and humid climates. Privatization is being strongly opposed by labour unions of workers in SWM. The solid waste management department is usually associated with low-status, and top officials frequently transfer out of it after a short stint, which hampers continuity in management.

Major changes toward decentralization are happening in some countries (e.g., in India). The trend to privatization also has implications for planning and management. Abrupt privatization, without careful arrangement of contracts and sound monitoring criteria, has led to problems, although some functions such as repair, and maintenance of vehicles etc., have been successfully privatized. One of the concerns is how well these private arrangements serve the poor sections.

A major challenge for MSW managers in this region (with the exception of Israel) is how the needs and views of underprivileged communities (e.g., squatter settlements) can be expressed, understood, and incorporated into decision making. As long as squatter areas are treated as illegal and denied services, it is difficult for the solid waste authorities to arrange for effective interface between the MSWM system and the informal planning of the settlements. Such cooperation is being achieved, however, through the mediation of NGOs in several countries, for example, Orangi Pilot Project in Karachi, PROUD in Mumbai, Society for Clean Environment and United Way in Baroda etc. Citizens' environmental organizations are on the increase in the region. In general, the role of NGOs and local communities can be extremely helpful in experimenting with waste reduction through neighborhood composting and the promotion of more recycling. Even in middle-class areas in the subcontinent, local groups are organizing to improve street cleanliness and to facilitate more efficient waste collection (e.g., the Civic Exnora street groups in India).

A recent development is computer modeling to aid administration and planning for MSWM, which is undertaken in national institutes such as The National Environmental Engineering Research Institute in Nagpur, India. The usefulness of the models, however, depends on the reliability of the basic data for the place where conclusions are to be applied. Expertise and funds delivered by international agencies and donors have been of great benefit in organizing the full range of MSW services and decisions in the less affluent countries. But, there are complaints that pressure from international loan agencies and equipment vendors has led to hasty or poorly conceived privatization or the adoption of inappropriate equipment and procedures. This situation should improve with the increase in understanding of solid waste issues worldwide. Perhaps the greatest impediment to improving planning in MSWM in this region is lack of (i) knowledge of waste quantities and characteristics and factors that affect their variations; (ii) generators' attitudes, behaviors and needs; (iii) the actual costs of different activities; (iv) staff performance; and (v) sound practices elsewhere.

(a) *Japan:* (Ref: Rachel et al 2009)

In Japan, most limited resource is the land suitable for urban purposes. The country is comprised of 6,800 islands, and 61% of its surface is covered by mountains. These features complicate the transport of waste and make it difficult to find sites for new landfills. Japan's high population density is another factor contributing to waste generation. In 2001, the country had a population of 127 million, which works out a population density of 341people/km². This compares to average densities of 29 people/km² in the US, 192 in Italy, and 233 in Germany. The demographic data also suggests that the Japanese population is highly concentrated in urban areas and that the population within these urban areas is generally denser than in many other countries. For example, in 2000, 60% of the area within Tokyo city limits had a population density > 15,000 people /km². The same criterion is met by less than 20% of the area within New York or Los Angeles, and approximately 20% of the area within Paris city.

As a result of these geographic and demographic conditions, land in and around urban areas is in high demand, making the siting of new landfills both difficult and extremely costly. To compensate, the Japanese waste management infrastructure has relied primarily on incineration to reduce the quantity of waste that is landfilled. Today, for example, nearly 70% of Japan's municipal solid waste (MSW) is incinerated.

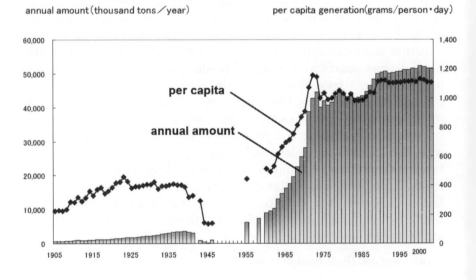

Fig.10.1 Waste generation in Japan (Source: Rachel et al 2009)

The majority of substances composing MSW include paper, plastic, putrescible/organic matter, glass, textiles, metal and rubber. Of the four main types of MSW management – landfilling, incineration, composting and anaerobic digestion – incineration plays a major role in Japan for the reasons mentioned. On the other hand, landfilling and recycling are very common methods of managing waste in other developed countries in Europe, North and Latin Americas. The trend of the annual quantities of waste and per capita generated for Japan is shown in Fig. 10.1

According to the Japan MOE 2004 Report, the remaining capacity of the final disposal sites is 152.61 million cubic meters and the remaining life time is less than 6 years. However, in order to tackle the challenges normally encountered by waste generation in Japan, "integrated waste management" the 3Rs method (Reduce, Reuse, and Recycle) is adopted as the best and most preferable way to deal with solid waste.

In Japan, a variety of waste collection approaches are used, including curbside and drop off collection, roadside bins, and pneumatic collection systems in approximately 10 urban centers (e.g., Ajiahama, Tenjinkawa, and Itami). Some municipalities offer curbside collection that is supplemented by drop-off facilities. In these cases, the materials collected typically include paper and paper packaging, glass and PET bottles, and aluminum and steel cans. In a few programmes, residents are encouraged to separate the remaining fraction of waste into combustibles and non-combustibles. The combustible materials are delivered to an incinerator while the non-combustibles are landfilled directly. Other programmes may collect waste and recyclables together, and then separate some recyclable materials such as aluminum and steel, at a mixed-waste MRF, transfer station, or prior to incineration. Over the period 1989 to 1998, of the total MSW collected, the fraction that is landfilled is decreased from 21.6 to 7.5% (nearly 64%), the recycling has increased from 4.5 to 14.6% (nearly 317%), and the incineration part increased from 73.9 to 77.9%. All the recycling activities are subjected to both 'Containers and Packaging Recycling Law' and the 'Home Appliances Recycling Law'.

The un-interrupted Incineration facilities are typically much larger and provide energy recovery through the generation of hot water or electricity compared to other ones. The emissions from these facilities are much more cost-effective to control because of economies of scale. Concerns about dioxin emissions from incineration facilities resulted in the passage of the "Law Concerning Special Measures against Dioxins" in 1999.

The treatment of waste is usually performed under the provisions of 'Waste Disposal and Public Cleansing Law' (MOE 2008).

(b)   Thailand: (sources: PCD, Thailand and MNRE, Thailand)

Thailand is a fast developing country in south East Asia. The status of waste management is shown in the Fig. 10.2. Out of 15.04 million tons of waste generated, 12.04 million tons (84%) is collected. The recyclable potential of MSW is high, as much as 80%. But the actual recycled material is around 22% (3.1 million tons).

MSW Collection and Disposal: Most cities employ compaction-type trucks to collect solid waste generated in their areas. Generally, around 80-90% of MSW is periodically collected. However, in rural areas, collection services are not widely covered and open dumping and burning are typical practices for disposal of MSW.

Only about 37% of solid waste collected is properly disposed through 'sanitary landfills', and around 63% is disposed improperly, i.e., open dumping and open burning (PCD Thailand). There are only 97 properly designed disposal facilities (91 sanitary landfills, 3 incinerators, and 3 integrated-system facilities) under operation, serving about 480 local administrations throughout Thailand (MNRE, Thailand). See Fig.10.2. Twenty more are under construction (PCD, Thailand)

STATUS OF WASTE MANAGEMENT

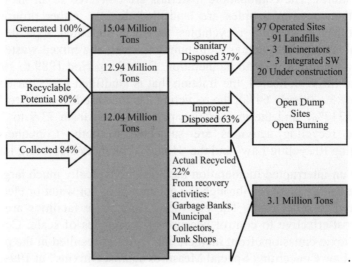

Fig. 10.2 Status of waste management (Source: PCD, Thailand)

Recycling: Nearly 90% of MSW is potentially compostable and recyclable materials. Therefore, a number of incentive campaigns with the cooperation of public and private sectors as well as NGOs have been undertaken to encourage the recycling activities that help to reduce the waste volume from sources. Recycling systems existing at disposal facilities are used by collection crews and scavengers.

There is provision to charge service fees for *Collection and Disposal* under rules, but they are under revision to comply with the MSW management approaches contemplated under National Plan.

Organizations Responsible for MSW Management: Local administrations have the responsibility to handle MSWM within their defined areas, while Central government, Ministry of Natural Resources and Environment (MNRE), takes care of policy making and framing guidelines and technical assistance, while Ministry of Interior coordinates among local administrations.

Waste management Policy: Two major National waste management policies are in place to achieve proper waste management: 1. To promote 3Rs hierarchy (Reduce, Reuse and Recycle), *and* 2. To encourage local administrations to establish central solid waste disposal facilities with an integrated approach of using appropriate technology, and beneficially utilizing waste through composting and energy recovery (MNRE, Thailand).

The Government has come up with a National Plan to improve the waste management situation with the following goals: (i) to ensure that not less than 50% of MSW will be disposed by the year 2009 and up to 100% by 2017, and (ii) to increase efficient disposal facilities by not less than 50% of all 38 provinces by 2009, and up to 100% by 2013 (PCD, Thailand).

To achieve the goals, National Plan and Policy provides the following strategies: *Social strategies,* to promote participation between public sectors, private sectors and citizens, and to create public awareness to reduce waste and to increase the utilization of organic wastes and recyclable wastes; *Economic strategies,* to encourage investments by private sectors for utilizing clean technologies for goods production, and waste treatment and disposal management, and to levy tax, if necessary, for reducing waste generated at the manufacturing stage; *Legal strategies,* to establish laws and revise existing laws and regulations as well as to emphasize on law enforcement in order to bring effectiveness in various steps of waste management; and *Supportive strategies,* to support R & D of appropriate technologies for producing environmental friendly products as well as products from recyclable materials (PCD, Thailand).

The National Plan envisages an Integrated Waste management system, shown schematically (PCD, Thailand) in Fig.10.3.

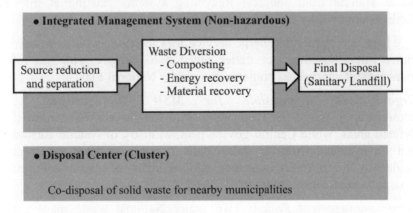

Fig.10.3 Integrated Management System (PCD, Thailand)

Area Clustering Approach for Establishing Central MSW Management Facilities is a part of integrated management system. The purpose is to encourage local administrations to come together to establish central disposal facilities with suitable technologies. This will reduce the disposal cost which is based on the amount of MSW generated for the local administrations. There are approximately 300 clusters formed throughout the country. It is estimated that about 28 clusters generate more than 250 tons MSW per day; and other clusters less than 250 tons per day (MNRE, Thailand).

Electricity from WTE systems for Domestic Wastewater Treatment: Royal Thailand Government has a policy to utilize electricity generated from MSW for running the domestic wastewater treatment plants. Hence, the excess electricity from incineration or biogas plants is supplied to local administrations in order to operate their wastewater treatment plants; sometimes, the electricity is sold to Provincial Electricity Authority at reasonable prices.

10.2 Africa

(Sources: Mwesigye et al, 2009, Rachel et al 2009)

Waste management problems in Africa are varied and complex. Waste is typically disposed off without consideration for environmental and human health impacts. The indiscriminate and improper dumping of MSW onto any available space, known as open-dumps, is increasing and is compounded by poverty, population explosion, decreasing standards of living, poor governance, and low level of environmental awareness. Disposal of hazardous waste along with non-hazardous waste without segregation is common practice. MSW management has been an inflexible problem in recent times beyond the capacity of most municipal/state governments. The problem is expected to aggravate with significant increase in waste generation as a result of industrialization, urbanization and modernization of agriculture in Africa.

The standard of waste management in the region is low; it suffers from limited technological and economic resources as well as poor budget. Added to it, the people especially the poor are reluctant to pay as they consider the waste disposal as a welfare service. These problems are worse in African countries afflicted by conflict and political instability, for instance, Côte d'Ivoire, Sudan, Somalia and Liberia. Such situations encourage illegal trans-boundary traffic of hazardous wastes. For example, there was illegal dumping of dangerous wastes from Estonia and Netherlands into Côte d'Ivoire in August 2006. The toxic waste pumped into Nigeria, Benin, Togo, Sierra Leone, Guinea, Zimbabwe and other African countries in the 1980s by dishonest waste traders from developed countries led to the adoption of the Bamako Convention on the Control of Trans-boundary Movement of Hazardous Waste in Africa in 1991.

The legal and institutional framework for the environmentally sound management of waste across Africa is either lacking or inadequate. All the countries have not ratified the Multilateral Environmental Agreements on wastes and chemicals (MEAs), particularly the Basel, Stockholm, and Rotterdam Conventions. Comprehensive national waste legislation is lacking although several countries have sketchy legislation on hazardous waste management.

Improper waste disposal has resulted in poor hygiene, lack of access to clean water and sanitation by the urban poor. The Continent urgently needs infrastructural, institutional, legal reforms and attitudinal changes. It also needs to adopt Environmentally sound management (ESM) of wastes including waste minimization focusing on the promotion of the 3Rs – Reduce, Reuse and Recycle; Waste to Wealth Initiatives towards poverty reduction and alleviation; Corporate Social Responsibility by producers of wastes; and involvement of Public-Private Partnerships.

The international community needs to support transfer and diffusion of knowledge and technology, and promote investments for implementing environmentally sound waste management practices, and to strengthen their national human and institutional capacities and to create awareness about integrated waste management practices. The scale of necessary investments is beyond the capacity of African countries.

Assessment of waste management capacities in some countries in Africa: A Regional Needs Assessment was conducted in June/July 2001 covering the English speaking African countries and, based on this exercise, capacity building activities were initiated. A rating scheme was used based on the factors: (a) Priority given to waste management, (b) Skills in waste management, (c) Financial resources, (d) Facilities and infrastructure, (e) Monitoring and information, (f) Training activity, (g) Project activity, (h) Institutional network, (i) Regulatory framework, (j) Legislative enforcement, and (k) Administrative system. Fig.10.4 shows the status of waste management (rating of waste management capacity) in the Region in 2001, and updated in 2004.

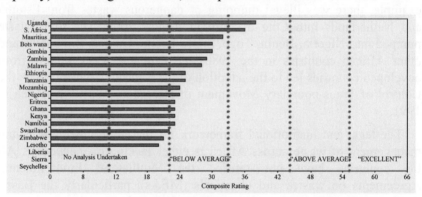

Fig.10.4 Rating of waste management capacity

(a) ***Nigeria:*** Nigeria is the most populous country in Africa and over the past 50 years, has recorded third largest urban growth rate in the world at 5.51% annually (UNWUP 1999). Nearly ten percent of the population (~21 million people) lives below the national poverty line (World Bank 1996).

The magnitude of the solid waste problem in Nigeria is hard to comprehend. The garbage 'dumps' are located on the side of the highways, at the borders of cities and slums. Since the garbage is not contained, it spreads into the road, blocking traffic. When refuse accumulates, households and businesses pile it in the center of major roads and burn it (Emily Walling et al 2004); hence a fair percentage of the trash never makes it to informal dumps.

The Federal Environment Protection Agency (FEPA) was established in 1988 to control the growing problems of waste management and pollution in Nigeria (Onibokun and Kumuyi 2003). FEPA prepared Vision 2010 document in an attempt to address environmental problems in the country. Regarding SWM, the goal is to 'achieve not less than 80 percent effective management of the volume of municipal solid waste generated at all levels and ensure environmentally sound management' (Vision 2010, 2003). Strategies to achieve this goal include education and awareness programs, developing collaborative approaches to integrative management of MSW, strengthening existing laws and ensuring compliance, and encouraging local and private sector participation. But the prevailing poverty and government corruption has prevented effective implementation of these plans. In addition, there is little to hold the government or the public accountable to the regulations developed by FEPA and Vision 2010 (Bankole 2004).

The composition of MSW in Nigeria in 2007 is given in the Table 10.1.

Table 10.1 Composition of MSW in Nigeria (Source: Sha Ato et al 2007)

Waste source	Waste category (%)							
	Putrescibles	**Plastics**	**Paper**	**Metals**	**Glass**	**Textiles**	**Fines**	**Ohters**
LD	57.5	6.10	4.30	2.50	2.30	2.90	21.0	3.40
MD	53.7	7.10	4.10	2.01	1.70	2.40	27.1	1.70
HD	36.4	8.04	2.59	1.75	0.86	3.67	41.0	5.73
COMM	27.9	10.20	10.90	3.40	6.90	1.20	36.4	3.10
INS	44.8	5.90	8.90	0.90	1.20	0.30	36.4	3.10

LD = low density; MD = medium density; HD = High density; COMM = commercial; INS = institutional.

In Nigeria, no organization seems to be willing to take responsibility for regulation of waste management. For example, in Ibadan, the waste management never rests with a single authority since the late 1980s (Onibokun and Kumuyi 2003).

Since the local governments could not collect funds for solid waste disposal (Onibokun and Kumuyi 2003), private companies were contracted for waste disposal. However, these companies are often as much ineffective as the State (Onibokun 1999). In Lagos, the sixth largest city in the world, around 20 to 25 percent of city's budget is allocated to waste management. However, even with proper garbage-collecting trucks, the extremely dense streets of Lagos make it impossible for the trucks to maneuver through to collect the excessive amounts of garbage that are produced. While in the five other mega-cities of the world, with over ten million people, over forty trips are made per day from the city to the dump site, only two trips are possible each day in Lagos (UNESCO 2003).

The organic matter content, pH, particle size distribution, bulk density, total porosity and the hydraulic conductivity, and available heavy metals (zinc, copper, iron and lead) of soils from an open dump site are greatly affected by the large volume of wastes dumped on such locations. Waste–amended-soils (Open dump sites soils) have high organic matter content (Anikwe 2002).

Causes for MSW problems: The Nigerian government is beset by lack of adequate policies and human resources, insufficient facilities, and financial difficulties (Agunwamba 1998). Lack of a system for accountability in the government for the lapses adds to the problem. For example, it is not uncommon for the government to withhold employees' wages for several months at a time if finances are tight (Emily Walling et al 2004). People feel entitled to waste collection services, and do not believe that waste management is part of each individual's responsibility (Emily Walling et al 2004).

The government has recently come up with national regulations for solid waste control, characterized by strong governmental involvement, development of an administrative infrastructure to regulate pollution, establishment of pollution control measures as a national priority, and "end-of-the-pipe" management (Mazmanian and Kraft 2001). However, corruption and inadequate finances promise that these regulations will be equally as ineffective as earlier attempts. So, regulatory flexibility, management of pollution through market based and collaborative incentives, introduction of pollution prevention, and a shift to an oversight capacity of local and regional governments have been attempted. Sustainable development approach incorporating pollution prevention at the individual, business, and industrial levels, institution-building, and attempts to balance human and natural system needs would also lead to sound MSWM system (Mazmanian and Kraft 2001).

In Nigeria, like in many developing countries, potential to develop market incentives exists. Thousands of people make their living by scavenging recyclable materials from open waste dumps (Kasseva and Mbuligue 2000). Their job is extremely hazardous; they get less pay and are being exploited. Significant increase in diseases contracted by landfill scavengers, and AIDS and other blood diseases from hazardous medical wastes are reported (Kasseva and Mbuligue 2000). These scavengers and other entrepreneurs, however, potentially provide an important source of social capital and programs, and the success depends on their economic well-being.

Programmes to develop locally based waste management services, and to strengthen cooperation among neighborhoods, municipal employees, and public/private organizations; and to train and educate young entrepreneurial volunteers to collect and transport household garbage to intermediary points, from which it is transported to the final dumpsite (Doan 1998) could be used to encourage source separation and recycling and composting as has been done in Ivory Coast in 1990s. Agunwamba (1998) estimates that in Nigeria, efficient recycling and composting programmes that include creating markets and market incentives, could save 18.6% in waste management costs and 57.7% in landfill avoidance costs. There is already high demand for scavenged materials (Kasseva and Mbuligue 2000); since MSW has a large proportion of organic material, composting could be an effective option to reduce waste volume.

There has been lack of international involvement in social and environmental issues in Nigeria. Recently several environmental organizations are formed to attract funds for management projects to the country (Vision 2010 1996). Additionally, community identification plays an important social role in Nigeria, despite its being composed of a large number of different ethnic groups

(Emily Walling et al 2004). The community and organizational participation and strengthening institutions can be achieved only in long-term, and there may be significant setbacks throughout the process.

Given training and reasonable solutions, Nigerian's citizens would likely be very enthusiastic to devote time and effort managing MSW. It is believed that education at all levels is an important beginning for a practical, effective and lasting municipal solid waste management.

(b) ***Tanzania: City of Dar es Salaam***: The status of MSWM in the city of Dar es Salaam, and five other municipalities have been studied during 1992-1999 and 1999-2004 respectively. The results of these studies are summarized.

Before 1992, the Dar es Salaam City Council (DCC) could not offer an effective solid waste collection service for the city. While solid waste generated at that time amounted to 1400 tonnes/ a day, the DCC had the capacity to collect between 30 and 60 tonnes (2% to 4%) of this amount. The city environment was characterized by large amounts of dumped garbage in public open spaces, on streets and major roads and in open drains, resulting in flooded roads, ground water pollution, soil contamination, and escalating outbreaks of communicable diseases like cholera, diarrhea and dysentery. The situation was particularly serious in the central business district. Like any other African city, DCC suffers from lack of equipment and financial resources to purchase spare parts and fuel for the fleet, mixed signals on political will, un-focused City leadership and short of an official disposal site.

Under the auspices of the Sustainable Cities Programme (a UN Habitat programme) a working group on solid waste management was formed, with the Dar es Salaam City Council as a lead partner. The objectives adopted by the working group on SWM was to improve the cleanliness of the city through increase of collection and disposal of waste, to create sustainable income generating activities for CBOs and small trade enterprises involved in waste collection and recycling, and reduce the waste volumes by encouraging recycling, reuse and composting.

In order to achieve these objectives, the following strategies were formulated:

- Conducting emergency city-clean-up campaigns

- Involvement of the political, administrative and city/municipality decision making machinery from the grassroots

- Private sector involvement in solid waste management

- Community involvement in solid waste management through awareness campaigns and pilot demonstration projects

- Improved management of refuse disposal sites and

- Promotion of recycling, reuse and composting

The strategies are being implemented on the ground through sustainable projects involving:

- Provision of communal waste storage facilities

- Procurement of solid waste collection equipment

- Development of a sanitary landfill and landfill gas extraction

- Creation of public awareness campaigns and,

- Promotion of community based solid waste collection.

With the exception of a sanitary landfill development project, implementation of other projects has been successful to the extent of achieving over 55% of the expected results. Successes achieved in implementing demonstration projects are scaled up to work for the entire city and later replicated to other five Municipalities in the country from 1999 to 2004. Some of the important observations made during this period are:

Involvement of the Private sector: The involvement of the private sector and local communities in solid waste management activities have created employment opportunities to a substantial number of jobless city residents, mostly, unemployed women and youths.

Income generation: Solid waste management activities have been offering income generation opportunities. Income generated is not only from wage payments but also from selling recovered materials from solid waste. Re-used solid waste creates items like plastic bags, plastic containers, knives, spoons, frying pans, gutters, etc., which are on sale in major markets in the city. Dar es Salaam City Council and the five Tanzania Municipalities are also having a good source of income from 'Refuse collection charges' collected from the residents.

Practical and attitudinal changes: There are increasing signs of waste segregation at source and storing in dust bins to a large extent, sorting at communal waste collection points and a large number of organized groups of people involving in solid waste recycling. Increasingly, people no longer regard SWM activities as useless but rather beneficial that can be utilised to generate income and alleviate poverty (Source: Dar es Salaam City Council, Tanzania)

Kenya: Nairobi city

There is not much literature on solid waste management (SWM) services in Kenya with the exception of Nairobi (Ikiara et al., 2004).

Solid wastes in Nairobi are due to wide range of industrial, service and manufacturing activities. High-volume of solid wastes is generated by the chemical, petroleum, metals, wood, paper, leather, textile and transportation industries. The smaller generators of waste include auto and equipment repair shops, electroplaters, construction firms, dry cleaners and pesticide applicators.

Waste quantities: There are no statistics for total production of wastes in Kenya. A study on MSW in Nairobi produced the waste generation statistics shown in Table 10.2. The difference in calorific value (CV) between low and higher income households is particularly noticeable in the data.

The total MSW production in Nairobi was projected as follows: Shops and restaurants: 94 t/day; Houses: 1285 t/day; Markets: 82 t/day; Road sweepings: 69 t/day

Table 10.2 Waste generation statistics

Source	Quantity	Density	CV (KCal/Kg)
Restaurants	6.79 kg/day	0.28	1630
Other commercial	1.39 kg/day	0.26	1692
High income households	0.654 kg/person/day	0.30	1233
Middle income households	0.595 kg/person/day	0.26	1349
Low income households	0.552 kg/person/day	0.28	630
Markets	2.425 kg/day	0.38	1427
Road sweepings	48.3 kg/km	0.23	n/a
Average		0.28	1032

Waste composition: The same study estimated the composition of MSW, given in Table 10.3.

Table 10.3 Composition of MSW

Material	Average (%)	High income (%)	Low income (%)
Food	51.5	50	57
Paper	17.3	17	16
Textiles	2.7	3	2
Plastic	11.8	14	12
Grass/wood	6.7	8	2
Leather	0.9	1	1
Rubber	1.5	1	2
Glass	2.3	2	2
Cans	1.7	2	1
Other metal	0.9	1	0
Others	2.7	7	4

Most beverages are sold in returnable bottles, which accounts for the low proportion of glass and cans. The use of cans is stated to be on the increase. It is noticed that the difference in composition between high and low income is not very great. From actual observation, however, it appears that in many MSW samples, the proportion of dust is substantially greater than shown in the above analysis, which should therefore be treated with caution.

The chemical analysis provides the following data:

Moisture - 64.2%; Ash - 8.9%; Combustible - 26.8%; C - 49.33%; H - 5.45% N - 1.22%; S - 0.14%; Cl - 0.21%; O - 43.75%

Storage and collection: Municipalities are responsible for managing MSW, including commercial wastes. The private sector handles industrial wastes.

Industrial wastes constitute about 23 per cent of the total solid wastes generated in the city. The collection and disposal of industrial waste in Nairobi is done by industries themselves. Though its disposal is done at a Municipal dumpsite, the industries have the responsibility to collect and dispose the waste at the designated dumping site.

It is estimated in 1997 by JICA study that only about 25% of the waste in Nairobi is collected. An alternative study, undertaken by UNCHS in 1997, has shown that 90% is collected by the Nairobi City Council

(NCC). In the mid 1980s, the appalling NCC performance and demand for municipal solid waste management services attracted private sector providers. It is now estimated that there are at least 60 private companies engaged in solid waste collection services in the city (JICA, 1998). Private companies serve 45-73% of the households, 32% of the institutions, 50% of the industries and 16.7% of the commercial enterprises. About 81% of the households served by private companies live in the high and middle-income areas (largely the western part) of the city. The majority of the private companies are either small family ventures or a hybrid between a community based organization (CBO) and a private firm. Even NCC, which has the social responsibility of providing SWM services to all citizens, concentrates its efforts on residential areas and institutions that can afford private service at the expense of areas inhabited by the poor. In Nairobi, the private sector offers a more reliable service, for which a fee of Ks 200/household/month is charged. According to the JICA study, the NCC collects 80 t/d of which 91% is from stations and 9% from door-to-door. Private contractors collect around 115 t/ day.

Individual household waste containers are present at higher and middle income households only. Low income households make use of communal containers or dumping stations - where waste is hand loaded into vehicles. Some dumping stations are constructed of concrete but others are just informal piles, which are sometimes burnt.

The vehicles may be tractor/trailers, open tippers, roll-ons or compaction vehicles, depending on the size of the city. Typical payloads are 2-3 tonnes. Vehicles are not covered and plastic/paper blows away during transit. A study shows that, in 1996, 50% of the Nairobi City Council refuse collection vehicles were non-operational at any one time.

The extent and nature of the solid waste management problems can be summarized as follows:

First, the collection ratio, that is, the proportion of the solid waste collected to generated, is low; as low as 25 per cent.

Second, there is marked inequality in the geographical service distribution. The Western part of the city is well serviced by the private parties and the NCC while the Eastern part is hardly serviced. High-income and some middle-income residential areas together with commercial areas are well serviced by private companies and the NCC. Small private firms are increasingly servicing some of the relatively better-off low-income areas. The core low-income areas (slums and other unplanned settlements) where 55-60% of Nairobi residents live receives no waste collection service save for localized interventions by community-based organizations (CBOs). The 1998 JICA study found 26% of households in high-income areas,16% of those in middle-income areas, 75% of those in low-income areas, and 74% of the surrounding area do not receive any service. Not surprisingly, thus, residents in low-income areas dissatisfied with waste collection services, are aware of the health risks associated with the problem, and are willing to pay for improved services in spite of their low incomes.

Third, there is widespread indiscriminate dumping in illegal sites and waste pickers litter the city with unusable waste materials without control.

Fourth, there is only one official dumpsite (NCC-owned and operated), which is full and located in a densely populated part of the city, 7.5 km from the central business district along a road with heavy traffic. Moreover, waste pickers and dealers 'control' this dumpsite, forcing the NCC and private companies to 'bribe' to access the dump.

Fifth, the city has no transfer facilities.

Sixth, solid wastes in the city are not segregated, with the exception of unstructured reuse of some waste materials at the household level. The private contractors that collect waste do not process waste in any way and dump at Dandora dumpsite which is littered with all types of wastes from hospital wastes, manufacturing/industry wastes, paper and biodegradable materials. To cut costs, many generators of solid wastes have taken to combustion at the site, which causes air pollution problems. The bulk of these wastes contain plastics, which when burnt generate carcinogenic vinyl chloride monomers and dioxins.

A JICA study (1998) revealed that the residents around the dump site suffered from smoke, smell, and broken glasses. Respiratory and stomach problems among children are common.

Waste minimisation and recycling: In the major cities, Nairobi and Mombasa, paper, cans, glass and plastic bottles are collected for recycling by private businesses or individuals. Such services are not found in the smaller towns. Recycling of materials such as papers, tyres, plastics, used clothes, and metals, is becoming increasingly popular. Organic wastes are also increasingly being recycled to produce compost products. For example, community-based organizations (CBOs) managed by women are recycling market waste from Korogocho Market to produce organic manure for sale. However, the percentage of solid waste recovery is only

8% of the recyclable and 5% of the compostables. There is recovery going on in the industries but the rate is unknown. The groups involved in these efforts are facing a number of problems such as land to conduct the composting, and lack of a stable market for the recovered materials, especially for wastepaper and compost. The self-help activities of the Mukuru project earned Kshs 1.55 million in 1996 from the recovery of 1,018 tons of materials per year. This income was not sufficient for the project's 60 members and for investment to improve efficiency.

About 6,000 tonnes of waste oils are recycled in Kenya, out of a total of 27,000 tonnes produced from vehicles. Waste oils are also used as fuel and for wood preservation. The recycling process, however, produces acid tars. The Oil Industry Waste Management Committee, however, expects cement kilns to use the oils as secondary fuel. The agrochemical industry have also used cement kilns, and in 2005, has negotiated with Lomé IV financing to incinerate the existing stockpile of about 100 tonnes. GIFAP, the international trade association, which has its African headquarters for the Safe Use Programme in Nairobi, has arranged this.

A survey was conducted as part of this study at the Dandora dumpsite where scavengers recover recyclable materials from municipal solid waste. The scavengers were found to be recovering more than 30 different types of materials, with the major ones being ferrous metals (aluminium and copper). While there is considerable potential in recycling, there is a problem of recyclables being contaminated. In addition, there is no policy on recycling in the country. This has led to the importing of waste materials by recycling companies and to the exploitation of waste pickers by middlemen and recycling firms. Industry operators encourage the setting up of recycling schemes (such as for aluminium cans, bottles, and polythene materials) to improve environmental conditions while also generating incomes to the poor.

Disposal of waste: There are no controlled landfills in Kenya and complete reliance is placed on open uncontrolled and burning. The main MSW dump site serving Nairobi is located at Dandora. This is totally uncontrolled and burning. Many scavengers are always present; and one can find tannery sludges, hospital waste including used syringes and other industrial wastes. The situation in smaller towns is similar, although the proportion of waste collected may be even lower.

Problem of plastics in Kenya: As elsewhere in the world, the problem of overuse, misuse and indiscriminate and inadvertent littering of plastic bags is serious in Nairobi. Because the plastic bags are either free or inexpensive there is widespread use and because most bags are thin and highly fragile, re-use is minimal. According to one of the leading supermarket chains in Kenya, approximately 8 million bags are given out by the supermarkets alone every month and two times as much in the informal sector in Kenya.

Legislation and Enforcement: There is currently no specific legislation on waste management. Before the Environmental Management and Co-ordination Act was enacted in 1999, Kenya was relying on the Public Health Act (cap 242) and the Local Government Act. These acts empower LAs to establish and maintain MSW management services and require them to provide the services. The Acts, however, neither set standards for the service nor insist on waste reduction or recycling. In addition, the Acts do not classify waste into municipal, industrial and hazardous types or allocate responsibility over each type.

There is little enforcement of waste management standards. Whatever enforcement exists, that is undertaken by local authorities under the Public Health Act. For example, a waste incinerator at a shoe factory has been closed for production of smoke; a factory discharging heavy metals causing a sewage works to cease functioning is also closed until a treatment plant is constructed.

The Ministry of Environmental Conservation (National Environmental Secretariat) covers pollution control (wastes), EIAs, resource management, planning and education.

The dump sites are selected by the municipal authorities and there are no rules for the location and operation of dump sites. The EIA procedure will come into effect when it is introduced.

Under the new legislation, it is expected that monitoring of discharges to ground and surface water would be undertaken by the Ministry of Land Reclamation, Regional and Water Development (MLRRWD) which may be delegated to the local authorities ultimately.

The World Bank funded the preparation of the National Environmental Action Plan, which contains some aspects related to SWM.

CBOs, RAs, farmers, informal agents: With NCC's awful performance and the failure of private service to extend into low-income and unplanned settlement areas, community-based initiatives in waste collection, transport, storage, trading and recycling started to emerge in 1992. There are now a number of CBOs, including charitable organizations, ethnic associations, welfare societies, village committees, self-help groups, and residential (or neighbourhood) associations (RAs). Majority of the CBOs are engaged in waste composting although the main activity of about 44 % of them is neighbourhood cleaning (Ikiara et al., 2004). One-third of CBOs are involved in waste picking. Despite individual and localized performances, the community in general plays a small waste management role.

NGOs and international organizations support CBOs through training, marketing and provision of tools and equipment, among other ways. 55.6 % of the CBOs report having been sponsored or facilitated by local and international NGOs and United Nations agencies like the UNFPA and UNCHS (HABITAT) (Ikiara et al., 2004). Important NGOs include Foundation for Sustainable Development in Africa (FSDA), Uvumbuzi Club and Undugu Society of Kenya. Other institutions offering assistance to CBOs in Nairobi include the National Council of Churches of Kenya (NCCK), the private sector, Norwegian aid institutions, and the Japan International Cooperation Agency (JICA). Donor agencies play a direct role and also an indirect one, by funding the NGOs that assist CBOs (King, 1996). Neighbourhood or Residential Associations (RAs) have emerged in many middle and high-income residential areas to organize provision of failed public infrastructure services. It is estimated that there are over 200 registered RAs in the city, engaged in improvement of security, roads, and cleanliness. They are contracting, organizing, and monitoring private SW collection service. The pioneering RA, Karen and Langata District Association (KARENGATA) and the Nairobi Central Business District Association (NCBDA) have emerged as highly organized, resourceful, and influential groups. Through a memorandum of understanding with the NCC, NCBDA has not only donated garbage storage bins for use in the CBD but also engaged in policing (security) and road and public toilet rehabilitation projects. There is now visible improvement in security and in availability of clean public toilets and the storage of solid waste in the CBD.

RAs in Nairobi have formed two umbrella associations 'We Can Do It' and Kenya Alliance of Residential Associations (KARA), which lobby for improved services, facilitate formation of new RAs, and provide technical assistance to potential RAs.

Farmers are also becoming important actors in Nairobi's SWM sector. The increasing number of urban and peri-urban farmers collects poultry waste, green vegetable waste, and cow dung as well as food waste from hotels, markets and other institutions, and transport it to use either as animal feed or as organic fertilizer. The actual amount of waste removed from the municipal waste stream through this route is not known.

Many informal agents (waste pickers, traders and dealers, itinerant buyers, informal dump service providers and informal recycling enterprises) are also involved in Nairobi's SWM sector, albeit as a secondary activity (Ikiara et al., 2004). These actors are involved in all SWM domains, including waste collection, separation, storage, re-use, recovery, recycling, trading, transport, disposal, and littering. They reduce the waste that has to be disposed, more significantly, in non-serviced areas inhabited by the urban poor. Like urban farmers, the actual contribution of these informal actors to SWM in the city and other parts of the country is not known.

Poor performance of MSWM: The poor SWM performance in Nairobi is attributable to many factors. Expansion of urban, agricultural and industrial activities has generated vast amounts of solid and liquid wastes that pollute the environment and destroy resources. The problems are mainly due to lack of appropriate planning, inadequate political will and governance, poor technology, weak enforcement of existing legislation, the absence of economic and fiscal incentives to promote good practices, and lack of analytical data concerning volumes and compositions of waste substances.

Administration of Nairobi is chaotic, with the NCC and the Central Government (particularly the Ministry of Local Government and the Provincial Administration in the Office of the President) often clashing and duplicating roles. Moreover, as the policymakers (NCC councillors) are generally not knowledgeable, the mismanagement and corruption have become the hallmarks of the NCC. The by-laws related to prohibiting illegal disposal of waste, specifying storage and collection responsibilities for SW generators, and indicating the Council's right to collect SWM charges is not effectively implemented. The Central Government also fails to oversee the performance effectively.

This dysfunctional local administrative system has led to the inefficiency of NCC operations - unprecedented deterioration of physical infrastructure, lack of transfer facilities, widespread indiscriminate waste dumping, lack of system-wide co-ordination and regulation of stakeholders, absence of strong and effective partnerships between the NCC and other SWM partners, lack of policy support for waste re-use and recycling, community's indifference to involve in SWM, prevalence of casual littering due to lack of public education and non-enforcement of NCC bylaws (Ikiara et al., 2004). Rapid population growth and urbanization (like other cities in developing world) add to the problem. *Public concern about waste management:* People in Nairobi are concerned by the inadequate management of MSW and the continuing decline in standards. In Nairobi, a survey showed that 36% of respondents thought that the problem of garbage collection was very serious and a further 22% saw the problem as moderate. If people are to be expected to pay for a service, however, the quality – particularly of collection - must be dramatically improved. It is unlikely that low income groups are prepared for the service.

10.3 Latin America

(source: The World Bank 2008)

Latin American countries (LAC) have considered solid waste management more crucial for the improvement of MSW services.

The Continent is highly urbanized with 78% of its 518 million populations living in cities; 114 cities have population more than 500,000, housing 225 million inhabitants and generating 98 million tons of waste per year. There are thousands of small and medium size cities also, with 209 million population generating 56 million tons of waste annually. The MSWgeneration status in cities of different sizes is given in the Table 10.4.

Table 10.4 Estimated MSW generation in LAC (2005)

City Size (million)	Total Population (million) [no. of cities]	MSW Generation Rates (kg/cap/day)		Total MSW Generated (million tons/year) [% of total]	
		Domestic	Municipal	Domestic	Municipal
>1	183 [55]	1.04	1.25	69 [55%]	83 [54%]
0.5 to 1	45 [59]	0.69	0.98	11 [9%]	15 [10%]
0.2 to 0.5	58	0.68	0.88	14 [11%]	19 [12]
< 0.2	151	0.56	0.68	31 [25%]	37 [24%]
Totals	434	0.79	0.97	125	154

MSWM Practices and Problems: Municipalities are accountable for SWM services throughout the Continent. *Waste collection* is generally satisfactory in the large cities: typically 85% of waste is collected in capital cities and large metropolitan areas. However, in the poor peri-urban areas of large cities, collection services are often deficient. Small to medium cities have lower collection levels and efficiencies which on average are about 69%.

Waste disposal is generally lacking; only 23% of MSW collected is disposed in sanitary landfills, 24% goes to controlled landfills, and the rest to open dumps or water bodies. In capital cities and metropolitan areas, about 60% of MSW collected is disposed in sanitary landfills. In small and medium cities open dumping predominates. Overall 60% of all MSW generated in LAC ends up in unknown disposal sites. However, surveys in Colombia, Chile and Mexico show that many sanitary landfills *do not meet* basic standards for sanitary operations, and many do not have the necessary EIA approval or environmental operating license.

Service financing is very poor; average cost recovery is less than half of actual recurrent costs of service provision. Most cities, especially small and medium cities, have little knowledge of the actual costs of service provision. The average rate collected is US$2.49 per household. The efficiency of service provision is often poor. The reasons for the high cost of poor quality service are often excessive employment, and low labour and vehicle productivity.

The *Service Costs* for different components of MSWM in LAC are:

Collection: US$ 15-40 per ton
Street sweeping: US$ 10-20 per km
Transfer: US$ 8-15 per ton
Disposal: US$ 4-15 per ton

LAC Average: US$ 29 per ton collected, transported and 'adequately' disposed.

The estimated costs in low-, middle- and high-income countries for different components of MSWM are given in Table 10.5. There is a very large difference between high-income countries and low-income countries, ranging between 10 to 20 times.

Table 10.5 Estimated Costs of Adequate MSWM

Contries →	Low-income	Middle-income	High-income
Average Waste Generation	200 Kg/cap/yr	300 Kg/cap/yr	600 Kg/cap/yr
Average per-Capita income	370 US$/cap/yr	2,400 US$/cap/yr	22,000 US$/cap/yr
Collection	10-30 US$/t	30-70 US$/t	70-120 US$/t
Transfer	3-8 US$/t	5-15 US$/t	15-20 US$/t
Final Disposal	3-10 US4/t	8-15 US$.t	15-50 US$/t
Total Cost	16-48 US$/t	43-100 US$/t	105-190 US$/t
Total Cost Per-Capita	3-10 US$/cap/yr	12-30 US$/cap/yr	60-114 US$/cap/yr

Private Sector Participation: Private sector participation is drawn in all countries to varying degree in all phases of the waste management – recycling, collection, transport, and disposal – and all enterprises ranging from large-scale multinationals to small-scale enterprises are involved. The efficiency of these private sector enterprises varies significantly, although the experience generally is good. Big cities started contracting out in 1970s which has spread to intermediate cities in 2000s. Half of urban population in Latin American countries is now served by private operators (WB 2008). In cities, for example, such as Sao Paulo, Buenos Aires, Bogota, Santo Domingo, and La Paz, waste collection is done almost completely by private enterprises (UNEP). In Brazil in 1998, 40 firms collect 65% of urban waste nationwide (up from 40% in 1982).

Due to lack of institutional capacity and structure for arranging service contracts, awarding concessions, and monitoring compliance with contract conditions, the supervision/regulation is feeble in municipalties that engaged private parties for improving services. As a result, performance standards under such arrangements are neither satisfactorily established nor adhered to, and the expected levels of efficiency are seldom achieved.

There is lack of competition, transparency and accountability at municipal level, and there is need to ensure them.

Private sector/small-scale initiatives are also happening in low-income and difficult-access areas. Such small-scale waste collection enterprises under the sponsorship of NGOs or technical cooperation organizations are operating in Bolivia, Colombia, Costa Rica, Ecuador, Panama, and Peru. The government either pays to these organisations for their services or allows them to collect fees directly from the households in their areas (UNEP). The role for micro-enterprises in MSWM practices has been established.

Institutional failures: A number of LAC cities have come up with solid waste management master plans, though very few have implemented. The solid waste management programmes are largely ad-hoc, heavily influenced by the political environment of the time. Legal and regulatory framework is often dispersed, overlapping and incoherent. The existing legislation, in general, does'nt take into account the economic reality of municipalities, resulting in non-compliance of the provisions. Further, lack of inter-municipal coordination (both rural and metropolitan) leads to inefficiencies and loss of economies of scale. Another important factor is failure to plan the system strategically; for example, politicians and planners fail to recognize the importance of NIMBY and the need for positive public involvement to deal with it. Added to it, enforcement is ineffective.

Disposal methods: The final disposal is appropriately based on sanitary landfill; there has been an increase in the number of sanitary landfills and controlled landfills, but open dumping still exists and is common. The private sector is also involved to a limited extent, in operation of landfills. For example, in Buenos Aires, as well as in some of the large cities in Brazil, landfills are operated by private parties. Privately operated landfills are also located in cities of Columbia, Ecuador, Mexico, Panama, Paraguay, and Venezuela. Most manual landfills are operated by small-scale enterprises.

Driven by Carbon financing, Landfill Gas-to-Energy (LFGTE) system has been quite successful in Argentina, Brazil, Chile, Mexico, and Uruguay (WB 2008).

The Latin American countries have gained considerable experience with composting, but the status is mostly disappointing.

No large-scale incineration (waste-to-energy) is preferred due to economics, as well as high moisture content and low calorific value of the waste generated.

Waste minimization and recycling efforts have been recent. On average, the estimated MSW recycled is 3% in LAC. However some countries are doing better: Mexico recycles 10% of waste stream; Paper and cardboard are recycled in Brazil (44%), Colombia (57%), Chile (50%), and Ecuador (40%).

Brazil recycles 87% of aluminum cans, 70% of steel cans, 35 percent of PET containers, and 45% of glass bottles (WB 2008).

Source separation and separate collection is on the increase: 20% of municipalities in Colombia and 5% of municipalities in Brazil have been conducting source separation and separate collection.

Like in other developing countries, recycling is predominantly performed by informal sector: PAHO estimates there are 500,000 wastepickers in LAC, 29% women and 42% children. Wastepickers face high health and accident risks, and live in conditions of extreme poverty. Many of them work at dumpsites, impeding attempts to operate as sanitary landfills. Social programmes are needed to improve their living and working conditions. In addition, the recycling activities need to be moved from dumpsites to waste sources by properly organizing cooperatives and microenterprises. Such microenterprises and cooperatives are successfully organized for informal wastepickers in Brazil, Colombia, Mexico, and Peru.

Sector setting in client countries: Strategies are required to effectively conduct MSWM services. Some common elements of these strategies could be:

- Regional landfill construction including LFGTE;
- Closure and/or remediation of open dumps;
- Strengthening national and local institutions including the private sector;
- Development of local/regional integrated MSWM strategies;
- Promotion of waste minimization and recycling;
- Social inclusion of wastepickers;
- Public communication and outreach.

Regional Strategy: MSWM Projects in LAC should give priority to: expanding collection to poor neighborhoods and settlements, improving final disposal, and promoting waste minimization and recycling. To achieve these, the authorities should focus on six key issues: (1) Strategic planning for integrated waste management, (2) Better institutional arrangements, (3) More efficient operations, (4) More effective financial management, (5) Improved environmental protection, and (6) Waste minimization and recycling strategies.

Compatibility with sustainable development: Landfills contribute significantly to greenhouse gases, leading to global warming. Therefore, the landfill construction and operation have to be compatible with the concept of sustainable development. For the countries in Latin America, it is an important opportunity to introduce LFGTE projects as mitigating measure, taking advantage of emerging markets for carbon emission reductions. Incidentally, the indirect benefit is that LFGTE projects can help finance properly operated sanitary landfills which is a must for the LFGTE projects. But, this may result in displacement of wastepickers currently working at open dumpsites. The planners have to come up with comprehensive socio-economic integration strategies so that sustainable sanitary landfills will be in place and the lives of wastepickers are taken care of.

In terms of policy, the region needs to start working on various fronts: (i) in the medium term LFG treatment (burning) has to be mandatory for security and sanitary reasons, even if this means loosing the CDM potential, (ii) Minimization practices, that is, reduction, re-use, recycle, and composting need to be included in the clients agenda, for sanitary, environmental and economic reasons. These actions will trigger additional eligible CDM (Clean Development Mechanism) activities that in the medium term will replace LFG revenues from CDM. The World Bank is currently developing a recycling methodology which is expected to have a significant impact in the sector, mainly because of the social implications that may come with it (informal work of waste pickers).

The Bank can help the cities in LAC by:

- Assisting countries in establishing national MSWM policies and programmes with the elements described above,
- Ensuring that projects are designed within an integrated MSWM strategic planning framework,
- Targeting immediate financing (IBRD, IFC, CFU) toward improved landfill disposal and the introduction of LFGTE components to help ensure operational viability, and
- Support local and national efforts to expand waste minimization and recycling, and improve the lot of wastepickers.

Chile: Santiago city (source: Paula Estevez 2003): Studies of the solid waste issues in Chile are relatively new and recent. Chile has been one of the few countries in Latin America which has witnessed tremendous economic growth in the last two decades; this growth is accompanied by an increased industrial activity and a significant and uncontrolled rise in the quantity of waste, creating huge social and environmental costs. Nearly 40% of the Chilean population lives in Santiago Metropolitan Region. During 2001, the annual amount of MSW produced in Santiago was 2,267,743 metric tons, and is projected to grow to 3,693,914 metric tons by 2011 (CONAMA 2002).

Until 1990 all the MSW produced in Santiago was disposed in 'garbage dumps.' Due to policies framed during 1990s to control this problem, the entire MSW collected in Santiago is deposited <u>currently</u> in authorized sanitary landfills. However, none of this waste is recycled or processed; therefore, current landfills will be filled within the next 20 to 40 years. Land in Santiago is scarce because of its high population, the large and increasing spread of urban areas, and its geographical location, making it difficult to find space for new landfills. Moreover, landfills have been opposed strongly by the politicians, people and NGOs. The political, geographical and environmental challenges do not make the landfills a sustainable alternative for MSW management. Therefore, there was an attempt to assess the use of relevant waste-to-energy technologies as a possible answer to Santiago's current MSW management problems incorporating environmental and economic considerations (Estevez 2003). Santiago Metropolitan Region with 6 million inhabitants represents nearly 40% of the Chilean population (INE 2003). The city produces 1.1 kg of garbage per capita daily. As seen in Table 10.6, during 2001 the annual amount of MSW produced in Santiago was 2,267,743 metric tons. On a year-to-year basis, volume is growing at 5% (CONAMA 2002).

Table 10.6 MSW annual production in Santiago

Year	Metric tons/year	Metric tons/month
2001	2,267,743	188,979
2002	2,381,130	198,428
2003*	2,500,187	208,349
2004*	2,625,196	218,766
2005*	2,756,456	229,705
2006*	2,894,279	241,190
2007*	3,038,993	253,249
2008*	3,190,942	265,912
2009*	3,350,489	279,207
2010*	3,518,014	293,168
2011*	3,693,914	307,826

projected, (Source: CONAMA, 2002)

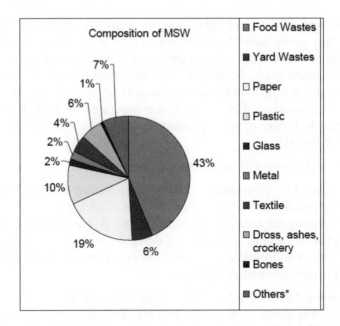

Fig. 10.5 MSW composition in Santiago city (source: CONAMA 2002)

Santiago is divided into 44 municipalities, responsible for the collection, transport and final disposal of municipal solid waste. The Environmental Health Department (SESMA) oversees and inspects the operation and management of all the solid waste treatment or disposal facilities, and also ensures the compliance of health standards and regulations. The National Environmental Commission (CONAMA) is responsible for conducting environmental assessment, to approve landfills or other projects regarding the final disposal of MSW, and for imposing penalties for the noncompliance of environmental regulations. The Santiago Regional Government (Intendencia Metropolitana) acts as coordinator, facilitator and, if required, a mediator between these bodies. About half of all residential solid waste generated in Santiago is organic, while paper accounts for 18.8%, plastic 10.3% and textiles 4.3%. Metals and glass make up a smaller percentage, 2.3% and 1.6% respectively, Fig.10.5 (source: CONAMA 2002). The Municipalities in Santiago have contracted all the waste management services to two private sector companies – EMERES (Empresa Metropolitana de Tratamiento de Residuos Solidos) to operate in southern part, and KDM (Kiasa Demarco S.A.), a subsidiary of the U.S. based company Kenbourne in the northern part of Santiago.

Methodology: The waste produced at the household level is left in black plastic bags in the street for collection. There is no source separation at the origin of waste. This waste is collected 3 times a week by trucks. The trucks, depending on the distance of the municipality to the landfill, transport the waste directly to the landfill or to one of two transfer stations. The waste in this station is not separated or treated, it is only transferred to bigger or special trucks that take the waste to the landfill for final disposal.

There are only three authorized landfills in Santiago:

(a) *Loma Los Colorados:* This landfill, managed by KDM S.A is located in the Municipality of Til-Til (63,5 km north of Santiago) covering an area of 600 hectares and is expected to reach final official capacity in 2046. It is designed to receive 150,000 metric tons of solid waste per month coming from the Municipalities in the northern part of Santiago.

(b) *Santa Marta:* This landfill, managed by EMERES S.A., is located 12 km south of Santiago in Talagante. It started operations in April 2002 and was designed to receive 60,000 final metric tons of solid waste per month. This landfill covers an area of 296 hectares and it is expected to reach final capacity in 2022. It serves a population of 1,212,896 inhabitants from the southern part of Santiago.

(c) *Santiago Poniente:* This landfill, managed by EMERES S.A., has started operations in October 2002 and is designed to receive 40,000 tons of MSW per month, serving people in the eastern and central Municipalities of Santiago.

As recycling is not obligatory in Chile, there is little recycling perception among the citizens, and recycling is minimal, sporadic and accomplished in an informal way. It is estimated that 9% of the total amount of MSW generate in Santiago is recycled.

The recovery and commercialization of recyclable material is done manually by informal sector – street cardboard collectors (cartoneros) and scavengers (cachureros) – who as individuals recover small volumes of paper, glass and aluminum cans from homes and businesses. Another informal commercial sector buys the collected material and sells it to recycling companies which are few.

There are small pilot projects but volumes are insignificant. Still, some government authorities are trying to raise recycling consciousness through the use of collecting containers, organizing household compost projects, encouraging recycline in public offices and universities, and conducting educational programs in schools.

A National Policy for Municipal Solid Waste Management was approved in 1997 to develop an economically viable Integrated SWM System that minimizes environmental impact and eliminates harmful human health effects.

The National Policy establishes a basic strategy that focuses on the priority objectives regarding MSW, namely, to prevent MSW creation; if not possible, to minimize its creation; MSW treatment; and disposal of MSW that couldn't be treated.

Since landfills is not a sustainable waste disposal facility for Santiago for reasons already explained, one promising method to reduce waste volume is by burning waste through Waste-to-Energy technology. The WTE systems with energy recovery could address Santiago's long term needs.

Recent study that included financial and technical issues establishes that the most appropriate technology for Santiago is the mass burn plant with manual pre-sorting of some recyclable materials such as metals, glass and papers before combustion (Estevez 2003).

The current mass burn systems have been operating successfully and reliably and are widely considered as a proven technology. In this category the Martin Grate technology is the most widely used one because it is simple, easier and less expensive to install than RDF burning. Another advantage of mass burning is that it offers ample flexibility for the kind of feedstock supplied, e.g., one can co-fire other fuels such as waste tires or sewage sludge residues from waste water treatment plants.

The project evaluation demonstrates that a WTE Plant for Santiago, with a capacity of 1,200 metric tons/day, would be able to generate more income than the investment (Estevez 2003)

10.4 Europe

(*Source:* UNEP)

Europe has many industrialized countries. In cities of these countries, since resources and skills are available and planners are quite well informed, short-, medium-and long-term plans for waste management are common.

Many countries have in general adopted and implemented EC policy on waste. Thus, the key directives on incineration, landfill and packaging all form part of the legislative drivers for waste management. Other measures include the setting of targets for recycling, recovery (usually including recycling and energy recovery) and landfill diversion; the targets are either aspirational or have been introduced as statutory requirements. The general emphasis for waste management is to increase diversion from landfill and maximise all forms of recycling and recovery. Various policy measures, including economic instruments, have been introduced or proposed, e.g., tax on disposal to landfill, support for recycling markets, etc. Not only has the individual Member State's policy developed over recent years with regard to Energy from Waste, but also the EC Directives have had a major impact on the operation and technical requirements of EfW plant across the EU. The Waste Incineration Directive 2000 imposes stringent emissions and other environmental controls on EfW plant, and became law in the UK in 2002 (CIWM 2003). These directives may be too expensive for Eastern European countries to pursue fully.

Generally, the prospects for energy recovery in Europe can be categorised into two main groupings: (a) Countries that have significant EfW capacity already in place and have potential for further expansion of that capacity, e.g., Germany, Netherlands and Sweden, which already treat between 25 and 41% of municipal waste by EfW but are likely to see only a small expansion in capacity over the short term, and (b) Countries with relatively low existing EfW capacity include Norway, Finland and the UK – and these countries are likely to see significant growth in capacity over the next 5 to 10 years. The opportunities in Norway and Finland are likely to be based around small-scale (less than 100 000 tpa) facilities and co-firing of waste-derived fuels within existing biomass facilities. The southern European countries, Spain, Italy, and Portugal, will also need to increase energy recovery though, for various reasons including cost, the pace of development may be less than that experienced in the northern European countries. UK has a low usage of EfW relative to the majority of other European Countries. This is a reflection of the economic, legislative and policy drivers in the UK (CIWM 2003).

Fig.10.6 shows the domestic waste management in EU countries from 1999-2000 indicating the fractions recycled, incinerated, composted, landfilled and others (Wagner, L 2007).

Treatment of Municipal Waste in Europe 2001

Fig. 10.6 Domestic waste management in EU countries from 1999-2000 indicating the percentages recycled, incinerated and landfilled (courtesy of Dublin Waste to Energy project).

Western Europe has the distinction of endorsing and largely implementing the integrated waste management system than any other region in the world. It is mandatory for the Western European governments to design their waste management systems around the well-known waste management hierarchy, with waste prevention given the highest priority, followed by reuse, recycling, materials recovery, energy recovery, and disposal as the last option. However, while most Northern European countries attach higher priority to materials recovery than energy recovery, France assigns equal weight to all, to keep materials out of landfills.

Western European governments usually arrange financing for waste management at national level, which ensures that not only the national policy priorities become incorporated into solid waste management systems, but all aspects of the system are financed together. As a consequence of this integrated approach to waste management, Europe has more experience with waste prevention than other regions; and recycling and materials recovery is well supported in Northern Europe. It is not the same in the southern EU countries and in the transition economies of Eastern Europe. The time-honored social democratic nature of European national and regional governments expects the public sector to be the prime mover in waste management. However, in recent years, governments are increasingly turning to the private sector to deliver waste management services. The formerly state-owned enterprises in Eastern Europe are under transformation, mostly in the course of involving private sector to a large extent.

Waste management planning and policy decisions in Europe are generally done at the level of national ministries, which respond to political pressure for environmental protection and the need for clean air and water. Sometimes, these decisions conflict with the more routine responsibilities of local authorities, who have to manage the flow of waste on a daily basis.

Regarding decision making, countries like France have highly centralized decision-making processes; while those like the UK leave most decisions to the local authorities. Decisions on specific local programmes may be broadly consultative as they are in The Netherlands; or responsive to adversarial citizen action as in Great Britain; or less attentive to citizen input as they are in Spain. In Northern Europe, implementation and monitoring tend to occur within the framework of a generally consensus-oriented culture, where noncompliance is the exception, rather than the rule.

National institutes, Technical universities or other academic institutions not only define research programmes but respond to suggestions and proposals from consulting and independent organizations to investigate particular problems or monitor the success of new programmes.

Integrating Energy from Waste into the community and the environment: There are many examples of good practice in several European countries and in the UK regarding the integration of WTE plant into the community. These show that plants can be designed to accommodate the needs of a particular town or community and can then provide lower cost district heating to that same community (as well as electricity) thus 'closing the loop' and utilising the residual waste emanating from a locality in a positive and beneficial manner. Additionally, facilities are now being designed in a variety of imaginative ways to make a positive contribution to the environment. In Vienna, one of its Energy-from-Waste (EfW) plants is an architectural feature of the city. In many other European cities, EfW is integrated into the district heating infrastructure. For example, Paris has three large EfW facilities with extensive district heating systems on the Peripherique, which supply around one third of central Paris' heat requirement (CIWM 2003).

Survey by Austrian consultancy firm, TBU: *(ref: Cesar Preda 2006)*
The objective of the survey on MSW management in Europe has been to
gather and convey relevant and comprehensive information on status,
achievements and shortcomings in the area of MSW management across
Europe to stimulate suitable policy initiatives where necessary.

It is undertaken in 47 Member countries comprising 'Council of
Europe' founded in 1949.

Of the 24 countries responded on the issue of waste collection, ten
countries, Austria, Andorra, Belgium, Germany, Liechtenstein, Lithuania,
Luxembourg, Monaco, Switzerland and UK report 100% performance.
The remaining countries mostly from Eastern and South-eastern Europe
report below 100%: The Czech republic, Denmark, Finland and Slovenia
report >95%; Azerbaijan, Estonia, Greece, Hungary and Sweden report
90-95%; Bulgaria, Romania and Poland, 80-85%; and Ireland 77%.
Montenegro reports the minimum of 50%.

Regarding national waste plan, 26 Member countries have *a National
level agency* dealing with MSW issues, and 29 Member countries have
developed a national waste plan. Five countries have no national plan:
two are miniature countries; two are from Caucasia (Georgia and
Azerbaijan) where issues related to the environment attract much less
political attention, and the fifth, Turkey where MSW strategy is being
prepared as part of a National Environmental Strategy. The Survey has
also covered the quantities of MSW- recyclables and compostables –
collected, the range of typical disposal costs, data on landfill levies, and
time targets within the national plan for the reduction of MSW and so on.
Based on the information brought out in the survey, the Council of
Europe has adopted a Resolution 1543 on 16th March 2007. The
resolution urges Member countries to develop an integrated approach to
MSWM in order to contribute to sustainable urban development in
Europe, in particular by:

(a) Ensuring compliance with occupational health and safety
standards during the collection, processing and landfilling of all
types of waste, in particular by banning any bare-handed
operations and any recovery of waste from landfills without
proper protection and regular health checks for the persons
involved,

(b) Establishing regular waste collection systems for all urban,
suburban and rural areas and including in the relevant legislation
phased targets for the provision of municipal solid waste collection

systems in accordance with the requirements of European Union directive No.1999/31/EC on the landfill of waste,

(c) Enforcing compliance with strict standards for landfilling, for instance: landfills must be fenced and protected; waste accepted at landfills must be recorded; waste placed in landfills must regularly be covered with suitable materials in order to reduce odours, windblown litter and vermin; adjacent groundwater must be monitored,

(d) Depending on local hydrogeology, suitable measures for groundwater protection (such as landfill liners and leachate collection etc) must be put in place;

(e) Planning waste management through the development of strategies including gradual reduction/phasing out of the landfilling of specific waste streams, given their recyclability and/or the impacts related to their disposal (e.g., biodegradable waste);

(f) Assigning municipalities responsibility for managing waste from households, businesses, institutions, and construction and demolition activities within their territory and enabling municipalities that are too small to provide the relevant services to set-up inter-municipal consortia for SWM;

(g) Facilitating cooperation between European towns and cities to allow information exchanges so that the best solutions in terms both of administrative management of MSW and of processing technologies disseminated and used Europe-wide;

(h) Encouraging R&D in the field of solid waste processing and recycling.

(a) **Switzerland:** (source: The Swiss Confederation – Waste management)

As early as 1960s, Switzerland became a pioneer in solid waste management by rigorously installing treatment and incineration plants with stringent emission standards. Today it can be acknowledged that Switzerland has succeeded in moving from basic waste removal to an environmentally friendly process of waste disposal and recycling. Now, incineration plants in the country are efficient power plants which produce clean heat and electricity.

Municipal solid waste generation in Switzerland has been on the increase year by year. By 2007 they had reached 720 kg/ person. Today, half of all MSW generated are collected separately and recovered – a ratio that has more than doubled over the past 20 years. The recycling was also used everywhere in the country, and Swiss recycling rates are among the highest in the world. The remaining wastes are incinerated in clean processes which generate electricity and heat, meeting some 2 % of the country's final energy requirements.

Waste disposal facilities: The requirements for disposal facilities are specified in the Technical Ordinance on Waste (TVA) released by the Government. It specifies stringent requirements for waste that is to be landfilled. Today, three different types of landfill sites are used for different types of waste in Switzerland:

Landfills for inert materials: only rock-like wastes may be disposed of, from which virtually no pollutants will be leached out by rainwater. These include materials such as construction waste (concrete, bricks, glass, and road rubble) and uncontaminated soil that cannot be used elsewhere. At suitable locations, landfills for inert materials do not require any special sealing.

Landfills for stabilized residues: are designed for the disposal of materials of known composition, with high concentrations of heavy metals and only a small organic component, and which cannot release either gases or substances readily soluble in water. Typical materials include solidified fly ash and flue gas cleaning residues from municipal waste incinerators, and vitrified treatment residues. These sites are subject to more stringent requirements than the above. Impermeable linings are required for the base and sides of the landfill, and leachate is to be collected and, if necessary, treated.

Bioreactor landfills: chemical and biological processes are expected to occur. At these sites, drainage controls are required. In addition, any gases emitted are to be captured and treated. Given the unpredictable composition of their contents, bioreactor landfills are at greatest risk of requiring expensive remediation at a later date. Certain types of waste (e.g. incinerator slag) are required to be disposed of in separate compartments, isolated from other types of waste. If these wastes were intermixed, heavy metals would be leached out in much greater quantities as a result of the relatively low pH of incinerator slag. Compartments for residual wastes have also been established at numerous bioreactor landfill sites.

(ref: Separate collections in Switzerland:
http://www.bafu.admin.ch/abfall/01472/index.html?lang=en)

Environmentally sound management of solid (nonhazardous) wastes and sewage: In 2007, about 5.5 million tonnes of MSW were generated, which equates to approx. 720 kg per inhabitant. The percentage of all MSW collected separately was 51 % or 2.8 million tonnes. In 1989, the peak year to date for MSW incineration, the figure was only 27 %. Since then, the volume of segregated MSW has more than doubled from 160 to 370 kg/person/year.

The level of MSW incineration has remained relatively stable in recent years despite population growth, averaging 2.6 million tonnes per annum. The per capita volume of refuse for disposal fell from 440 to 350 kg per year. Financing waste disposal on the polluter-pays principle (e.g. Switzerland's refuse-bag levy) has contributed to progress in this area.

The success of separate collections is also reflected in the composition of the household waste left for regular refuse collection. Changing consumption patterns are making a significant difference. Goods made of natural products such as wood, leather or metal are being replaced by composite products majority of which contain plastic cannot be separated. Biogenic waste from the kitchen or garden as well as food waste account for 27 % of incinerated waste, the largest category by weight. Paper and card come next, accounting for 20 %, while composite products and composite packaging weigh in at 18 % and plastics at 15 %.

In recent years, the Swiss Confederation's waste management policy has significantly reduced the level of environmental pressure caused by waste management, despite continuous growth in the total volume of MSW generation. This trend can be attributed to the introduction of high waste management standards, to a highly effective infrastructure, and to a financing system that makes the waste producers responsible for the costs of disposal.

Causality principle: The causality principle implies that anyone harming the environment must bear the costs. In 2001 the private sector – i.e., companies, households and farmers – committed 530 million francs to waste management, to which 1.5 billion francs were added from public expenditure. Out of this, a little less than 1.1 billion francs was passed on to responsible parties through taxes. The remaining 418 million francs, funded through tax receipts, represent a shortfall: this amount still to be charged to concerned parties for the causality principle to be fully enforced in Switzerland.

The environmentally friendly disposal of municipal waste in Switzerland costs only 30 centimes per person and per day. The huge investment made to introduce the separate collection for new incineration plants did not increase this amount because the plants were able to rapidly market the heat, electricity and metal they produced. Today the costs per person and per day are lower than at the end of the 1980's.

Waste management facilities: Switzerland has a well-developed network of waste management facilities. Virtually every region is equipped with the infrastructure required in order to dispose of its own wastes. This helps to minimize transport costs and vehicle emissions. Since the introduction of the landfilling ban on 1st January 2000, all non recycled combustible waste in Switzerland must be incinerated in appropriate plants and end up in one of the country's 28 municipal solid waste (MSW) incinerators. Since the plant "Thun MSWI" came on stream in 2004, a total incineration capacity of 3.29 million tonnes has been available in the country which is sufficient to allow the landfilling of combustible waste to be dispensed with from now on.

Hazardous wastes: Hazardous waste accounts for about 6 % of all waste. Each year around 1.2 million tonnes hazardous wastes are consigned to special reprocessing, or disposed of within the country, or exported in line with the provisions of the Basel Convention. The remediation of disused hazardous waste landfill sites will cost the Swiss economy well over 1 billion francs. In 2005, 43 % of the hazardous waste was incinerated, 22% was landfilled after appropriate pretreatment, 23% underwent chemical/physical treatment and 12% was directly recycled. Chemical/physical treatment takes place mainly in Switzerland. This approach is applied to polluted wastewater, soil from contaminated site remediation and emulsions.

Renewable energy potential: Since 1997, the Confederation has been awarding grants to support the development of innovations in environmental technologies. Production of renewable energy from waste is an essential component in waste management. Whether it is diesel fuel production from plastic waste or used edible oils, or biogas from organic waste or sewage sludge fermentation, the search for all possible technological solutions has attracted more attention since the increase in energy prices. The reclaiming of metallic elements (copper, zinc, nickel) in slag or ashes from electro filters also benefits from the price escalation on the minerals market.

Electricity, heat and fertilizers production with biomass: In Switzerland, some 1.3 million tonnes of biogenic wastes are generated every year. 740,000 tonnes are processed in the country's 333 composting and anaerobic digestion plants with an annual capacity in excess of 100 tonnes/ year, while 300,000 tonnes are reckoned to be recycled in private gardens and on neighbourhood compost heaps. Nevertheless, a further 250,000 tonnes or so still finds its way into the municipal solid waste incinerators along with the normal domestic refuse. Biogenic wastes are useful for the production of electricity, heat and fertilizer. During the last 10 years the production of industrial and agricultural waste production from electricity, heat, gas with biogenic waste from farms, and industrial factories has increased six-fold.

In order to ensure that sectors that need biomass are not deprived because of the increasing interest in biomass for energy and heat production, the four federal agencies concerned developed a common strategy on biomass usage. This strategy is based on the cascade classification for the use of biomass. The production of high value added products such as food and construction materials should remain a top priority. Synergies should be checked and applied consequently. For example waste and byproducts from the food industry can be used for animal feed. Wastages coming from the husbandry of animals can be used for energy production in biogas plants, while other organic wastes can be used for digestion and the production of fertilizer for agriculture. The energy produced in biogas plants can support the digestion process and heating required by industry.

Future optimization: There is opportunity for waste management structures to be optimized locally. For example, the collection of waste can be regionalized and measures can be taken to standardize the collection systems. To increase the efficiency of waste management, it is also necessary to focus on product design and improve social and environmental criteria all along the life cycle of goods and services.

Greece: Greece is in a relatively early stage of its waste management infrastructure growth. Economic development, intense urbanization and change in consumption patterns have resulted in an increase of solid waste generation. The quantity of municipal waste generated in Greece increased by 42.5% from 1995 to 2002 (Table 10.7); still it is far below the average generation rate of 500/capita/year in many European countries (Commission of the European Communities, 2006).

Table 10.7 MSW Production in Greece (kg/capita/year), (source: EEA 2005)

1995	1996	1997	1998	1999	2000	2001	2002	2003
306	344	372	388	405	421	430	436	441

With regard to municipal solid waste composition (Table 10.8), very few analyses of urban wastes have been carried out, resulting in problems for monitoring the composition changes through time, season and economic activities (Agapitidis & Frantzis, 1998).

Table 10.8 Solid waste composition in Greece (Agapitidis & Frantzis 1998)

Material type	Athens 1985	Thessaloniki 1987	Rhodos 1989	Iraklio 1987	Naxos 1993	Average
Putrecibles	59.8%	51.7%	43%	52.5%	48%	47%
Paper	19.5%	17.7%	17%	17.2%	20.9%	20%
Metals	3.8%	5.9%	10%	2.8%	3.1%	4.5%
Plastics	7%	7.2%	10%	14.3%	10.3%	8.5%
Glass	2.6%	4.1%	14%	1.4%	4.2%	4.5%
Other	7.45%	13.4%	6%	11.7%	13.5%	15%

The 'uncontrolled' dumps where waste is thrown indiscreminately are spread across the country – off cliffs, on banks of torrents, rivers and stream beds, on coast, in the immediate vicinity of springs used for water supply, abandoned quarries, forested areas, and archaeological sites - flouting both national and EU legislations. As a result, the environment has deteriorated steadily resulting in pollution of ground water, soil and air, problems of public health, aesthetics and ecological vulnerability of the region. Further, their location in those sites made them unqualified to be licensed for an upgradation. Greek authorities acknowledged in 2005 that at least 1125 illegal or uncontrolled waste dumps were operational. According to the Ministry of Environment, the number of uncontrolled dumps decreased from 3500 to 1450 approximately in the year 2002 and tends to decrease further to 500 in 2007.

In an effort to confront the problem, 43 sanitary landfills to serve 55% of the population were constructed; and 24 were under construction to serve 18% of the population in 2007. In Attica and Thessaloniki prefectures, 93% and 71% of the population respectively is served by a single sanitary landfill per prefecture.

In addition, 12 waste transfer stations in Athens region and 3 in other regions of the country are in operation.

Recovery and recycling: The quantity of packaging waste generation in Greece has increased from 68 kg/ capita in 1997 to 94 kg/capita in 2002 against the objective of the legislation. Greek government, in order to improve the waste management services, has encouraged municipalities to take initiatives to reduce packaging waste, and the private companies to extensively undertake recycling of paper packaging. Since 2004, permits are provided to organizations started by industries and commercial units, to execute reuse and recycling programmes; and a large number of municipalities have entered into agreements with them.

The household solid wastes are separated at home into multi material recyclables and refuse. The commingled recyclables are collected and transferred to a material recovery facility. In 2007, only four materials recovery facilities are in existance. A mechanical material recovery facility, constructed at Ano Liosia Landfill, has not set in efficient operation. The packaging waste recycled in Greece during 2005-06 is about 14% of the total packaging waste generated, which has reached to 25% in 2007. The fraction of recyclables has varied significantly: 65% of paper, 10% of metals, 19 % of glass and 3 % of plastic.

Greece is in the beginning of the solid waste management effort, and much has to be done both quantitatively and qualitatively (EEA, 2005). It has been proposed to make collection more consistent and reliable, and to provide wider information on the recycling projects, and to identify a secondary material market for the sorted materials.

The public pay the MSWM charges to the municipalities based on the location of the residence and the total population of the area, and not on the quantities of waste produced. On the contrary the best proposition shall be to pay the charges as determined by the waste quantities produced in order to help those participating in recycling programmes to pay less.

Biodegradable waste disposal: Greece would take more time than set by the directive to reduce the waste going to the landfill. Provision has been made for reducing biodegradable wastes reaching landfill to 75% by 2010, 50% by 2013, and 35% by 2020.

Biochemical treatment plants and/or energy recovery plants are planned where economics prove feasible.

Three compost plants for commingled refuses have been constructed. Two are operating less efficiently while the third is not operational.

There is not any incineration plant under elaboration or construction in 2007.

Legal provisions: The legislative framework for waste management in Greece is mainly based on the EU legislation. A number of Acts were issued which were supposed to correspond to the Waste Framework Directive (Council Directive) but bureaucratic obstacles come in the way. They were very complex resulting in considerable delays in formulating waste management planning.

In an attempt to rationalize, the National Policy Plan (NPP) was formulated. The Policy goals are the prevention or reduction of waste production, the recovery of waste by means of recycling, re-use or reclamation, the closure and restoration of all uncontrolled dumps until 2008 and the establishment of an adequate network of disposal facilities choosing the best available technology. The planning of disposal capacity has to be carried out on the regional level, and the Regions are legally obliged to issue periodical Regional Waste Plans. The pertinent Prefecture provides permits for waste handling, collection and transport as well as processing and disposal facilities, though the local authorities (Municipalities, associations) are competent to execute the works and to develop and maintain a reliable, efficient and cost effective system of solid waste collection and disposal. In addition a number of Ministerial Decisions were issued setting the technical specifications for the design, operation and maintenance of sanitary landfills as well as recycling programmes. Despite the legal provisions in place, inappropriate waste disposal and management practices persist resulting, as already mentioned, in the degradation of surface and groundwater, air pollution and forest fires (EEA, 2005).

The investigations of Greek Ombudsman (GO) on complaints concerning uncontrolled disposal of wastes, deviations from the approved environmental provisions, as well as systematic lapses in the selection and approval of the siting of solid waste treatment facilities (The Greek

Ombudsman, 2003, 2004) reveal that these problems have been compounded due to lack of comprehensive MSWM; such an arrangement should include:

- Application of a nationwide and long term planning
- Effective and appropriate legislative and institutional framework
- Effective coordination and participation of local authorities
- Citizens mobilization in terms of awareness and public involvement and participation
- Proper training and support of the use of innovative technologies

10.5 North America

Local governments in North America have primary responsibility for managing MSW, with some involvement by state or provincial authorities and less by the federal government. Canada's approach is even more decentralized than that of the US; the Canadian federal government has few legislative mandates regarding MSWM, except regarding hazardous wastes.

Local public works departments manage solid waste services, although health or the environment departments are often involved. In addition, parties who have an interest in local MSWM decisions have access to the political process, including elected officials, the news media, business interests, and citizens' organizations. In some states and provinces, due to economic and environmental pressures of waste disposal, the responsibility for waste management has shifted from the local to the state/provincial level. In US, the issues are sometimes taken to the federal level due to serious contentious concerns.

Plans and incentives: Most states have developed comprehensive legislation, innovative approaches, and highly skilled solid waste staff. The solid waste plans developed by majority of states define the goals and agenda for regional waste management action. These plans and supporting law often place necessities on the resources and programmes of the local community and suggest suitable programme approaches. Some laws require local governments to set up recycling programmes that attain specific levels of recycling while other laws impose recycling tasks on industries and businesses.

States and provinces also encourage local waste management approaches, but the indicators of programme must see the funding available. Legislation often contains provisions for grants, matching funds for feasibility studies, technical assistance, programme development and implementation, training aspects, public education, educational curriculum materials, household hazardous waste and special waste programmes, marketing and service directories, and information networks for both public and private waste managers.

The US Environmental Protection Agency (EPA) and many states have established a hierarchy for waste management that ranks options in terms of their desirability and relative role in an integrated waste management system: reduction and reuse, recycling and composting, waste-to- energy incineration, and landfill disposal. This planning scheme depends on local geographic, environmental, social, and economic conditions. The State of Oregon, for example, has established strict priorities for local planners, resulting in several highly successful integrated waste management strategies on the county level and one of the most successful recycling programmes in the US. Recycling plans are mandatory in some states, while others make them available to local communities on a voluntary basis. Some states encourage a regional approach in order to better coordinate waste management efforts.

Tax and other incentive-based policies and deposit-refund systems are used in particular jurisdictions. The beer industry in Ontario has been remarkably successful operating a deposit-refund system for years with more than a 90% capture rate. The soft drink industry across Canada has been unsuccessful in operating deposit-refund systems, largely due to the decentralized nature of its distribution. Product procurement guidelines mandating the purchase of paper, lubricating oil, retread tires, building insulation, and other products with a certain recycled content, as in US, are also used by the governments.

Ownership and management: Ownership and management of solid waste collection and disposal facilities varies from fully publicly owned and operated programmes, to government contracts with private firms, to freely operating private firms in an open market. The private or contractual systems are favoured because of increased system efficiency and service due to competition, less susceptibility to political influence, greater management flexibility, and lesser strain on government budgets. The advantages of a publicly-owned-operated system include its nonprofit character, government purchasing advantages, centralized operation, and standardized procedures.

Due to restricted local government expenditure, municipalities have increasingly turned to private ownership and operation of solid waste disposal and collection services. Private ownership also transfers much of the technical, financial, and potential cleanup risks to the private sector. In addition, many municipalities prefer privatization because they do not have ready access to the necessary team of operators, engineers, and maintenance personnel required for running a facility.

Some municipalities prefer public ownership of solid waste facilities because public ownership gives community officials more control over facility development and operation. Another alternative is some form of joint public/private ownership and operation agreement. This option has the advantage of enabling the risks and costs of facility design, construction, and operation to be allocated between a community and its contractors in a way that is tailored to local needs and circumstances. Many municipalities choose facilities that are publicly owned, but privately operated, often by the same firms that designed and built the facilities. To encourage efficient MSW operations, some cities, such as Phoenix, Arizona have turned to privately operated service in some areas, while maintaining municipally operated service in others. Some of these cities have even adopted competitive bidding between the public works department and private contractors.

Case study: New York City: (source: Themelis 2002) Since 1950s, New York City (NYC) disposed most of its solid wastes in the giant (about 20 million square meters) Fresh Kills landfill in Staten Island. But, it was closed in 2001 (re-opened temporarily after the September 11 attack on the World Trade Towers). In 2002, New York City having a population of 8 million, has generated about 12,000 metric tons per working day of residential wastes (collected by the City) and nearly an equal amount of commercial and institutional wastes, collected by private contractors. The Department of Sanitation (DOS) of NYC collects waste in three streams, separated at the household level: (1) Recyclable paper ('clear' bags), (2) recyclable 'metal-glass-plastics' (MGP, 'blue' bags) and (3) all other wastes ('black' bags). The four million metric tons of 'black' bag MSW collected annually by NYC are disposed as follows (Fig.10.7):

(a) *Recycling*: In recent years, NYC has launched a campaign to increase recycling to the present level of about 700,000 metric tons. The paper stream consists of mixed paper, newspapers, magazines, and corrugated cardboard and represents about 65% of the recyclables collected by NYC. Most of this stream is used in paper recycling plants in Staten Island (Visy Paper) and elsewhere. The residue from the paper stream (12-15% of the paper stream) consists of plastics (mostly from plastic bags) and some unusable paper.

Although this material is combustible and has a relatively high heating value, after compacting into 0.7- ton bales, it is sent to landfills.

The MGP stream also goes to a sorting operation where steel cans (about 8%), iron and steel parts (18%), aluminum cans and foil (1%) are sorted out manually and mechanically (e.g., using electromagnets). A small fraction of recyclable plastics (5%; mixed color HDPE, natural HDPE, PET) and clear glass (4%) are also recovered. The residues of the MGP stream consist of a large amount of broken glass mixed with small particles of plastic, metal and dirt (about 40% of the stream) and plastic bags (about 10%). The glass residue is used as 'day cover' in landfills. The plastic residue is baled and sent to landfills. The sorting, baling, and further treatment or disposal of the various products of the paper and MGP streams are contracted by the City to several private parties.

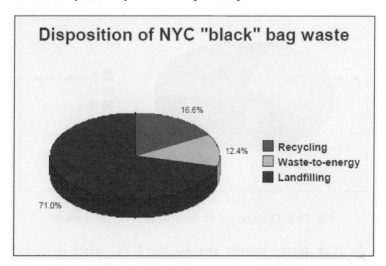

Fig. 10.7 Disposition of NYC solid waste generated

(b) ***Waste-to-Energy***: About 500,000 tons of 'black bag' waste go to two Waste-to-Energy plants, one in New Jersey (Essex County WTE) and the other in New York (Hempstead WTE).

(c) *Landfilling*: The remaining 2.8 million metric tons of 'black bag' waste are transported, mostly by truck (> 90% of waste) to Pennsylvania, Virginia, and New Jersey (Fig.10.8) where landfills are located.

Transportation of MSW: Prior to closing Fresh Kills, the NYC's DOS trucks travelled relatively short distances within the City to unload at marine transfer terminals from where the waste was transported by barges to the Fresh Kills landfill in Staten Island. Presently, six of the thirteen transfer stations are located outside New York City and the DOS trucks must travel distances up to 60 km to unload. It has been estimated that as many as one thousand trucks cross to New Jersey each working day over the existing two bridges and two tunnels (Columbia Earth Institute 2002, Fresh Kills report).

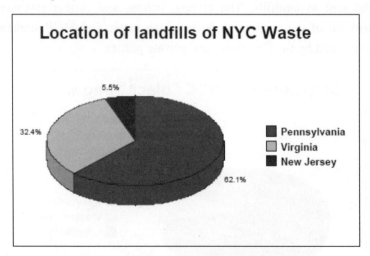

Fig. 10.8 Distribution of landfills of NYC solid waste

At the New Jersey transfer stations, the NYC waste is loaded onto 20-ton tractor trailers that transport it to landfills in Pennsylvania, Virginia, and New Jersey. The total distance traveled daily by the NYC DOS 10-ton trucks crossing to NJ is estimated at 64,000 km per working day. The distance traveled by the 20-ton trucks (average of 480 km per round trip) is estimated at 216,000 km each day. Wang et al (2000) have estimated fuel consumption for various types of heavy trucks. On the basis of the above data, the fuel consumption for transporting NYC MSW to other states by truck for landfilling is estimated at 40 million liters per year (about ten million gallons; Columbia Earth Institute 2002).

Ideal disposition of NYC MSW materials: MSW consists of many materials with entirely different characteristics which should go to different processing places. Since metals and glass are not combustible or compostable, 'recycling' is the most appropriate for them. Most of the collected paper and some plastics (e.g., PET and PE) are sorted out and recycled: Visy Paper on Staten Island (300,000 tons per year) is an example of a modern, efficient plant that operates fully on recycled feedstock. The non-recyclable paper, plastics and fibers contain useful energy; therefore, they can be burnt in a properly designed combustion chamber to generate steam which can be utilized to produce electricity. Finally, the materials to be landfilled are inorganic compounds such as non-recyclable glass and ashes from the Waste-to-Energy power plant.

Table 10.9 shows the classification of NYC MSW under four categories of recyclable, combustible, compostable, and landfillable. The ideal disposition of Table 10.9 is not easily realizable because of social, economic and market factors. For example, New York City citizens are already asked to separate three streams: Paper (clear bag), plastic, metal and glass (PMG; in blue bags) and trash (black bags). Despite an intensive campaign by the Recycling Bureau of NYC-DOS, the recycling rates in some areas of NYC are low; that is one reason why the present rate of city-wide recycling is less than one half of the projected maximum (Table 10.9).

Table 10.9 also shows that the maximum compostable fraction is 19%. However, separating and composting the 'wet' fraction requires the development of a regional market for nearly 0.5 million tons of compost product. In the absence of a 'wet-dry' system of collection, the compostable fraction will remain mixed with the other materials in the black bag stream. Therefore, the two alternatives for the black bag stream are combustion or landfilling. Table 10.9 shows that ideally, only about 6% of the NYC MSW needs to be landfilled compared to the present 71%.

Table 10.9 Classification of NYC MSW by most appropriate method of disposal (in thousands of short tons/year; numbers in parenthesis show assumed maximum % recyclable each material; Themelis et al., 2002)

	Collected Short tons	% of total MSW	Recyclable short tons	Combustible short tons	Compost-able short tons	Landfill-able short tons
Cardboard (90%)	229	5.1%	206	23		2
Newsprint (90%)	446	9.9%	401	43		5
All other paper (90%)	869	19.2%	434	414		41
Plastic bags/film (50%)	252	5.6%	126	126		13
All other plastics (50%)	193	4.3%	97	96		10
Wood, textiles, leather, rubber (20%)	608	13.4%	122	486		49
Food and plant wastes (0%)	879	19.4%			879	88
Disposable diapers (0%)	178	3.9%		178		18
Miscellan. Organics (0%)	409	9.0%		409		41
Glass (90%)	234	5.2%	210			24
Aluminum scrap (100%)	42	0.9%	42			
Ferrous scrap (100%)	185	4.1%	185			
Total:	4,524	100.0%	1,823	1,775	879	291
Fraction of NYC MSW	100.0%		40.3%	39.2%	19.4%	6.4%

*Landfillable is assumed to consist of 10% ash from all combustible streams and 10% of non-recyclable glass.

Waste Generation & Management Data by Country

Region /Country	MSW 1, 2 Generation Rate IPCC - 1996 values 4 (tonnes/cap/yr)	MSW 1, 2, 3 Generation Rate Year 2000 (tonnes/cap/yr)	Fraction of MSW disposed to SWDS IPCC-1996 values 4	Fraction of MSW disposed to SWDS	Fraction of MSW incinerated	Fraction of MSW composted	Fraction of other MSW management, unspecified 5	Source
Asia								
Eastern Asia	**0.41**	**0.37**	**0.38**	**0.55**	**0.26**	**0.01**	**0.18**	
China		0.27		0.97	0.02	0.01		1
Japan	0.41	0.47	0.38	0.25	0.72	0.02	0.01	2, 31
Rep. of Korea		0.38		0.42	0.04		0.54	3
Southern and Central Asia	**0.12**	**0.21**	**0.60**	**0.74**	**-**	**0.05**	**0.21**	
Bangladesh		0.18		0.95			0.05	4
India	0.12	0.17	0.60	0.70		0.20	0.10	4
Nepal		0.18		0.40			0.60	4
Sri Lanka		0.32		0.90			0.10	4
South-eastern Asia		**0.27**		**0.59**	**0.09**	**0.05**	**0.27**	

Contd…

	C1	C2	C3	C4	C5	C6	C7	Ref
Indonesia		0.28		0.80	0.05	0.10	0.05	4
Lao PDR		0.25		0.40			0.60	4
Malaysia		0.30		0.70	0.05	0.10	0.15	4
Myanmar		0.16		0.60			0.40	4
Philippines		0.19		0.62		0.10	0.28	4, 5
Singapore		0.40		0.20	0.58		0.22	6
Thailand		0.40		0.80	0.05	0.10	0.05	4
Vietnam		0.20		0.60			0.40	4
Africa								
Africa 6		**0.29**		**0.69**			**0.31**	
Egypt				0.70			0.30	4
Sudan		0.29		0.82			0.18	7
South Africa			1.00	0.90			0.10	4
Nigeria				0.40			0.60	4
Europe								
Eastern Europe		**0.38**		**0.9**	**0.04**	**0.01**	**0.02**	
Bulgaria		0.52		1.00	0.00	0.00	0.00	8
Croatia				1.00	0.00	0.00	0.00	8
Czech Republic		0.33		0.75	0.14	0.04	0.06	8
Estonia		0.44		0.98	0.00	0.00	0.02	8
Hungary		0.45		0.92	0.08	0.00	0.00	8
Latvia		0.27		0.92	0.04	0.02	0.02	8
Lithuania		0.31		1.00	0.00	0.00	0.00	8
Poland		0.32		0.98	0.00	0.02	0.00	8
Romania		0.36		1.00	0.00	0.00	0.00	8
Russian Federation	0.32	0.34	0.94	0.71	0.19	0.00	0.10	9
Slovakia		0.32		1.00	0.00	0.00	0.00	8
Slovenia		0.51		0.90	0.00	0.08	0.02	8
Northern Europe		**0.64**		**0.47**	**0.24**	**0.08**	**0.20**	
Denmark	0.46	0.67	0.2	0.10	0.53	0.16	0.22	8
Finland	0.62	0.50	0.77	0.61	0.1	0.07	0.22	8
Iceland		1.00		0.86	0.06	0.01	0.06	8
Norway	0.51	0.62	0.75	0.55	0.15	0.09	0.22	8
Sweden	0.37	0.43	0.44	0.23	0.39	0.10	0.29	8
Southern Europe		**0.52**		**0.85**	**0.05**	**0.05**	**0.05**	
Cyprus		0.68		1.00	0.00	0.00	0.00	8
Greece	0.31	0.41	0.93	0.91	0.00	0.01	0.08	8
Italy	0.34	0.50	0.88	0.70	0.07	0.14	0.09	8
Malta		0.48		1.00	0.00	0.00	0.00	8

Contd…

Portugal	0.33	0.47	0.86	0.69	0.19	0.05	0.07	8
Spain	0.36	0.60	0.85	0.68	0.07	0.16	0.09	8
Turkey		0.50		0.99	0.00	0.01	0.00	8
Western Europe	**0.45**	**0.56**	**0.57**	**0.47**	**0.22**	**0.15**	**0.15**	
Austria	0.34	0.58	0.4	0.30	0.10	0.37	0.23	8
Belgium	0.40	0.47	0.43	0.17	0.32	0.23	0.28	8
France	0.47	0.53	0.46	0.43	0.33	0.12	0.13	8
Germany	0.36	0.61	0.66	0.30	0.24	0.17	0.29	8
Ireland	0.31	0.60	1.0	0.89	0.00	0.01	0.11	8
Luxemburg	0.49	0.66	0.35	0.27	0.55	0.18	0.00	8
Netherlands	0.58	0.62	0.67	0.11	0.36	0.28	0.25	8
Switzerland	0.40	0.40	0.23	1.00	0.00	0.00	0.00	8
UK	0.69	0.57	0.90	0.82	0.07	0.03	0.08	8
Central, South America and Caribbean states								
Caribbean		**0.49**		**0.83**	**0.02**		**0.15**	
Bahamas		0.95		0.7			0.3	10
Cuba		0.21		0.90			0.1	11
Dominican Republic		0.25		0.90	0.06		0.04	12
St. Lucia		0.55		0.83			0.17	13
Central America		**0.21**		**0.50**			**0.50**	
Costa Rica		0.17						14, 15
Guatemala		0.22		0.40			0.60	16, 17, 18
Honduras		0.15		0.40			0.60	4
Nicaragua		0.28		0.70			0.30	4
South America								
South America		**0.26**		**0.54**	**0.01**	**0.003**	**0.46**	
Argentina		0.28		0.59			0.41	4
Bolivia		0.16		0.70			0.30	19
Brazil		0.18		0.80	0.05	0.03	0.12	20, 21
Chile				0.40			0.60	4
Colombia		0.26		0.31			0.69	22
Ecuador		0.22		0.40			0.60	23
Paraguay (Asuncion)		0.44		0.40			0.60	24
Peru		0.20		0.53			0.47	4, 25
Uruguay		0.26		0.72			0.28	26, 27
Venezuela		0.33		0.50			0.50	28

Contd…

Municipal Solid Waste Management

North America								
North America	0.70	0.65	0.69	0.58	0.06	0.06	0.29	
Canada	0.66	0.49	0.75	0.71	0.04	0.19	0.06	29, 30, 31
Mexico		0.31		0.49			0.51	32, 33
USA	0.73	1.14	0.62	0.55	0.14		0.31	34
Oceania								
Oceania	0.47	0.69	1.00	0.85			0.15	
Australia	0.46	0.69	1.00	1.00				4, 31
New Zealand	0.49		1.00	0.70			0.30	4

MSW generation and management data for some countries whose data are available are given above (source: IPCC 2006).

[1]Data are based on weight of wet waste.

[2]To obtain the total waste generation in the country, the per-capita values should be multiplied with the population whose waste is collected. In many countries, especially developing countries, this encompasses only urban population.

[3]The data are default data for the year 2000, although for some countries the year for which the data are applicable was not given in the reference, or data for the year 2000 were not available. The year for which the data are collected is given below with source of the data, where available.

[4]Values shown in this column are the ones included in the *1996 IPCC Guidelines*.

[5]Other, unspecified, includes data on recycling for some countries.

[6]A regional average is given for the whole of Africa as data are not available for more detailed regions within Africa.

(Source: 2006 IPCC Guidelines for National Greenhouse gas Inventories; Vol.5, Chap.2: Waste generation, composition, and management Data).

1. Urban Construction Statistics Yearbook of China – Year 2000 (2001). Ministry of Chinese.Construction. Chinese Construction Industry Publication Company
2. OECD Environment Directorate, OECD Environmental Data 2002, Waste. Ministry of Environment, Japan (1992-2003): Waste of Japan, http://www.env.go.jp/recycle/waste/ippan.html.
3. 1. '97 National Status of Solid Waste Generation and Treatment , the Ministry of Env, Korea,1998.
   2. '96 National Status of Solid Waste Generation and Treatment , the Ministry of Env, Korea, 1997.
   3. Korea Environmental Yearbook, the Ministry of Environment, Korea, 1990.
4. Doorn and Barlaz, 1995, Estimate of global methane emissions from landfills and open dumps, EPA-600/R-95-019, Office of Research & Development, Washington DC, USA.
5. Shimura et al. (2001).
6. 2001 National Environmental Agency, Singapore (www.nea.gov.sg.) and www.acrr.org/resourcecities/waste_resources/europe_waste.htm.
7. Ministry of Environment and Physical Development, Higher Council for Environment and Natural Resources, Sudan (2003), Sudan's First National Communications under the UNFCC
8. 2000 Eurostat (2005). Waste Generated and Treated in Europe. Data 1995-2003. European Commission -.Eurostat, Luxemburg. 131p
9. Problems of waste management in Russia: Not-for-Profit Partnership "Waste Management – Strategic Ecological Initiative" http://www.sagepub.com/journalsProdEditBoards.nav?prodId=Journal201691

10   The Bahamas Environment, Science and Technology Commission (2001). Commonwealth of the Bahamas. First National Communication on Climate Change. Nassu, New Providence, April 2001, 121pp.

11   1990   OPS/OMS (1997). Análisis Sectorial de Residuos Sólidos en Cuba. Serie Análisis 1. Sectoriales No. 13, Organización Panamericana de la Salud, 206 pp., 2. López, C., et al. (2002). República de Cuba. Inventario Nacional de Emisiones y Absorciones de Gases de Invernadero (colectivo de autores). Reporte para el Año 1996/Actualización para los Años 1990 y 1994. CD-ROM Vol. 01. Instituto de Meteorologia-AMA-CITMA. La Habana, 320 pp. ISBN: 959-02-0352-3.

12   Secretaria de Estado de Medio Ambiente y Recursos Naturales (2004). República Dominicana. Primera Comunicación Nacional a la Convención Marco de Naciones Unidas sobre Cambio Climático. UNEP/GEF, Santo Domingo, Marzo de 2004, 163 pp.

13   1990   Ministry of Planning, Development, Environment and Housing (2001). Saint Lucias's Initial National Communication on Climate Change, UNEP/GEF, 306 pp.

14   Lammers, P. E. M., J. F. Feenstra, A. A. Olstroom (1998). Country/Region-Specific Emission Factors in National Greenhouse Gas Inventories. UNEP/Institute for Environmental Studies Vrije Universiteit, 112 pp.

15   Ministerio de Recursos Naturales, Energía y Minas (1995). Inventario Nacional de Fuentes y Sumideros de Gases con Efecto Invernadero en Costa Rica. MRNEM, Instituto Meteorológico Nacional, San José, Septiembre 1995.

16   Ministerio de Ambiente y Recursos Naturales (2001). República de Guatemala. Primera Comunicación Nacional sobre Cambio Climático..

17   JICA (Agencia Japonesa de Cooperación Internacional) (1991). Estudio sobre el Manejo de los Desechos Sólidos en el Area Metropolitana de la Ciudad de Guatemala. Volumen 1.

18   Guatemala de la Asunción, diciembre 2001, 127 p.,OPS/OMS (1995). Análisis Sectorial de Residuos Sólidos en Guatemala, Diciembre 1995, 183 pp.

19   1990   Fondo Nacional de Desarrollo (FNDR). Cantidad de RSM dispuestos en RSA-años 1996 y 1997, La Paz, Bolivia., 2. Ministerio de Desarrollo Sostenible y Medio Ambiente/Secretaria Nacional de Recursos Naturales y Medio Ambiente (1997). Inventariación de Emisiones de Gases de Efecto Invernadero. Bolivia – 1990. MDSMA/SNRNMA/SMA/PNCC/U.S. CSP, La Paz, 1997.

20   Ministry of Science and Technology, Brazil (2002). First Brazilian Inventory of Anthropogenic Greenhouse Gas Emissions. Background Reports. Methane Emissions from Waste Treatment and Disposal. CETESB. 1990 and 1994, Brazilia, DF, 85 pp.

21   CETESB (1992). Companhia de Tecnologia de Saneamiento Ambiental. Programa de gerenciamiento de residuos sólidos domiciliares e de services de saúde. PROLIXO, CETESB; Sao Paulo, 29 pp., IBGE: Instituto Brasileiro de Geografia e Estadística. http://www.ibge.gov.br/home/estadistica/populacao/atlassaneamiento/pdf/mappag59.pdf in November 2004.

22   1990   Ministerio de Medio Ambiente/IDEAM (1999). República de Colombia. Inventario Nacional de Fuentes y Sumideros de Gases de Efecto Invernadero. 1990. Módulo Residuos, Santa Fe de Bogotá, DC, Marzo de 1999, 14 pp.

23   BID/OPS/OMS (1997). Diagnóstico de la Situación del Manejo de los Residuos Sólidos Municipales en América Latina y el Caribe., Doorn and Barlaz, 1995, Estimate of global methane emissions from landfills and open dumps, EPA-600/R-95-019, Office of Research & Development, Washington DC, USA.

24   1990   MAG/SSERNMA/DOA – PNUD/UNITAR (1999). Paraguay: Inventario Nacional de Gases de Efecto Invernadero por Fuentes y Sumideros. Año 1990. Proyecto PAR GLO/95/G31. Asunción, Noviembre 1999, 90 pp.

25   1990   Estudios CEPIS-OPS y/o Estudio Sectorial de Residuos Sólidos del Perú. Ditesa/OPS., Lammers, P. E. M., J. F.
       1994   Feenstra, A. A. Olstroom (1998). Country/Region-Specific Emission Factors in National Greenhouse Gas
       1998   Inventories. UNEP/Institute for Environmental Studies Vrije Universiteit, 112 pp.

26   Ministerio de Vivienda, Ordenamiento Territorial y Medio Ambiente/Dirección Nacional de Medio Ambiente/Unidad de Cambio Climático (1998). Uruguay. Inventario Nacional de Emisiones Netas de Gases de Efecto Invernadero 1994/Estudio Comparativo de Emisiones Netas de Gases de Efecto Invernadero para 1990 y 1994. Montevideo, Noviembre de 1998, 363pp.

27   OPS/OMS (1996). Análisis Sectorial de Residuos Só,Ministerio de Vivienda, Ordenamiento Territorial y Medio Ambiente/Dirección Nacional de Medio Ambiente/Unidad de Cambio Climático (2004). Uruguay. Segunda Comunicación a la CMNUCC. 330p. lidos en Uruguay. Plan Regional de Inversiones en Medio Ambiente y Salud, Marzo 1996.

28   2000   Ministerio del Ambiente y de los Recursos Naturales Renovables. Ministerio de Energía y Minas (1996). Venezuela. Inventario de Emisiones de Gases de Efecto Invernadero. Año 1990. GEF/UNEP/U.S CSP.

29   1992   Organization for Economic Cooperation and Development (OECD) http://www.oecd.org/dataoecd/11/15/24111692.PDF

30   The Fraser Institute, Environmental Indicators, 4th Edition (2000).
http://oldfraser.lexi.net/publications/critical_issues/2000/env_indic/section_05.html.

31   UNFCCC Secretariat, Working paper No.3 (g) (2000). Expert report, prepared for the UNFCCC secretariat, 20 February 2000.

32   1992   http://www.oecd.org/dataoecd/11/15/24111692.PDF.

33   INE/SMARN (2000). Inventario Nacional de Emisiones de Gases de Invernadero 1994-1998, Ciudad de Mexico, Octubre 2000, 461 p.

34   Waste generation from: BioCycle (January 2004). "14th Annual BioCycle Nationwide Survey: The State of Garbage in America", Waste disposition from: BioCycle (December 2001). "13th Annual BioCycle Nationwide Survey: The State of Garbage in America"; Personal Communication: Elizabeth Scheele, U.S. EPA.

Waste-to-Energy Facilities in USA

(Source: Themelis 2006)

The most recent survey by BioCycle Journal and Columbia University (Simmons et al 2006, Themelis and Kaufman 2004) showed that the generation of MSW increased from 369 million short tons (1.1 short tons = 1 tonne) in 2002 to 388 million tons in 2004, i.e., at the rate of 2.5% per year. Landfilling accounted for 249 million tons or 64% of the MSW generated. The MSW generation per capita remained at 1.3 tons/year (3.2 kg/day), by far the highest in the world. A comparison of the BioCycle/EEC data for 2002 and 2004 data (Table) shows that in the intervening two years, recycling plus composting increased by 11.8 million tons, landfilling by 6.3 million tons and WTE by 0.5 million tons.

Table MSW generation and disposal in US in 2002 and 2004

	MSW Generated	Recycled or composted	Waste-to-Energy	Landfilled
2004, million tons	387.9	110.4	28.9	248.6
2004, percent	100%	28.5%	7.4%	64.1%
2002, million tons	369.4	98.6	28.4	236.8
2002, percent	100%	26.7%	7.7%	65.6%

BioCycle/EEC surveys (BioCycle, Jan. 2004 and April 2006)

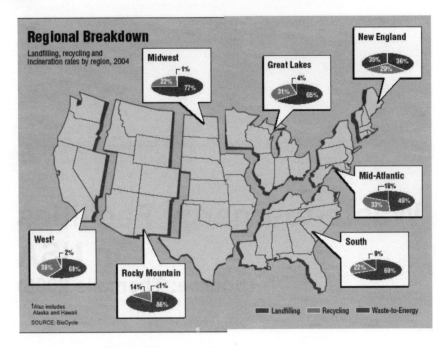

Fig. Breakdown of disposal of MSW by region
(Biocycle/ EEC SOG, April 2006)

The figure shows that many of the WTE facilities are located in the East, and most of the recycling is in the coastal states of US.

Thermal treatment facilities installed in the 21st century are based mostly on the grate combustion of 'as received' MSW. US facilities follow this type of treatment and on an industrial scale, the dominant WTE technology is grate technology, because of its simplicity and relatively low capital cost. The majority of the facilities is grate combustion ("mass burned as received" or RDF), and represent over 80% of the total capacity of WTE in the US. Three dominant technologies – those developed by Martin, Von Roll, and Keppel-Seghers – are grate technologies. In terms of novel technologies, gasification (JFE), direct smelting (JFE, Nippon Steel), fluidized bed (Ebara) and circulating fluidized bed (Zhejiang University) are in operation around the world, while some of them are under investigation and discussion for possible implementation in the WTE facilities that will be constructed in the US (Themelis, 2003, 2007, Psomopoulos et al., 2009). One of the most successful types of facilities is the RDF-type process of the SEMASS facility in Rochester, Massachusetts, USA, developed by Energy Answers Corp. This facility is discussed in chapter 6.

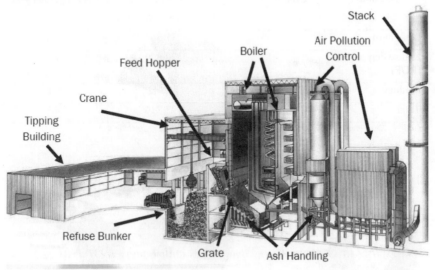

Mass burn facility for 'as received MSW'
(Source: WTERT Brochure)

The WTE facilities in US can therefore be classified broadly into three categories:

1. Mass burn plants generate electricity and/or steam from garbage by feeding *MSW as received* into large furnaces dedicated solely to burning garbage and producing power. Typical one is shown above.

2. Refuse-derived fuel (RDF) plants shred the MSW; recover some recyclable materials, and combust the homogenized fuel in a combustion chamber. The RDF producing facility may be next to the furnace or at another location.

3. Modular waste-to-energy plants are similar to mass burn facilities but are smaller and typically pre-fabricated off site and assembled where they are needed. See Table for details. In the last column, the figures represent percentage.

Table Operating WTE plants in US

Technology	Number of plants	Capacity, tons/day	Capacity, tons/year
Mass burn	65	71,354	22.1
Refuse derived fuel (RDF)	15	20,020	6.3
Modular	9	1,342	0.4
Total	89	92,716	28.8

. J. V. L. Kiser and M. Zannes, Integrated Waste Services Association, April 2004

(Taken from Themelis 2006)

Fig. Distribution of WTE plants; the numbers are shown in the brackets.

There are 89 waste-to-energy power plants operating in 27 states serving 31 million people. They are fed by 29 million tons of MSW and have a generating capacity of 2,700 megawatts of electricity. They recover 0.7 million tons of ferrous and non-ferrous metals; also, three million tons of MSW are used in place of soil or stone aggregate in the maintenance of landfills.

Taking into account the electricity generated and the methane emissions avoided has led several independent studies to conclude that for each tonne of MSW diverted from landfilling to WTE, GHG emissions are reduced by an estimated 1.1 to 1.3 tons of carbon dioxide.

Therefore, in addition to the energy benefits, the combustion of MSW in WTE facilities reduces US greenhouse gas emissions by about 26 million tonnes of carbon dioxide. In theTable, the air emissions of WTE and fossil-fuelled power plants are compared.

Table: WTE and Fossil fuel Power plants (O'Brien and Swana 2006)

Fuel	Air emissions (kg/MW h)		
	Carbon dioxide (CO_2)	Sulphur dioxide (SO_2)	Nitrogen oxides
MSW	379.66	0.36	2.45
Coal	1020.13	5.90	2.72
Oil	758.41	5.44	1.81
Natural gas	514.83	0.04	0.77

In addition to methane, landfill gas contains several volatile organic compounds and chlorinated hydrocarbons.

Potential for clean energy: In 2004, 28.9 million tons of garbage were combusted in waste-to-energy power plants in US and generated a net of 13.5 billion kilowatt-hours of electricity, greater than all other renewable sources of energy, with the exception of hydroelectric and geothermal power.

Emissions and public health issues: In the past, all high temperature processes, including metal smelting, cement production, coal-fired power plants and incinerators were the sources of enormous emissions to the atmosphere. In particular, incinerators were the major sources of toxic organic compounds, dioxins and furans, and mercury as already discussed.

However, in the last fifteen years and at the cost of about one billion dollars, the 89 WTE facilities operating in the US have implemented air pollution control systems that has led EPA to recognize them publicly as a source of power 'with less environmental impact than almost any other source of electricity' *(www.wte.org/epaletter.html).*

In 1995, the EPA adopted new emissions standards for WTE facilities pursuant to the Clean Air Act. Their Maximum Achievable Control Technology (MACT) regulations dictated that waste-to-energy facilities should comply with new Clean Air Act standards. MACT includes dry scrubbers, fabric filter bag houses, activated carbon injection and other measures that were implemented at the cost of over one billion dollars. Waste-to-energy facilities now represent less than 1% of the US emissions of dioxins and mercury.

Decrease in dioxin and mercury emissions: WTE plants have decreased dioxin emissions since 1987 by a factor of 1,000 to about 12 grams TEQ (toxic equivalent) total (See Figure). The major source of dioxin emissions, as reported by EPA, is backyard garbage burning that emits close to 600 grams annually, and about 1000 grams TEQ annually by thousands of landfill fires as reported by Federal Emergency Management Administration (FEMA, May 2002)

(*www.fireox-international.com/fire/FEMA-LandfillFires.pdf*).

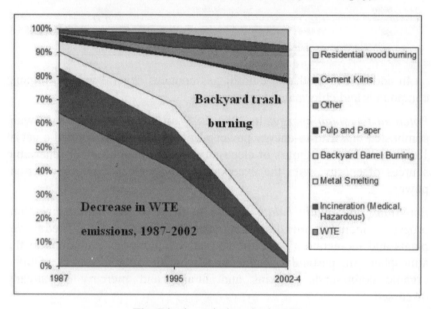

Fig. Dioxin emissions in the US

(Source: P. Deriziotis, MS thesis,
Columbia University, 2003; Data by EPA)

Due to the phasing out of most applications of mercury, the use of mercury in U.S. processes and products decreased to less than 360 tonnes by 2002. Also, many communities have put in place strong recycling programs that keep older mercury-containing products out of the MSW sent to WTE facilities. This trend and the installation of MACT pollution control systems have reduced the U.S. WTE emissions to the atmosphere from 80 tonnes of mercury in the eighties to less than one ton in 2002 (Figure). The major sources of mercury in the atmosphere now are the coal-fired powerplants.

The only remaining WTE emissions of concern are nitrogen oxides whose WTE emissions correspond to only 0.22% of the total US NOx emissions.

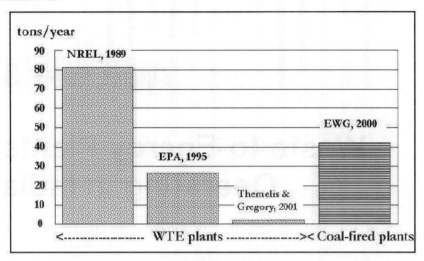

Fig. Reduction of mercury emissions from WTE plants in US
(Source: Themelis and Gregory 2002)

WTERT Council: To support R & D aspects of WTE facilities, Waste-to-Energy Research and Technology Council (WTERT) was co-founded in 2002 by the Earth Engineering Center of Columbia University (www.columbia.edu/cu/earth) and the Integrated Waste Services Association (IWSA) (www.wte.org) which represents most of the waste-to-energy facilities in the US. The mission of WTERT is to increase the recovery of materials and energy from used solids and, in particular, to advance both the economic and environmental performance of waste-to-energy technologies worldwide (Themelis 2006).

Waste-to-Energy Plants Operating in India

A few examples operating through PPP are already mentioned at relevant places. A few more important ones are detailed below:

1. *Bio-Methanation Plant, Lucknow*: The Lucknow Nagar Nigam (LNN) faced major threat from disposed wastes as its two landfills were overflowing. Lucknow produces around 1800 MT of MSW daily. Inability to identify a suitable area in proximity forced LNN policy managers to look for other alternatives. The city also faces a huge crisis in terms of energy requirements, especially for its industries, and an option of generating environment-friendly power from MSW was determined to be the most optimum solution to the problem. Studies estimate that MSW in Lucknow has the potential of generating 1000 MW of power while industrial waste has the potential of generating 700 MW of power. The LNN, therefore, invited firms for development and execution of a WTE power generation and bio-fertiliser producing plant. The facility is designed to handle a minimum of 300 TPD of municipal waste and uses the bio-methanation process for conversion of waste to energy with help of a BIMA digester, a technology that is being used in over 50 WTE plants worldwide. The estimated project cost is Rs. 76 crore (LNN, 2002).

Stakeholders/Partnerships and Financial Outlay: The LNN invited Chennai-based Enkem Engineers to be the project promoters. Enkem India Ltd. floated a Special Purpose Vehicle (SPV) called Bio Energy for the project. ENTEC, an Austria-based firm, provided the project technology and digester for production of methane. M/s C.G.E.A. Asia Holdings, Singapore, is responsible for O&M.

The proposed financing plan and the source of funding is as follows:

Funding Type	Source	Amount (Crore Rupees)
1. Promoter Equity	SPV – Asia Bio Energy	20
2. Government Subsidy	MNES (@30 million per MW)	15
3. Equipment Supplier	Supplied on Operating Lease basis	11
4. Debt	Contract-based lending from IDFC	20
5. Credit	Deferred credit being provided by equipment supplier	10
Total		76

The following have taken an equity stake in the project: (i) Enkem Engineers Pvt. Ltd., Chennai, (ii) Entec Environment Technology, UgmBH, Austria, (iii) Innovative Umwelltechnik Ges.mbH, Austria, (iv) Jurong Engineering Ltd., Singapore, (v) CGEA Asia Holdings (P) Ltd., France, (vi) Larsen and Toubro, India, (vii) IDFC

Other stakeholders include LNN, NEDA, LDA, UPPCL, UPPCB and GoUP. The GoUP has given a guarantee against any default in payment of electricity charges. The solid waste is provided by LNN while the electricity is purchased by UPPCL.

Description of the Project: This project was designed as *the first solid waste power project in India* which the MNES identified as a full-scale national demonstration plant. Although the project was initiated in 1998, the project got delayed because of finalisation of land transfers, government guarantee, identification of financiers and other related formalities, which could be completed only by August 2001. The plant construction was completed in August 2003 (see picture).

The project is being executed on a Build Operate and Transfer **(BOT)** basis. The land for the project was provided by the LNN on a lease for a period of 30 years. LNN was given a subsidy to the tune of Rs. 75 lakhs from MNES for providing this land. While the technical inputs are being provided by firms from Austria and Singapore, human resource for execution was provided by Indian firms. As part of the original contract, LNN agreed to assure provision of 113 to 120 MT of MSW daily to the operator although the Plant has been designed to take a maximum of 300 MT daily. Depending on success of the initial phase, LNN was to increase supply to the Plant over a period of time.

*Broad features of the project as envisaged include:*300 MT of solid waste to be treated daily to provide biogas, 5 MW power will be generated by using biogas as fuel for five gas generators, 70 tonnes of organic fertiliser will be produced daily as bye product, there will be no toxic liquid or gas effluents from the Plant.

Cost Recovery: As per the original framework, LNN was to get 1% of the cost of power sold to UPPCL and 5% of the organic fertiliser produced, an amount estimated around Rs.35 to 40 lakhs annually.

Electricity is being sold to UPPCL at rate of Rs. 2.48/- per unit. As per estimates, the Plant was to produce around 5000 units/hour by beginning of 2004. Power generation over 1 MW was started by Dec 2003 and is now expected to reach 5 MW soon.

2. *Integrated SWM Project, Guwahati:*

   Hyderabad based Ramky Enviro Engineers Ltd has secured an integrated SWM project. The project will be developed on 60 acres of land on BOOT basis. The company signed a 20 year concession agreement with Guwahati Municipal Corporation on Oct 29, 2008.

The project being implemented under JNNURM. A grant of Rs.350 million for the Rs.1.02 billion project has been approved. The Municipal corporation has set up SPV, Guwahati Waste Management company Ltd, to implement the project. The project involves two segments – under the first segment, primary (door-to-door) collection of MSW, transportation to the processing site and disposal of MSW is being undertaken. The second segment involves setting up a plant with a capacity to convert 650 tpd of MSW into refuge derived fuel (RDF), a 50 tpd compost plant to produce manure as well as a 6 mega watt power plant to run on the RDF produced.

3. *Integrated SWM Project, Coimbatore:*

The Coimbatore City Municipal Corporation in Tamil Nadu is implementing an integrated SWM project under JNNURM. The project awarded to JV of UPL Environmental Engineers Ltd and Bharuch Enviro Infrastructure Ltd, is being implemented on BOOT basis and has a concession period of 20 years.

The 965.1 Million project is being implemented under two components. The first component, requiring an investment of Rs.269.1 million, involves distribution of bins, primary segregation of garbage, secondary storage in bins and transportation of waste to transfer stations. (Source: Nagpur and Lucknow: Solid Waste Mgt Part-II, Background Paper,12 th Finance Commission by IPE(P) Ltd).

The 696 million second component involves development, design, engineering, finance, construction implementation and O&M of transfer stations, transportation of MSW from transfer stations to waste processing and disposal sites, processing and disposal facility for treating 400 tonnes of waste generated from Coimbatore. The JV has formed an SPV, Coimbatore Integrated Company Ltd, to implement the second component of the project. The project is expected to be completed by Sept 2009.The capital grant of 70% is being funded under JNNURM and the private operator UPL will contribute the remaining 30% in the form of Equity and Debt. The CMC will pay to UPL total NPV for 20 years Rs.264.85 crores (Tipping fee)

4. *MSW Management Project: Navi Mumbai*

The NMMC is planning to set up an MSW processing unit with a capacity of 500 tpd under the JNNURM on PPP basis. The project aims at the bioremediation of old accumulated waste, acceleration in aerobic composting of waste using windrows under controlled conditions for converting these to organic compost, converting combustible waste to RDF and recovering dry recyclables like glass, metals and plastics.

The operational methodology under the project is divided into six steps. Beginning at waste segregation, the project will move on to composting, RDF manufacturing, plastic recycling, inert processing to civil bricks and sanitary landfills.

Two main products are expected to be generated from the MSW. These are compost/organic fertilizer and fuel pellets/RDF. Segregated waste like plastic will be sent to recycling units.

Of the total 500 tpd of waste received, about 125 tpd of RDF and 100 tpd of compost will be recovered. The remaining components will comprise about 25 tpd of recyclables, 125 tpd of rejects that will be sent to sanitary landfills and 25 tpd of inert processing paver blocks. The balance 100 tpd will comprise moisture.

The project will be implemented on PPP basis. In April 2008, the corporation entered into an agreement with the UK based firm Ecomethane for tapping landfill gas (LFG) and availing carbon credit under the clean development mechanism (CDM) technology.

The firm will be responsible for arranging all financial, legal and technical aspects of investment in the LFG project, design and contract effective LFG tapping, develop all CDM related documentation, undertake baseline study, monitor plans, register the project with the CDM executive board, and monitor emission reduction of the project in accordance with the United Nations Framework Convention on Climate Change procedures during the project's lifetime. The Corporation on its part will be responsible for providing electricity to the site.

The corporation is expected to receive an upfront amount of Rs 6.38 million for the project, and generate revenue of about Rs 80 million in the 10 years.

5. *MSW Management Project: Kollam*

Preliminary work on construction of a comprehensive MSW management project has started in Kollam district of Kerala. The project was secured by Jamshedpur Utility and Services Company Limited (JUSCO). An agreement for constructing the project was signed between JUSCO and the Kerala Sustainable Urban Development Project on July 29, 2008.

The Rs 650 million project is being funded by the Asian Development Bank. The project scope involves establishing an integrated solid waste processing and disposal facility, a new sanitary landfill capacity, closure of the existing waste dump site, erection of an electromechanical compost plant and establishment of a leachate treatment plant.

The plant will have a capacity to process and dispose of 100 tpd of waste. Currently, seepage of leachate into nearby drinking water wells and the Ashtamudi Lake, especially during the rainy season, is a major problem. The establishment of the landfill site is expected to obstruct the seepage. The project is expected to be made operational by February 2010-11

6. *MSW projects in Jaipur:*

Rs.20 crores plant has been set up on PPP basis by M/s Grasim at Langaria Was in Jaipur for conversion of MSW into RDF. The plant accepts about 300-400 MT unsegregated SW per day at plant site from Jaipur Municipal Corporation (JMC).

The rejects/inert material will be dumped at the adjoining sanitary landfill site which is being constructed by JMC.

The Project is on BOOT basis -built, own, operate and transfer of land only for 30 years. The corporation has allotted underdeveloped land equivalent to 25 acres to the private operator @ Rs.1 per sq metre on lease for 30 years. The operator M/s Grasim is supposed to pay revenue share of Rs.1,20,190 per annum to JMC. The RDF produced is being used in the cement plant of M/s Grasim so there is no problem of marketing.

JMC has a bio-medical waste treatment plant established on BOOT basis at village Rupari on the 4 acres land provided by JMC on lease basis.

JMC has worked as a facilitator to arrive at a negotiated rate of Rs.3.10 per day per bed to be paid by the hospitals/clinics to the service provider.

482 hospitals (Government/private), 8000 beds are being provided facility.

The common bio-medical waste treatment facility is treating about 4 tonnes of bio medical waste daily generated from hospitals/clinics.

Work on an additional plant at Agra road is in progress.

(Source: Guwahati, Coimbatore, Navi Mumbai and Kollam Projects-Indian Infrastructure Vol II,Issue 7, Feb 2009)

Compost plant for waste treatment: Another project on PPP basis for production of compost from MSW is coming up in Jaipur on DBOOT basis at Sewapura at a cost of 8 crores.

JMC will supply mixed waste of 250 MT per day at the plant site. Land has been provided to the private operator for a lease period of 30 years. 100 MT of compost is likely to be generated per day.LOA has been issued to the private operator

7. *Municipal Solid Waste Management Project in Asansol Urban Area*

Asansol MC generates about 700 tonnes of solid waste every day. A project of Rs.44 crores has been prepared and proposed under JNNURM. The five ULBs have to shell out 30% of the project cost. ADDA (Asansol Durgapura Municipal Corporation) have decided to enlist PSP in implementation. The private partner will set up and run processing plants at three sites and manage the landfill site.ADDA invited bids for selection of private operator on BOOT basis. 5 ULBs have to ensure 350 tonnes of Garage at the processing site every day. The consortium of Gujarat Enviro Protection Infrastructure Ltd and Hanjer Bio-tech Ltd has been selected. Tipping fee has been agreed as Rs. 85/tonne of MSW. The project is to be completed in 24 months. Door to door collection of waste at such a large scale has been conceived of for the first time. For sustainability of the project, ULB will collect: RS.5 to 25 per month per household and Rs.25 to 50 per month in commercial areas.

Composting Plants in India

No.	State	City	Facility manufacturer	Installed capacity
1.	Andhra Pradesh	Vijayawada	Excel Industries Ltd	125 TPD
2.		Thirumala	NA	NA
3.		Vizianagaram	NA	NA
4.	Assam	Kamarup	NA	NA
5.	Chhatisgarh	Dhamtari	NA	NA
6.		Rajnandgaon	NA	NA
7.		Jagdalpur	NA	NA
8.		Rakpur	NA	NA
9.		Korba	NA	NA
10.		Bhilai	NA	NA
11.		Durg	NA	NA
12.		Raigad	NA	NA
13.	Delhi (UT)	Delhi	Nature And Waste Inc India (BALSWA Plant)	500 TPD
14.		Delhi	Private Organo-PSOS Plant, (Tikri Plant)	150 TPD
15.		Delhi	MCD Plant, Okla	300 TPD
16.		Delhi	NDMC Plant, Okla	300 TPD
17.	Gujarat	Ahmedabad	Excel Industries Ltd, Ahmedabad	500 TPD
18.		Junagadh	NA	NA
19.		Rajkot	NA	NA
20.	Goa	Margao	M/s. Comets International Ltd	40 TPD
21.	Himachal Pradesh	Shimla	L&T	100 TPD
22.		Solan	Janseva Trust	50 TPD
23.		Sirmour	NA	NA
24.		Dharamshala	NA	NA
25.		Bilaspur	NA	NA
26.		Una	NA	NA
27.		Hamirpur	NA	NA
28.		Kangra	NA	NA
29.		Kullu	NA	NA
30.		Mandi	NA	NA
31.	Karnataka	Bangalore	Karnataka Compost Development Corporation	350 TPD
32.		Bangalore	Terra-Fersia Bio-Technologies Ltd	100 TPD
33.		Mysore	Vennar Organic Fertilizer Pvt. Ltd	200 TPD
34.		Mangalore	NA	NA
35.	Kerala	Thiruvananthapuram	POABS Envirotech Pvt. Ltd	300 TPD

37.		Adoor	NA	NA
38.		Atingal	NA	NA
39.		Chalakundy	NA	NA
40.	Madhya Pradesh	Bhopal	M. P. State Agro Industries	100 TPD
41.		Gwalior	NA	120 TPD
42.	Maharashtra	Nasik	M/s. Live Biotech	300 TPD
43.		Aurangabad	M/s. Satyam Bio-fertilizer Co. Ltd	300 TPD
44.		Thane	M/s. Leaf Biotech Ltd	300 TPD
45.	Meghalaya	Shillong	M/s. Anderson Biotech Pvt. Ltd	150 TPD
46.	Orissa	Puri	M/s. Krishi Rashyan, Kolkata	100 TPD
47.	Pondicherry	Pondicherry	Pondicherry Agro Services and Industries	100 TPD
48.	Tamil Nadu	Tiruppur	IVR Enviro Project (P) Ltd	100 TPD
49.		Nagercoil	NA	NA
50.	West Bengal	Kolkata	M/s. Eastern Organic Fertilizer P. Ltd	700 TPD

(Source: Asnani 2006)

Waste-to-Energy Status in China

The need for intelligent waste management has led to the concept of the 'hierarchy of waste management' that places the various means for dealing with MSW in order of environmental preference. Of the estimated one billion tons (907 million tonnes) of global 'post-recycling' MSW, close to 200 million tons (181 million tonnes) are processed in Waste-to-Energy (WTE) plants that recover the energy content of waste in the form of electricity or heat. The dominant WTE technology involves combustion of MSW on an inclined or horizontal grate. There are over 500 WTE plants of this type operating in 35 countries.

Most of the global urban MSW, i.e., over 800 million tons (725 million tonnes) is landfilled. The Earth Engineering Center of Columbia University has estimated that one square meter (about 10 square feet) is used up, forever, for every ten tons (nine tonnes) of MSW landfilled. True sustainable development requires that only inorganic residues be landfilled, as is already the practice in several countries. However, this would require us to considerably increase the present global WTE capacity of about 200 million tons (181 million tonnes) and this is a very costly proposition, especially for developing countries. Obviously, the need is greatest in large countries with rapidly growing cities, such as China and India, where existing dump sites are overfilled.

Waste management in China: China has the largest population (1.33 billion) on Earth and is experiencing rapid economic growth. This country has a GDP of $8.8 trillion in terms of Purchasing Power Parity (PPP), which is the third largest in the world after the EU and the US. However, its population is over four times that of the US so the actual per capita GDP is only $6.800 and corresponds to a fraction of the US GDP per capita.

Despite the relatively high capital cost of WTE, the central government of China has been very proactive with regard to increasing WTE capacity. One of the measures brought in provided a credit of about $30 per MWh of electricity generated by means of WTE rather than by using fossil fuels.

'Harmless treatment' of MSW in China: The term 'harmless treatment' in China means the disposal of MSW by recycling, composting, WTE and sanitary landfilling. The 'harmless treatment' rate is defined as the percentage of the weight of total MSW treated with these methods. The generation of MSW, and also the 'harmless treatment' fraction have been increasing over the past 30 years in China.

Table 1 shows the reported data from 2001 to 2007 and also the number of WTE plants and their total capacity. The Chinese WTE capacity has increased steadily from 2.2 million tons in 2001 to nearly 14 million tons by 2007. However, landfilling remains the dominant means of waste disposal in China.

Table 1 MSW generation, treatment and WTE capacity in China.

MSW generation million tons/y)	Fraction disposed by "Harmless Treatment", %	Number of WTE plants	Total WTE capacity, million tons/y
134.70	< 25	36	2.17
136.55	<30	45	3.39
148.57	50.8	47	3.7
155.09	52.1	54	4.49
155.77	51.7	67	7.91
148.41	52.2	69	11.38
152.14	62	66	14.35

Most WTE plants are located in eastern China, especially in the districts of the Changjiang and Pearl River Deltas. As of 2007, three provinces in these two districts, Guangdong, Zhejiang and Jiangsu had fifteen, fourteen and nine WTE plants, respectively. These plants constitute 64 % of the existing WTE capacity in China. This is explained by the relatively high economic development in these provinces.

China's 11th Five-Year Plan (2006-2011) is very ambitious, showing expected construction of many new WTE facilities across the country, shown in Table 2.

Table 2 New WTE plants planned for period 2006-2011

Region	Quantity	Capacity (10^6 tons/ yr)	Percentage (%)
East	56	15.03	67.7
Middle	9	2.4	10.8
West	10	3.13	14.1
North East	7	1.63	7.4
Total	82	22.2	100

WTE technologies used in China: Stoker Grate incinerator and Circulated Fluidized Bed (CFB) incinerator are the main types of technology used in WTE plants in China. According to a preliminary survey of 100 WTE plants in operation or under construction, most of the MSW incinerators are of the Grate combustion type ('mass burn'), and are based either on imported or domestic technologies. The CFB incinerators co-fire MSW with coal (up to 15 % coal by weight) and have been developed by Chinese academic research centers, such as Zhejiang University, Chinese Academy of Sciences (CAS), and Tsinghua University. Most of the new plants are based on the stoker grate design. There are 4 major local companies, Sanfeng Covanta, Shanghai Environmental, Everbright Shenzhen and Shenzhen Energy.

The capacity of the WTE plants built in earlier years was generally less than 800 tons/day (725 tonnes/ day). However, recent WTE plants are larger, typically over 1000 tons/day (907 tonnes/day). The capacity of a single line within a plant has also increased, from the 200 tons/day (181 tonnes/day) in early years to over 500 tons/day (453 tonnes/day) in recent years.

Air pollution control systems: Most of the air pollution control systems built in the Chinese WTE plants are similar to the predominant gas control systems in the US: a combination of semi-dry scrubber, activated carbon injection (to remove volatile metals and organic compounds) and fabric filter baghouses (to remove particulate matter). In some WTE plants, selective non-catalytic reduction is included to remove nitrogen oxides, such as, for example, the WTE plants under design for Guangzhou, Shantou, and Chongqing. A major problem that faced the western incinerators in the late 1980s was high emissions of dioxin. The US WTE plants in 1989 emitted a total of 10,000 grams of toxic equivalent dioxins (grams TEQ), corresponding to 100 nanograms TEQ per standard cubic meter of stack gas. This led to the US Environmental Protection Agency's regulation of Maximum Achievable Control Technology (MACT) that resulted in the retrofitting of about 90 WTE plants in the US and the closing of nearly 50 small plants. As of 2002, this retrofit resulted in decreasing WTE dioxin emissions by a factor of 1000, to less than 10 grams TEQ/Nm^3.

The emissions of polychlorinated dibenzo-p-dioxins and polychlorinated dibenzofurans (dioxins) from 19 MSW incinerators in China were investigated by the Chinese Academy of Science (CAS). Sixteen stoker grate and three circulating fluid bed incinerators with capacities from 150 to 500 tons/day (136-453 tonnes/day) were examined. The air pollution control systems of nine of the grate combustion WTE plants, consisted of semi-dry scrubber, activated carbon injection and fabric filter baghouse; the other seven plants did not use activated carbon injection. The results of this study showed that the dioxin emissions of these 19 MSW incinerators ranged from 0.042 to 2.461 nanograms TEQ /Nm^3; the average value was 0.423 ng TEQ/Nm^3. The dioxin emission levels of three MSW incinerators were higher than 1.0 ng TEQ/Nm^3, which is the emission standard in China. Only six MSW incinerators had dioxin emission levels below 0.1 ng TEQ/Nm^3, which is the emission limit in Europe, the US and other developed countries. Therefore, the average emissions of dioxins from Chinese incinerators ranged from being as low as European and US plants to being 24 times the western standard. Considering the significant amount of MSW generation in China, the dioxin emissions from some poorly-operated WTE have been a severe problem and caused an adverse public reaction against all WTE facilities.

The dioxin emission factors to the atmosphere from these 19 MSW incinerators were calculated to range from 0.169-10.72 μg TEQ for per ton MSW with an average 1.728 μg TEQ per ton MSW.

Broad Technical details of two typical WTE plants presently under operation are described here (Kalogirou 2010):

1. *Pudong Yuqiao Waste to Energy (WtE) Plant.*

   The Yuqiao WtE Plant in Pudong was put into operation in September 2002. The main technical details are the following:

   (i)   The Plant is equipped with three incineration lines, each with daily capacity of 330 tons, feeding with MSW from part of Pudong district.The total area of the Plant is 80,000 m^2 (20 acres); the plant operates for 7500 hours annually.

   (ii) The grate technology used is the SITY 2000 technology that is owned by Martin GmbH

   (iii) There are three gravity circulating water-wall boilers and each incineration-boiler line is coupled with a Flue Gas Cleaning System with semi-dry scrubber, bag filter and activated carbon injection system.

   (iv) Steam produced from two turbine-generators with each normal capacity of 8.5 MW; so the nominal capacity of the two turbines is 17 Megawatts.

   (v)   Bottom ash produced after incineration of MSW is reused as material for bricks after special treatment.

   (vi) Fly ash is transported to Shanghai Solid Waste Treatment Center for security special waste landfill.

   (vii) The thermodynamic characteristics of produced steam are 400°C and 40 bar pressure.

   (viii) The gate fee is around 240 RMB (28€)/ton

   (ix) The total investment was 670 million RMB

   (x)   The selling price for the electricity is 500 RMB/MWh

   (xi) Plant is equipped with continuous emission monitoring devices on flue gas emissions which calibrated according to the European standards (directive 2000/76).

2. *The Chnogqing Tongxing WtE Plant.*

   The Plant was first put into operation in March 28, 2005. The plant is using the Alstom SITY2000 design of Martin GmbH. Nearly all the equipment was fabricated locally to Martin specifications.The plant handles 50% of the waste generated in the Chongqing municipality.

The main technical details are the following:

- The company has three aspects: EPC contracts, Core technology and equipment & operation of the project.

- Chongqing Sanfeng Covanta Environmental Industry Co., Ltd has constructed and operated 25% of the WTEs in China.

- The technology adopted by Sanfeng Covanta is suitable for high water content, low heating value municipal waste which could burn steadily without pre-selection and auxiliary fuel.

- It has the capacity of 1200 ton per day with SITY2000 inclined reverse grate. It is the model project of the waste to energy for high water content and low heating value municipal solid waste.

- High efficiency incineration with residence time over 2 seconds, gas temperature above $850\,^{o}C$.

- The project used two sets of 58.39 t/h steam boiler, natural circulating system, $130\,^{o}C$ feeding water, $210\,^{o}C$ induced draft, 4.0 MPa steam pressure, $400\,^{o}C$ steam temperature.

- Waste heat from burning of waste is used for power generation from 220 kWh to 250 kWh from each ton of waste; and power pumped to grid is about 250.000 kWh supplying to 40.000 households.

- Previously, the dioxin limit in China was $1.0\ ng/Nm^3$ but now it has been reduced according to EE 2000/76 directive to $0.1\ ng/Nm^3$. Dioxin and furan emissions from this plant are around $0.05\ ng/Nm^3$.

- The bottom ash (25% of feed MSW) is used for road constructions, and fly ash (3% of the feed MSW) is made inert by on site solidification with cement.

- The bottom ash is used for building material. The waste water is recycled and after three levels treatmant, is used for watering the plant flowers.

- Gate fee is around 80 RMB (10 €/ton)

- The total investment was 370 million RMB

Since the beginning of the 21st century, China, more than any other developing country, is taking major steps to increase WTE capacity; it has increased its WTE capacity from 2 to 14 million tons of municipal solid wastes. This makes China the fourth largest user of waste-to-energy

(WTE), after the EU, Japan, and the US. There were 66 WTE plants in China by 2007; this is projected to increase to one hundred by 2012. Two thirds of these plants employ either imported or domestic versions of combustion on a moving grate; and the other third various forms of a home-developed technology, the circulating fluid bed reactor. There is preparation to increase to 140 WTE plants in the next 5 years (Kalogirou, 2010).

This study also examines in detail the environmental performance of Chinese WTE plants. Using as a yardstick the emission of dioxins from a group of 19 Chinese WTE plants, we found that seven operate below the EU dioxin standard (0.1 nanograms TEQ per standard cubic meter of stack gas) and 12 above this standard. The fact that several WTEs in China are able to control dioxin emissions to the very strict EU standard (which is 10 times lower than the present Chinese standard for dioxins) is very encouraging and indicates that Chinese operators and air pollution control systems can be as good as those in the west.

Fig. Fuzhou Hongmiaoling Energy-from-Waste Plant
(Credit:Sanfeng Coventa Environmental Co, Chongqing, China)

International Agreements and Commitments to Environmentally Sound Management of Waste

The three conventions related to waste, namely, Basel, Rotterdam and Stockholm Conventions are designated to protect human health and the environment from effects of hazardous chemicals and wastes.

Although legally separate ones, each governed by its respective Conference of Parties, all the three address the same fundamental challenge, namely, the environmentally sound management of hazardous products during their entire lifecycle from production to disposal, and wherever possible, their minimization and/or replacement with safer alternatives.

Waste Management:

The Basel Convention (Article 4) requires each Party to minimize waste generation and to ensure, to the extent possible, the availability of disposal facilities within its own territory. The objective of environmentally sound management of hazardous wastes underpins the Convention. At its fifth meeting in December 1999, the Conference of the

Parties adopted the Basel Declaration on Environmentally Sound Management.

The Stockholm Convention (Article 6) obliges Parties to develop strategies for identifying Persistent Organic Pollutants (POPs wastes), and to manage these in an environmentally sound manner. The POPs content of wastes is generally to be destroyed or irreversibly transformed. The Basel Convention Technical Working Group has developed technical guidelines on POPs wastes as part of its work programme and at the request of the Conference of Parties that adopted the Stockholm Convention.

Import/export controls:

The original Prior Informed Consent procedure of the Basel Convention (Article 4.1) was strengthened by the subsequent decisions of the Parties to prohibit the export of hazardous wastes from OECD to non-OECD countries (Decisions II/12 and III/1). The Basel Convention imposes strict conditions on the transboundary movement of hazardous wastes (Articles 4 and 6). Trade with non-parties is generally not permitted (Article 4.5).

The Rotterdam Convention (Articles 10 to 12) established a Prior Informed Consent Procedure based on the earlier voluntary guidelines. The Stockholm Convention (Article 3.2) restricts the import and export of POPs to cases where, for example, the purpose is not for environmentally sound disposal. It also requires that POPs should not be transported across international boundaries ignoring relevant international rules, standards and guidelines (Article6.1).

Environmental Releases:

The Stockholm Convention requires Parties to take measures to reduce or eliminate releases of POPs from intentional production and use (Article 3), unintentional production (Article 5) and stockpiles and wastes (Article 6). The principles of Best Available Techniques (BAT) and Best Environmental Practices (BEP) are to be further elaborated for and on behalf of the Conference of the Parties.

Hazard Communication:

Provision is made for the obligatory communication of hazard information under the Basel Convention (Article 4.2 f), the Rotterdam Convention (Article 5.1) and the Stockholm convention (Article 10).

Technical Assistance:

All three Conventions address the technical assistance needs of developing countries. The Basel Convention (Article 14) and the Stockholm Convention (Article 12) provide for regional centers for training and technology transfer, subject to views of Conferences of Parties. Basel has a Technical Cooperation Trust Fund, while Stockholm Convention (Articles 13 & 14) has a 'financial mechanism' operated by the Global Environment Facility (GEF) for the development of National Implementation Plans.

Types of Biogas Plants

Simple biogas plants are shown in the Figure. A: Floating-drum plant, B: Fixed-dome plant, C: Fixed-dome plant with separate gasholder. The gas pressure is kept constant by the floating gasholder. The unit can be operated as a continuous overflow type plant with no compensating tank. The use of an agitator is recommended. D: Balloon plant, E: Channel-type digester with folia and sunshade.

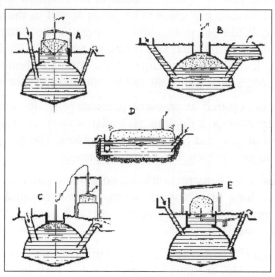

1. *Balloon Plant:*

   A balloon plant consists of a plastic or rubber digester bag, in the upper part of which the gas is stored. The inlet and outlet are attached direct to the skin of the balloon. When the gas space is full, the plant works like a fixed-dome plant, i.e., the balloon is not inflated; it is not very elastic.

   The fermentation slurry is agitated slightly by the movement of the balloon skin. This is favourable to the digestion process. Even difficult feed materials, such as water hyacinths, can be used in a balloon plant. The balloon material must be UV-resistant. Materials which have been used successfully include RMP (red mud plastic), Trevira and butyl.

   Advantages: Low cost, ease of transportation, low construction (important if the water table is high), high digester temperatures, uncomplicated cleaning, emptying and maintenance.

   Disadvantages: Short life (about five years), easily damaged, no employment locally, little scope for self-help.

   Balloon plants can be recommended wherever the balloon skin is not likely to be damaged and where the temperature is even and high. One variant of the balloon plant is the channel-type digester with folia and sun.

2. *Fixed dome Plant:*

   A fixed-dome plant (Figure) consists of an enclosed digester with a fixed, non-movable gas space. The gas is stored in the upper part of the digester. When gas production commences, the slurry is displaced into the compensating tank. Gas pressure increases with the volume of gas stored; therefore the volume of the digester should not exceed 20 m^3. If there is little gas in the holder, the gas pressure is low.

   If the gas is required at constant pressure (e.g., for engines), a gas pressure regulator or a floating gasholder is required. Engines require a great deal of gas, and hence large gasholders. The gas pressure then becomes too high if there is no floating gasholder.

   Advantages: Low construction cost, no moving parts, no rusting steel parts, hence long life (20 years or more), underground construction, affording protection from winter cold and saving space, creates employment locally.

   Disadvantages: Plants often not gas light (porosity and cracks), gas pressure fluctuates substantially and is often very high, low digester temperatures.

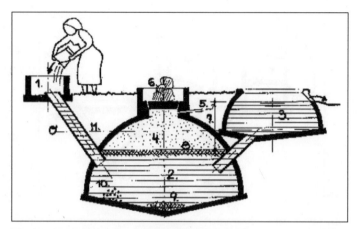

Fig. Fixed-dome plant: 1.Mixing tank with inlet pipe, 2. Digester,
3. Compensating and removal tank, 4. Gasholder, 5. Gaspipe, 6. Entry hatch,
with gas light seal and weighted, 7. Difference in level = gas pressure in cm WC,
8. Supernatant scum; broken up by varying level, 9. Accumulation of thick
sludge, 10. Accumulation of grit and stones,11. Zero line: filling height without
gas pressure.

Fixed-dome plants can be recommended only where construction can be supervised by experienced biogas technicians.

3. *Floating-Drum Plant:*

Floating-drum plant (Figure) consists of a digester and a moving gasholder. The gasholder floats either direct on the fermentation slurry or in a water jacket of its own. The gas collects in the gas drum, which thereby rises. If gas is drawn off, it falls again. The gas drum is prevented from tilting by a guide frame.

Advantages: Simple, easy operation, constant gas pressure, volume of stored gas visible directly, few mistakes in construction.

Disadvantages: High construction cost of floating-drum, many steel parts liable to corrosion, resulting in short life (up to 15 years; in tropical coastal regions about five years for the drum), and constant maintenance costs due to painting.

In spite of these disadvantages, floating-drum plants are always recommended. Water-jacket plants are universally applicable and easy to maintain. The drum won't stick, even if the substrate has high solids content.

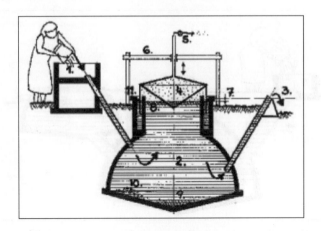

Fig: Floating-drum plant 1: Mixing tank with inlet pipe, 2: Digester, 3: Overflow on outlet pipe, 4: Gasholder with braces for breaking up surface scum, 5: Gas outlet with main cok, 6: Gas drum guide structure, 7: Difference in level = gas pressure in cm WC, 8: Floating scum in the case of fibrous feed material, 9: Accumulation of thick sludge, 10: Accumulation of grit and stones, 11: Water jacket with oil film.

Floating-drums made of glass-fibre reinforced plastic and highdensity polyethylene have been used successfully, but the construction cost is higher compared to the ones built with steel. Floating-drums made of wire-mesh-reinforced concrete are liable to hairline cracking and are intrinsically porous. They require a gas tight, elastic internal coating. PVC drums are unsuitable because they are not resistant to UV.

The floating gas drum can be replaced by a balloon above the digester. This reduces construction costs (channel type digester with folia), but in practice problems always arise with the attachment of the balloon at the edge. Such plants are still being tested under practical conditions.

(Source: Ludwig Sasse)

Nisarga-Runa Biogas Plant: The plant, developed by BARC, can use vegetable and fruit market waste, fruit and food processing industries waste, domestic and institutional kitchen waste, paper, garden waste, animal and abattoir waste etc. However, wastes such as coconut shells, egg shells, big bones, plastic/polythene, glass, metal, sand, silt, debris and building materials, wood, cloth/clothes, ropes, nylon threads, batteries, tyres/rubber, hazardous and chemical industries waste etc., cannot be treated and to be strictly avoided. Municipal authorities, therefore, have to ensure segregated waste before setting up the biogas plant.

Major components of the plants are a mixture/pulper (5 HP motor) for crushing the solid waste, pre-mix tank(s), pre-digester tank, air compressor, slow water heater, main digestion tank, gas delivery system, manure pits, tank for recycling for water and water pump and gas utilization system. The waste is homogenized in a mixer using water. This slurry enters the predigesting tank where aerobic thermophilic bacteria proliferate and convert part of this waste into organic acids. The slurry then enters the main tank where it undergoes mainly anaerobic degradation by a consortium of archaebacteria belonging to the Methanococcus group. These bacteria are naturally present in the alimentary canal of ruminant animals (cattle). They produce mainly methane from the cellulosic materials in the slurry. The undigested lignocellulosic and hemi-cellulosic materials then are passed on in the settling tank. After about a month, high quality manure can be dug out from the settling tanks. There is no odour to the manure at all. The organic contents are high and this can improve the quality of humus in soil, which in turn is responsible for the fertility.

As the gas is generated in the main tank, the dome is slowly lifted up. This gas is a mixture of methane (70–75 per cent), carbon dioxide (10–15 per cent) and water vapours (5–10 per cent). It is taken through GI pipeline to the lamp posts. Drains for condensed water vapour are provided on line. This gas burns with a blue flame and can be used for cooking. The gas generated in this plant is used for gas lights fitted around the plant. The potential use of this gas would be for cooking purposes. It can also be used to produce electricity in a dual fuel biogas–diesel engine. The manure generated is high quality and can be used for gardening and agricultural purposes. The plant can be installed at hotel premises, army/big establishment canteens (private/ government), residential schools/colleges, housing colonies, religious places/temple trusts, hospitals, hotels, sewage treatment plants etc. There are 5 such plants already in operation and about 5 others are proposed mostly in Maharashtra. The plant should be closer to source of waste being produced and the point of utilization of biogas power. The site should be free from underground cables; drainage pipes etc., and water table should be below 3 metres.

It is estimated that the life of the plant could be 20–30 years and payback period is estimated to be 4–5 years.

Table Cost Details of Nisarguna Biogas Plant

Treatment capacity (tonnes/day)	Installation cost	Monthly O & M (Rs in lakhs)	Methane generation charges (Rs)	Manure production (tonnes/day) (Cu m)
1	5-6	8000	100-120	0.1
2	9-10	12,000	200-240	0.2
4	20-22	15,000	400-480	0.3
5	28-30	22,000	500-600	0.5
6	65-70	50,000	1000-1200	2.5

Note: This is an approximate cost for biogas generation plant and may increase by 10–20 per cent depending on location, site, specific parameters, cost of materials, labour cost etc., in different states/ cities. Cost of additional infrastructure like office space, toilets, security, compound wall, flood control measures etc., and for power generation will be extra, if required (Asnani 2006).

Zero Waste Approach

Zero Waste programmes are the fastest and most cost effective ways that local governments can utilize to reduce climate change, protect health, create green jobs, and promote local sustainability. There are three overarching goals needed for sustainable resource management. 1. Producer responsibility: industrial production and design. 2. Community responsibility: consumption, discard use and disposal. 3. Political responsibility to bring both community and industrial responsibility together in a harmonious whole.

Zero Waste is a critical step to other necessary steps in the efforts to protect health, improve equity and reach sustainability. Zero Waste can be linked to sustainable agriculture, architecture, energy, industrial, economic and community development. Every single person in the world makes waste and as such is part of a non-sustainable society. However, everyone could be engaged in the necessary shift towards a sustainable society.

The only peer-reviewed internationally accepted *definition of Zero Waste* is that adopted by the Zero Waste International Alliance:

"Zero Waste is a goal that is ethical, economical, and efficient and visionary, to guide people in changing their lifestyles and practices to emulate sustainable natural cycles, where all discarded materials are designed to become resources for others to use. Zero Waste means designing and managing products and processes to systematically avoid and eliminate the volume and toxicity of waste and materials, conserve and recover all resources, and not burn or bury them. Implementing Zero Waste will eliminate all discharges to land, water or air that are a threat to planetary, human, animal or plant health." "If a product can't be reused, repaired, rebuilt, refurbished, refinished, resold, recycled or composted, then it should be restricted, redesigned, or removed from production."

Principles and Practical steps towards Zero Waste:

Demand decision makers manage resources not waste. Existing incinerators must be closed down and no new ones built. Landfill practices must be reformed to prevent all pollution of air and water including pre-processing all residues at landfills before burial to stabilize the organic fraction and prevent methane generation.

Landfills are a major source of greenhouse gases (particularly methane, which warms the atmosphere 23-72 times more quickly than carbon dioxide) as well as groundwater contamination. Incinerators and other burning and thermal treatment technologies such as biomass burners, gasification, pyrolysis, plasma arc, cement kilns and power plants using waste as fuel, are a direct and indirect source of greenhouse gases to the atmosphere and turn resources that should be reduced or recovered into toxic ashes that need to be disposed of safely. Neither landfills nor incinerators are an appropriate response to the challenge of peak oil, which will make any new incinerator impractical within its lifetime, as embedded energy and oil within products will become too costly to replace.

More energy can be saved, and global warming impacts decreased, by reducing waste, reusing products, recycling and composting than can be produced from burning discards or recovering landfill gases. Communities should fight any effort to introduce new incinerators, in any guise, and replace existing landfills and incinerators, with Zero Waste policies and programs, including EPR, resource recovery parks, reuse, recycling and composting facilities.

Integrated Solid Waste Management

Various concepts have been developed over the years to provide the basis for improving the management of solid waste in developing cities. Among these, the concept of Integrated Solid Waste Management (ISWM) provides a framework of developing a sustainable MSW management system, which has been very successful in various industrialized countries.

ISWM is defined as the selection and application of appropriate techniques, technologies and management programmes to achieve specific waste management objectives and goals (UNEP, 2001b). In the Integrated SWM, all aspects – technical, legislative, economic, social/cultural, institutional and environmental – along with community participation are considered, and the link between the stakeholders and MSWM aspects to develop an integrated approach is provided. Comprehensive in nature for managing MSW, ISWM is particularly suited to developing countries where SWM services are of poor quality and costs are high often with no effective means of recovering them.

UNEP (1996) laid out a series of questions to be answered when evaluating technologies and policies in the context of an Integrated MSW system (Zerbock 2003):

Is the proposed technology likely to accomplish its goals given the financial and human resources available?

What option is the most cost-effective in financial terms?

What are the environmental costs and benefits?

Is the project feasible given administrative capabilities?

Is the practice appropriate in the current social and cultural environment?

What sectors of society are likely to be impacted and in what way; are these impacts consistent with overall societal goals?

The answers may not always be immediately clear, but the process of research and evaluation of these criteria lends insight to the suitability of specific solutions to the situation as a whole. The conceptual framework and Externalities of ISWM are shown in Figrues (UNCHS 2000).

In an integrated approach, all elements of MSWM, namely, waste generation, source segregation, collection and transport of waste, recycling, resource recovery, existing disposal systems and their upgradation, and people's participation would be analysed and the problems identified.

The actual integration takes place in different operations and at various stages (Beukering *et. al.*, 1999) leading to sustainability such as: (a) using a range of collection and treatment options which include prevention, recycling, energy recovery and environmentally sound landfilling of solid waste; (b) involving all the stakeholders - waste generators (households, industries and agriculture), waste processors (formal and informal recyclers), NGOs and CBOs, government institutions (waste managers and urban planners), and financing agencies, and (c) relating the waste system with other systems concerning product design at the manufacturing stage which significantly impacts the recyclability of the product after its consumption (ARRPET 2004).

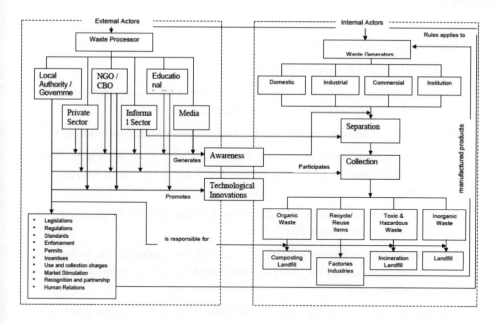

Conceptual framework of ISWM

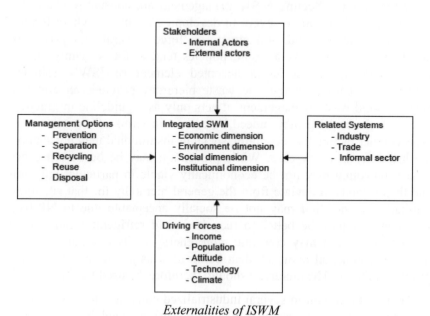

Externalities of ISWM
(Source: UNCHS 2000)

Basic Requirements of ISWM:

An ISWM should at least be (CREED, 2000):

(a) *Responsive:* balancing the local needs with wider institutional, technical and environmental constraints;

(b) *Equitable:* addressing the needs of all sectors of the community;

(c) *Empowering:* motivating and organizing local people to help them find solutions to problems at the local level by using indigenous ideas;

(d) *Decentralized:* into local authorities responsible for all problems of MSW over the whole city;

(e) Diverse: experimenting with a range of technologies or processes rather than attempting to find one single solution applicable to all situations; and

(f) Flexible: to allow developments and modifications in approaches and activities.

In practice, it is difficult to balance all these aspects at the same time, since the factors affecting MSW management are constantly changing. These problems are more severe in developing countries, where limited resources, budgets and infrastructure, force several comprises in developing a system. As a result, policies tend to focus mainly on the waste hierarchy which is an accepted element of ISWM (already explained in Ch.1). Though the waste hierarchy provides an effective basis of solid waste management, it acts only as a guideline in practice. Although the ranking may indeed be correct in terms of environmental aspects other factors like economy, social or institutional may not support the hierarchy in every case. While prevention may be best from every aspect, recycling may not be economically viable in particular situation and thus it tends to deviate from the general hierarchy for that situation. Similarly, incineration may not be socially acceptable due to NIMBY sentiments. It may be better to recycle an old refrigerator rather than reuse it because it may consume more energy in its old state creating more environmental damages than being used as a raw material in the primary industry. The hierarchy should, therefore, be used flexibly.

Though successful in several industrialized countries, it is not easy to plan and execute a successful ISWM framework in developing countries, since factors such as population growth, income levels, education and local awareness, that directly affect the waste generation change constantly in these countries. Further, lack of adequate finances, technical facilities and skilled manpower, lack of stringent laws and regulations and people's ignorance weaken the System. The policy makers, municipal authorities and the stakeholders in these countries must focus on making the existing system sustainable in order to achieve an ISWM (ARRPET 2004).

General Recommendations for ISWM development:

The practices should encompass the following aspects along with the basic requirements of an ISWM:

1. Public participation in the collection, segregation and disposal of garbage by forming eco-clubs or community based organizations.

2. Involvement of NGOs to work for various community-based solid waste management programmes that may improve social awareness, emphasize participation, create job opportunity for the needy, encourage small scale technology like composting and remove gender inequality.

3. Public-private partnerships leading to the privatization of some aspects of garbage collection, recovery and disposal. Relationships can be formalized by establishing MOUs, legal agreements between NGOs and user groups, and by ensuring enforcement of such agreements.

4. Even informal linkages should clearly define roles and responsibilities of all stakeholders. Provisions aimed at administrative restructuring of the local municipalities to enable them discharge their responsibilities more efficiently should be initiated. Improvement can be brought about by (i) motivating the municipal staff and improving their capacity by training, (ii) monitoring and supervision of waste management practices by the authorities, and (iii) introducing structural changes within the administration aimed at decentralizing authority and responsibilities.

Door to Door Refuse/Garbage Collection System in Surat Municipal Corporation – A project in Best Practice

Surat city is located on Mumbai-Ahmedabad rail corridor, and has a population of more than 4 million. The city is having a comprehensive system of MSW management and has pioneered many concepts in the field of Municipal Solid Waste Management. The waste is collected from the point of generation and is sent to the disposal site in a systematic manner.

On implementation of Door to Door garbage collection system, every citizen has been holding waste generated temporarily in dustbin till the garbage collection vehicle arrives. This has improved the overall scenario and surrounding environment compared to early days.

The city is provided with containers & dustbins to collect the waste being generated. Ward wise nuisance spots are identified from where MSW is collected for its disposal to final disposal site. The only problems noticed are (i) animals straying at each dustbin and nuisance spot, and

(ii) ragpickers collecting materials that can be recycled creating filthy appearance.

The main objective of this best practice is to comply with MSW Rules-2000. According to these Rules, the garbage has to be collected at doorstep of household; this has forced the residents to habituate themselves to store their garbage in a bin till vehicles of Door to Door collection system reach them. It was first started as Pilot project in one ward each of all zones utilising open tractors. Based on the results of the Pilot project, it was extended to three of seven zones through tender process. Three agencies were chosen and work was entrusted to them for a period up to February 2011. They deployed new vehicles of various categories (i.e. HGV/ MGV/LGV) with closed tops painted with green colour.

The other objectives are to improve hygiene of the city, and the environment by regular collection of waste from every house/shop on daily basis, reducing the menace of stray animals as well as the discharge of bad odour.

Strategies adopted:

The strategies adopted for the success of this system are the following: Selection of the right kind of vehicle based on width of road; Restricting the number of units in each route to 1,000 to 3,000; Strengthening the existing system of garbage collection; Providing with uniforms & identity cards to Drivers and "Swachchhta Mitra."; Allowing concession period of this project to seven years keeping in mind the useful life of vehicle; Equipping all garbage vehicles with proper alarm system to go to every door step regularly at scheduled time; Facilitating second shift for collecting waste from Commercial units during 4.00 pm to 11.00 pm daily in all zones while maintaining first shift collection timing, 7.00 am to 1.00 pm, for residential zone; Operating this system all the days in a year; Creating Public awareness on garbage management through campaign by the Contractor; Centralized complaint management system at Head office at Mugalsarai and Contractor's office with modern communication facilities; and Provision for segregated waste collection (Dry and Wet).

Roles of the Partners:

Handling of Door to Door Refuse/Garbage System is entirely carried out by Surat Municipal Corporation through its own budget, and there is no involvement of any financial collaboration or aid. However, initial investment on the procurement of vehicle is made by the contractors.

While inviting tenders, care is taken to include the clause for conducting awareness campaign for the Door to Door garbage collection system, because the Contractor to whom the work is entrusted is made responsible for the campaign. As the Contracting agency is paid for the work executed on weight basis, the agency is constantly aiming for improvement in the present practice of Door to Door garbage collection System. For example, conducting survey at regular intervals, make changes in TPM schedule for maximum coverage etc.

Procedure:

Normally Door to Door collection system vehicles reach to the concern ward office early in the morning every day to get confirmation regarding the route to be taken and receive information on complaints, if any. At each ward office, there is provision to make a complaint regarding noncoverage of any area / unit. Phone numbers of the supervisory staff is communicated to the area in order to reach in case of a problem. Frequent meetings with ward office are also held to effect improvements to the collection system.

Payment for the service:

Payments are made to the contractor on weight basis. Before arriving at the present practice, various options like lump sum base to cover city as a whole, to cover zone as a whole and payment on number of units covered were discussed.

As payment is made on weight basis, constant monitoring is required to avoid the malpractice in the collection of waste from the area which is not under the scope of agency. Public awareness plays main role to collect the waste in a segregated manner; hence, public awareness campaigns need to be taken up on mass-scale.

Cost of the Project:

The system of Door to Door garbage collection involves huge capital investment in procurement of vehicles for collection of garbage. It also requires proper manpower to run the system effectively and efficiently. Payment by the Municipal Corporation to the contracting agency for the garbage collected from doorsteps of residential and commercial units on weight basis has become viable for agency to run the system effectively. Municipal Corporation is incurring Rupees 180 million per annum for the collection of 625 Mt of garbage generated per day. The rate of payment to the contractor for first year of collection was kept as Rupees 630 per Mt; subsequently, an annual increase of 5% in the rate is given to compensate the inflation.

User charges are levied from 2007-08 which recover the cost of Door to Door collection system to some extent.

The contract period is fixed at 7 years.

Outcome of the Practice:

The Door to door collection system has realized a number of benefits as under:

Improvement in the over all environment because of public consciousness and habit of keeping waste in domestic bins;

Timely daily collection of waste from every house/shop;

Reduction in the number of stray animals and the odour around containers' spots;

Avoiding multiple handling of waste;

Reduction in number of containers;

Engaging extra sweepers/workers for carrying out sanitation work of new developing area in most effective manner;

Removal of old collection system through open tractors;

Improvement in the perception of citizens for environment cleanliness;

Improvement in the environment around the community containers;

Significant change in th people regarding health and hygiene parameters;

Reduction in the budget required for lifting and maintenance of containers due to decrease in the number of containers; and

Overall appreciation of the programme implemented by Surat Municipal Corporation by the citizens of Surat.

Replication of the System

This system, started in April 2004, has been running successfully and has become very unique. Representatives from various urban local bodies and city managers visited Surat to see the innovative system to replicate the same in their regions.

Representatives are allowed to study the system with site visits. Help is offered for documenting the project.

Impact: This best practice is discussed in several National seminars/work shops related to MSW management highlighting the unique features. There has been general appreciation in all these forums.

Centralised Co-digestion of Multiple substrates (CAD): Example of Denmark

Digestion of only manure yields a low biogas production due to the composition. The dry matter content in pig and cattle manure is usually ranges from 2 to 5 % with the vast majority of this dry matter being plant fibres. Co-digestion of manure and other organic feedstocks solves many practical problems. The high water content in manure ensures that the fermentation broth is diluted sufficiently to allow efficient mixing of substrate and microorganisms. Nutrient deficiency in single substrates is counteracted when co-digesting. Nitrogen, carbon, sulphur, and phosphorous have to be present in the blend in optimal proportions. Trace metals have to be present in adequate amounts in order for the microbial processes to perform satisfactorily.

In Denmark, a significant part of the produced amount of biogas arises from *manure co-digested with industrial organic waste*. Two types of biogas plants are in operation; the decentralized farm-scale plants treating manure from a single farm or a few farms, and the centralised co-digestion plants, normally operated as cooperatives or as private limited companies. A larger number of farmers supply manure to the centralised plant. Moreover, significant amounts of suitable organic residues are added to the process in order to enhance the biogas yield and thus strengthen the economics. The economic performance of the centralised biogas plants to a large extent dependent on the availability of high quality organic residues.

At present, 21 centralised co-digestion plants and approximately 60 farm-scale plants are in operation in Denmark. Together, they treat 1.5 million tonnes of manure and 0,3 million tonnes of industrial organic waste annually.

The Danish Centralised Co-digestion concept involves the agricultural sector, the energy production-end distribution sector, the food industry and agro-industry sector. The result is an optimized and integrated biological production system, *a bio refinery*. The centralised biogas plant concept is depicted in Figure.

The high-value products from the biogas plant constitute organic fertiliser and biogas.

Some of the features of the centralised Danish concept are that nutrients contained in pig and cattle manure produced by agricultural activities can be re-distributed among crop cultivators. i.e., farmers having many livestock units and too few hectares of farmland to apply the manure on according to Danish law, can re-distribute nutrients (nitrogen and phosphorous) via the centralised biogas plant to crop farmers having plenty of land, but no livestock units. Moreover, the anaerobic digestion process effectively reduces the offensive odor traditionally associated with raw manure and it also eliminates pathogens, weeds, and deceases.

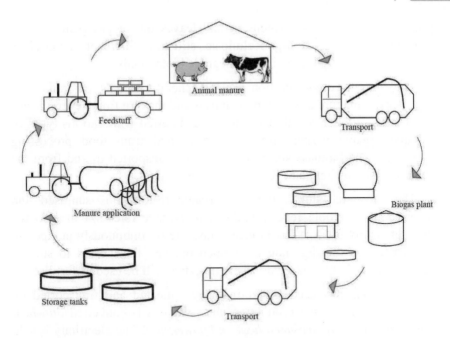

The composition of the co-digestion substrate consisting of manure and other organic substrates is set by Danish legislation. Minimum 75 % of the biomass has to be manure. A sector analysis performed by Aarhus School of Business in 2005, estimated the global market for manure handling to be worth DKK 740 billion (~EUR 100 billion). The simple monosubstrate configuration based on manure is not economically viable under Danish conditions.

Organic substrates resulting in higher biogas yield *have* to be added to the biogas reactor in order to boost the gas production and ensure an economically feasible process. These additional substrates can, for instance, be the organic fraction of source-sorted municipal waste (OFMSW), organic industrial by-products (fats, oils, spirits etc.), and energy crops (various silages, whole crops etc.).

A recent Danish socio-economic study investigated different scenarios, where centralised biogas plants were operated *without* addition of industrial organic resources. The aim was to analyse the feasibility of digesting solely farmyard manure. It was concluded that a combination of separation of the manure at the farms and centralised physico-chemical pre-treatment of the lignocellulosic fibre fraction could lead to a doubling of the practical biogas yield. The pretreatment technologies included wet-oxidation and pressure cooking. As of today, a large portion of the biogas potential in manure is recalcitrant and leaves the biogas plant via the effluent; the complex lignocellulosic structure of plant material is difficult, practically impossible, to degrade biologically.

The centralised co-digestion plant is generally situated centrally, in a high density manure area. Animal slurries and manure from several farms around are supplied to the plant, to be co-digested with various types of suitable organic wastes from agriculture and from food processing industries. The biomass substrate is usually transported to and from the centralized digestion plant (CAD) in vacuum trucks.

The substrate (slurry, manure, organic residues) is sanitised and digested in anaerobic reactor tanks. The average retention time in the digesters is of 15 days. The biomass substrate is continuously pumped in the reactor, as the digestate is pumped out and transported to storage tanks located next to the fields where digestate will be used as fertiliser.

The biogas produced is continuously collected and transported by pipelines to the energy production unit, where it is converted *into heat and electricity in a combined heat and power unit.* The electricity is sold to the grid and the heat is used at the biogas plant as process heat, while the main part is sold to heat consumers (housing or industry).

Benefits: The centralised co-digestion is a multifunctional technology providing renewable energy and benefits for the agriculture and environment. The environmental friendly renewable energy production is used to substitute fossil fuels and thereby increases energy security, reducing dependence on imported fossil fuels. The codigestion of manure helps the farming sector to handle and to redistribute the surplus of manure in other areas, where it could be used in environmentally friendly ways. Co-digestion provides an economically attractive and sustainable management of organic wastes and improves the fertilizer value of the animal manure and slurries. CO_2 emissions and losses of nitrogen to water systems are reduced and the establishment and operation of the biogas plant leads to creation of new local jobs and supports the rural economies.

Digested manure is a valuable fertilizer:

In Denmark, digestion of slurry is recognized to contribute to a better utilization of the slurry as fertilizer as documented from a large number of field trials. It is also evident that digestion reduces the smell problems after spreading the slurry.

The most important advantages are:

Energy sector: Energy production and CO_2 neutral

Agriculture sector: Improved utilisation of nitrogen from animal manure, balanced phosphorus/potassium ratio in slurry, homogeneous and light-fluid slurry, reduced transportation of slurry, possible to get large amounts of slurry with a full declaration of contents, and slurry free from weed seeds and disease germs.

The environment: reduced nitrogen leaching, reduced odour problems, reduced greenhouse gas emissions, and controlled recycling of waste.

Biogas plants contribute to a better utilization of nutrients in the agriculture.
Photo courtesy: Torkild Birkmose, DAAS

The physical and chemical process taking place in the biogas plant changes the fertilizing effect of the slurry in the field. The high content of ammonium is advantageous to the crops. In other words, it is often possible to replace nitrogen from commercial fertiliser by digested slurry and thus save money.

The thin, low-viscosity digested slurry seeps relatively quickly into the soil. This reduces the normally very high risk of ammonia volatilization. Trials have shown that the ammonia evaporation from surface applied digested slurry actually is lower than from surface applied pig slurry.

(Source: PROBIOGAS 2007)

Glossary

Aerobic: A biochemical process occurring in the presence of oxygen.

Acid gases: A general term used to cover sulfur dioxide, hydrogen chloride, hydrogen fluoride, and nitrogen oxides.

Air Classification: A process in which a stream of air is used to separate mixed material according to the size, density, and aerodynamic drag of the pieces

Algal Bloom: Increase of algae (usually aquatic plants) in surface waters. Algal blooms are associated with nutrient-rich run-off from composting facilities or landfills.

Anaerobic: A biochemical process occurring in the absence of oxygen.

Ash: Non-combustible residue resulting from a thermal process, classified as bottom ash (heavy and falls to bottom of combustion chamber).

Autoclaving: Sterilization via a pressurized, high-temperature steam process.

Baghouse: An emission control device that consists of an array of fabric filters through which gases pass in an incinerator to prevent particles from passing into the atmosphere.

Baler: A machine used to compress recyclables into bundles to reduce volume. Balers are often used on newspaper, plastics, and corrugated cardboard.

Basel Convention: An international agreement on the control of trans boundary movements of hazardous wastes and their disposal, drawn up in March 1989 in Basel, Switzerland

Biodegradable material: Waste material which is capable of being broken down by microorganisms into simple, stable compounds such as carbon dioxide and water. Most organic wastes, such as food wastes and paper, are biodegradable.

Biodiesel: Chemically, they are called 'methyl esters of fatty acids'; can be used as a fuel in a diesel engine.

Boiler Ash: Particulate matter deposited in the heat recovery system by the flue gas.

Buffer Zone: A buffer zone established between a composting facility and neighboring residents to minimize odor problems.

Bulking Agent: A material used to add volume to another material to make it more porous to air flow. For example, water treatment sludge may act as a bulking agent when mixed with municipal solid waste.

Calorific Value: The heat value liberated when a unit mass of substance is burned as fuel under standard conditions

Chemocar: A special vehicle for the collection of toxic and hazardous wastes from residences, shops, and institutions

Class A Solid Waste Facility: A commercial solid waste facility which handles an aggregate of between ten thousand (10,000) and thirty thousand (30,000) tons of solid waste per month.

Class B Solid Waste Facility: A commercial solid waste facility which receives, or is expected to receive, an average daily quantity of mixed solid waste equal to or exceeding one hundred (100) tons each working day, or serves, or is expected to serve a population equal to or exceeding forty thousand (40,000) persons, but which does not receive or is expected to receive solid waste exceeding an aggregate of ten thousand (10,000) tons per month.

Class C Solid Waste Facility: A commercial solid waste facility which receives, or is expected to receive, an average daily quantity of mixed solid waste of less than one hundred (100) tons each working day, and serves, or is expected to serve a population of less than forty thousand (40,000) persons.

Class D Solid Waste Facility: Any commercial solid waste facility for the disposal of only construction/demolition waste and does not include the legitimate beneficial reuse of clean waste concrete/masonry substances for the purpose of structural fill or road base material.

Co-incineration: The combustion of more than one category of fuel in the same plant.

Co-composting: Simultaneous composting of two or more diverse waste streams.

Cogeneration: Production of both electricity and heat from one facility, from the same fuel source.

Combustion: Burning of materials in an incinerator in MSWM

Commercial Waste: Waste materials originating in wholesale, retail, institutional, or service establishments such as office buildings, stores, markets, theaters, hotels and warehouses.

Compactor: Power-driven device used to compress materials to a smaller volume.

Composite liner: A liner system for a land-fill consisting of an engineered soil layer and a synthetic sheet of material.

Compost: The relatively stable decomposed organic material resulting from the composting process also referred to as humus.

Composting: Biological decomposition of solid organic materials by bacteria, fungi, and other organisms under aerobic conditions into a soil-like product.

Cullet: Clean, generally color-sorted, crushed glass used to make new glass produced.

Curbside Collection: Programmes where recyclable materials are collected at the curb, often from special containers, to be brought to various processing facilities.

Decomposition: Breaking down into component parts or basic elements.

Detinning: Recovering tin from "tin" cans by a chemical process which makes the remaining steel more easily recycled.

Dioxins: Heterocyclic hydrocarbons that occur as toxic impurities, especially in herbicides or when trash is burned.

Disposal: The final handling of solid waste following collection, processing, or incineration. Disposal most often means placement of wastes in a dump or a landfill.

District heating scheme: The heating of multiple community premises from a dedicated heat stream

Diversion rate: The amount of waste material diverted for recycling, composting, or reuse and away from landfilling or incineration.

E-Cycling (electronics recycling): The reuse or recycling of end-of-life electronic materials.

Emissions: gases released into the atmosphere.

Energy recovery: The process of extracting useful energy from waste, typically from the heat produced by incineration or via methane gas from landfills.

Environmental impact assessment (EIA): An evaluation designed to identify and predict the impact of an action or a project on the environment and human health and well-being. Can include risk assessment as a component, along with economic and land use assessment.

Environmental risk assessment (EnRA): An evaluation of the interactions of agents, humans, and ecological resources - typically evaluating the probabilities and magnitudes of harm that could come from environmental contaminants.

Ester: An organic compound of the form, R-O-R (R is short hand for carbon and hydrogen compound in organic chemistry).

Esterification: A chemical reaction in which an alcohol and an acid form an ester

Farm Dump: Refers to the placement of farm waste such as old equipment, household garbage, fence posts and wire, etc., on the farmer's property in an open pile.

Ferrous Metals: Metals that are derived from iron. They can be removed using large magnets at separation facilities.

Flow Control: A legal or economic means by which waste is directed to particular destinations.

Fluidized-bed incinerator: A type of incinerator in which the stoker grate is replaced by a bed of limestone or sand that can withstand high temperatures. The heating of the bed and the high air velocities used cause the bed to bubble, which gives rise to the term fluidized.

Fly ash: The highly toxic particulate matter captured from the flue gas of an incinerator by the air pollution control system, see Ash.

Garbage: A general term for discarded products.

Gasification: The process by which a solid or liquid feedstock is converted to a gaseous product by partial oxidation under the application of heat

Gob: That portion of coal which is difficult to utilize in a conventional combustion chamber;

Ground Water: Water beneath the earth's surface that fills underground pockets (known as aquifers) and moves between soil particles and rock, supplying wells and springs.

Hammermill: A type of crusher or shredder used to break up waste materials into smaller pieces.

Hazardous waste: Waste that is reactive, toxic, corrosive, or otherwise dangerous to living things and/or the environment

Heavy metals: Metals of high density, such as mercury, lead, and cadmium, that are toxic to living organisms.

Household hazardous waste: Products used in residences, such as paints, cleaning compounds, that are toxic to living organisms and/or the environment

Humus: Organic materials resulting from decay of plant or animal matter (also referred to as compost).

Incineration: The process of burning (combustion) solid waste under controlled conditions to reduce its weight and volume, and often to produce energy

Incinerator: An enclosed device using controlled combustion to thermally breakdown solid waste, including refuse-derived fuel, to an ash residue.

Incinerator Ash: The remnants of solid waste after combustion, including non-combustibles (e.g., metals and soot)

Industrial Waste: Materials discarded from industrial units or derived from manufacturing processes.

Inorganic Waste: Waste composed of matter other than plant or animal (i.e., contains no carbon).

Integrated solid waste management: Coordinated use of a set of waste management methods, each of which can play a role in an overall plan management and disposal.

In-vessel composting: Composting in an enclosed vessel or drum with a controlled internal environment, mechanical mixing, and aeration.

Itinerant waste buyer: A person who moves around the streets buying reusable and recyclable materials.

Landfill: Any solid waste facility for the disposal of solid waste on or in the land for the purpose of permanent disposal.

Leachate: Liquid that has percolated through solid waste or another medium and has extracted, dissolved or suspended materials from it, which may include potentially harmful materials.

Leachate pond: A pond or tank constructed at a landfill to receive the leachate from the area. Usually the pond is designed to provide some treatment of the leachate, by allowing settlement of solids or by aeration to promote biological processes.

Liner: A protective layer, made of soil and/or synthetic materials, installed along the bottom and sides of a landfill to prevent or reduce the flow of leachate into the environment.

Mandatory Recycling: Programs which by law require consumers to separate trash so that some or all recyclable materials are not burned or dumped in landfills.

Market waste: Primarily organic waste, such as leaves, skins, and unsold food, discarded at or near food markets.

Mass-burn incinerator: A type of incinerator in which solid waste is burned without prior sorting or processing.

Materials Recovery Facility (MRF): Any solid waste facility at which source-separated materials or materials recovered through a mixed waste processing facility are manually or mechanically shredded or separated for purposes of reuse and recycling.

Mechanical Separation: The separation of waste into various components using mechanical means, such as cyclones, trommels, and screens.

Methane: An odorless, colorless, flammable and explosive gas produced by municipal solid waste undergoing anaerobic decomposition at landfills.

Microorganisms: Microscopically small living organisms that digest decomposable materials through metabolic activity. Microorganisms are active in the composting process.

Monofill: A sanitary landfill intended for one type of waste.

Municipal Solid Waste (MSW): Includes nonhazardous waste generated in households, commercial establishments, institutions and light industrial process wastes, agricultural wastes, mining waste and sewage sludge.

MSW Composting: (Municipal Solid Waste Composting) The controlled degradation of municipal solid waste including after some form of preprocessing to remove non-compostable inorganic materials.

Mulch: Ground or mixed yard wastes placed around plants to prevent evaporation of moisture and freezing of roots and to nourish the soil.

Night soil: Human excreta.

Open dump: An unplanned "landfill" that is typically no leachate control, no access control, no cover, no management, and many waste pickers.

Organic waste: Waste containing carbon, including paper, plastics, wood, food wastes, and yard wastes; the term is used to material directly derived from plant or animal sources which can generally be decomposed by microorganisms.

Pathogen: An organism capable of causing disease.

Pollution: The contamination of soil, water, or the atmosphere by the discharge of waste or other offensive materials.

Pollution Control Residuals: End products of the thermal process which includes hot combustion gases composed primarily of nitrogen, carbon dioxide, water vapor (flue gas) and noncombustible residue (ash).

Primary material: A commercial material produced from virgin materials used for manufacturing basic products. Examples include wood pulp, iron ore, and silica sand.

Putrescible: Organic matter partially decomposed by microorganisms and producing a foul smell.

Pyrolysis: Chemical decomposition of a substance at high temperatures in the absence of oxygen, resulting in various hydrocarbon gases and carbon-like residue.

Rag picker: See waste picker.

Recyclables: Materials that still have useful physical or chemical properties after serving their original purpose and that can, therefore, be reused or remanufactured into additional products.

Recycling: The process by which recovered products are transformed into new products.

Refuse: A term often used interchangeably with solid waste.

Refuse-derived fuel (RDF): Fuel produced from MSW that has undergone processing. Processing can include separation of recyclables and noncombustible materials, shredding, size reduction, and pelletizing

Resource Recovery: The extraction and utilization of materials and energy from the waste stream. The term is sometimes used synonymously with energy recovery.

Retention Basin: An area designed to retain run-off and prevent erosion and pollution.

Reuse: The use of a product more than once in its same form for the same purpose.

Rubbish: A general term for solid waste.

Sanitary landfill: An engineered method of disposing of solid waste on land that meets most of the standard specifications.

Scrubber: Emission control device in an incinerator, used primarily to control acid gases, but also to remove some heavy metals.

Secondary material: A material recovered from post-consumer wastes for use in place of a primary material in manufacturing a product.

Secure landfill: A disposal facility designed to permanently isolate wastes from the environment.

Septage: Sludge removed from a septic tank (a chamber that holds human excreta).

Sewage sludge: Semi-liquid residue that settles to the bottom of canals and pipes carrying sewage or industrial waste waters, or in the bottom of tanks used in treating waste waters.

Site remediation: Treatment of a contaminated site by removing contaminated solids or liquids or treating them on-site.

Source separation: Setting aside of compostable and recyclable materials from the waste stream before they are collected with other MSW, to facilitate reuse, recycling, and composting.

Soil Liner: Landfill liner composed of compacted soil used for the containment of leachate.

Source Reduction: The design, manufacture, acquisition and reuse of materials so as to minimize the quantity and/or toxicity of waste produced. Source reduction prevents waste either by redesigning products or by otherwise changing societal patterns of consumption, use and waste generation.

Special Waste: Refers to items that require special or separate handling, such as household hazardous wastes, bulky wastes, tires and used oil.

Tipping Floor: Unloading area for vehicles that deliver municipal solid waste to a transfer station or municipal waste combustion facility.

Tire Derived Fuel (TDF): A tire that is shredded and processed into a rubber chip ranging in size from 1 to 4 inches. Depending on the requirements of the users, TDF may also be processed energy content ranging from 14,000 to 15,500 BTU per pound.

Transesterification: A chemical reaction that turns an alcohol and an ester to a different alcohol and ester

Tub Grinder: Machine to grind or chip wood wastes for mulching, composting or size reduction.

Vectors: Organisms that carry disease causing pathogens. At landfills rodents, flies, and birds are the main vectors that spread pathogens beyond the landfill site.

Virgin Oil: Pure vegetable or animal oil.

Virgin materials: Any basic material for industrial processes that has not previously been used, for example, wood-pulp trees, iron ore, crude oil, bauxite.

Waste Oil: Animal or vegetable oil obtained from restaurants, hotel kitchens and fast food centres.

Waste picker: A person who picks out recyclables from mixed waste wherever it may be temporarily accessible or disposed of (also called rag picker).

Waste reduction: Reducing the amount of waste that is produced initially.

Waste stream: A term describing the total flow of solid waste from homes, businesses, institutions and manufacturing plants that must be recycled burned or disposed of in landfills; or any segment thereof, such as the "residential waste stream" or the "recyclable waste stream.

Waste-to-energy (WTE) plant: A facility that uses solid waste materials (processed or raw) to produce energy. WTE plants include incinerators that produce steam for heating or generate electricity; they also include facilities that convert landfill gas to electricity.

White Goods: Large household appliances such as refrigerators, stoves, air conditioners and washing machines.

Water table: Level below the earth's surface at which the ground becomes saturated with water.

Wetland: An area that is regularly wet or flooded and has a water table that stands at or above the land surface for at least part of the year.

Windrow: An elongated pile of aerobically composting materials.

Worm castings: The material produced from the digestive tracts of worms as they live in earth or compost piles. The castings are rich in nitrates, potassium, phosphorous, calcium, and magnesium.

Worm culture: A relatively cool, aerobic composting process that uses worms and microorganisms; also known as vermin culture.

Yard waste: Leaves, grass clippings, prunings and other natural organic matter discarded from yards and gardens. Yard wastes may also include stumps and brush, but these materials are not normally handled at composting facilities.

References

- Sources   for   Glossary: www.state.wv.us/swmb/07_plan_pdf./Appendix_G.pdf)

  UNEP-IETC, *International Source Book on Environmentally Sound Technologies for Municipal Solid Waste Management*, Technical Publication Series no.6, Osaka/Shiga: UNEP International Environmental Technology Centre, 1996 (pp. 421-427)

- Abbott, John et al (2003): Environmental and Health risks associated with the use of processed Incinerator bottom-ash in road construction, BREWEB

- Ahmed, R., van de Klundert, A, and Lardinois, 1. Rubber Waste, Urban Solid Waste Series, Vol. 3. Amsterdam,

- Agarwal, A. et al (2005): Municipal solid waste recycling and associated markets in Delhi, Ind. Journal of Resources, Conservation and Recycling 44 (1), 73–90

- Agarwal V.S, (2005): Solid Waste management – Case Study of Nagpur.

- Ahsan, N (1999): Solid waste management plan for Indian megacities. Indian Journal of Environmental Protection 19 (2), 90–95.

- Agunwamba, J. C. 1998. "Solid waste management in Nigeria: problems and issues." Environmental Management. 22(6): 849-856.

- Ambulkar, A.R., Shekdar, A.V (2004): Prospects of biomethanation technology in Indian context: a pragmatic approach, Journal of Resources, Conservation and Recycling 40 (2), 111–128.

- Anaerobic   Digestion: http://en.wikipedia.org/wiki/anaerobic_digestion

- Anaerobic Digestion, www.wasteresearch.co.uk,

- An introduction to anaerobic digestion, www.anaerobic-digestion.com,

- Anaerobic digestion, www.waste.nl,

- Anaerobic Digestion Page, www.remade.org.uk,

- Anaerobic digestion, www.biotank.co.uk,

- Anaerobic digestion feedstock classification, www.wisbiorefine.org,

- Anaerobic co-digestion of sewage sludge and rice straw, www.bvsde.ops-oms.org,

- Anaerobic digestion of classified municipal solid wastes, www.seas.ucla.edu,

- Anaerobic digestion, www.energy.ca.gov,

- Anaerobic Lagoons for Storage/Treatment of Livestock Manure, www.missouri.edu,

- Anaerobic Digestion (Methane Recovery) Works], www.eere.energy.gov,

- Anaerobic digestion briefing sheet, www.foe.co.uk,

- www.anaerobic-digestion.com,

- Anaerobic digestion, What is; at www-sop.inria.fr,

- Fact sheet; www.waste.nl, retrieved 19.08.07

- Anikwe, M.A.N., Nwobodo, K.C.A., (2002): Long term effect of MSW disposal on soil properties and productivity of sites used for urban agriculture in Abakaliki, Nigeria. Bioresource Technolog 83, 241-250

- Arlosoroff, S (1982):'WB/UNDP Integrated Resource recovery Project: recycling of waste in developing countries' Ed. By k. Curi, Plenum Press, New York, N.Y.

- ARRPET (Asian Regional Research Program in Environmental Technology) (2004): 'Municipal Solid Waste Management in Asia', Asian Institute of Technology, Bangkok,Thailand, Principal Investigators: C.Visvanathan & Josef Trankler

- Asnani, P U (2006) : 'Solid Waste Management' in India Infrastructure Report, p160-189

- Asnani, P U (2004): United States Asia Environmental Partnership Report, USAID. Centre for Environmental Planning and technology, Ahmedabad.

- Asnani, P U (2005): Technical Committee Report, West Bengal SWM Mission 2005, Govt of West Bengal, Kolkata.

- Bankole, Philip Olatunde. 2004. "Nigeria Environmental Profile: National Environmental Outlook", United Nations Environmental Program, at jhttp://www.unep.net/profile/index.cfm?tab=100&countrycode = NG

- Barrett, C, et al., (2001): Conserving tropical biodiversity amid weak institutions, Bioscience, 51(6): 497-502.

- Basic Facts: Municipal Solid Waste." 2003. United States Environmental Protection Agency. http://www.epa.gov/epaoswer/nonhw/muncpl/facts.htm.

- Basics of Landfills:How they are constructed and why they fail?, Environmental Research Foundation, Annapolis, MD., Web site: www.rachel.org

- Basic Information on Biogas, www.kolumbus.fi,

- Bathija Raju et al (1997): 'Slums – The magnitude of the Problem', The Mumbai Pages, at http://theory.tifr.res.in/bombay/history/slums.html.

- Beede, David N., and David E. Bloom, "The Economics of Municipal Solid Waste." The World Bank Research Observer, Vol. 10, No. 2, August, 1995, pp. 113-150

- Bejoy Davis: Solid waste Management in Mumbai, Understanding our Civic issues, The Bombat Community Public Trust (Ed: Nita Mukherjee).

- Benefits of anaerobic digestion, www.afbini.gov.uk,

- Beychok, M.R., *Process and environmental technology for producing SNG and liquid fuels*, U.S. EPA report EPA-660/2-75-011, May 1975

- Beychok, M.R., *Coal gasification for clean energy*, Energy Pipelines and Systems, March 1974

- Beychok, M.R., *Coal gasification and the Phenosolvan process*, American Chemical Society 168th National Meeting, Atlantic City, September 1974

- Bezboruah, A.N., Bhargava, D.S., 2003. Vermicomposting of municipal solid waste from a campus. Indian Journal of Environmental Protection, 23 (10), 1120–1136.

- Beukering, P.V, et al (1999): Analysing Urban solid waste in Developing countries: A Perspective on Bangalore, CREED and IIED, Working paper 24, March 1999.

- Bhide, A.D., Shekdar, A.V (1998): Solid waste management in Indian urban centers. International Solid Waste Association Times (ISWA) (1), 26–28.

- Bijoy Davis: 'Understanding our Civic Issues – SWM in Mumbai', YUVA, Mumbai.

- Bingemer, H G and Crutzen, P J (1987): J. Geophys. Res. 92(D2), p 2181-2187

- Biodisel Now (2002/2004): athttp://www.biodieselnow.com

- Biogas Bonanza for Third World Development, www.i-sis.org.uk,

- Biogas as a road transport fuel, www.nfuonline.com, retrieved 24.10.07

- Biogas production in Europe: EurObserver, 2007 Biogas barometer, available   at http://www.energies-renouvelables.org/observ-er/html/baroSom.asp

- Biomethanisation of municipal waste in France, Edited by BIO Intelligence Service 3/3, February 2008, http://ec.europa.eu/environment/etap

- Blight, G E et al (1996): Some problems of waste management in developing nations, J.Solid Waste Technology and Management, 23, p19-27

- Bosdogianni, A (2007): MSWM in Greece – Legislation – Implementation Problems, Proc. Sardinia 2007, Eleventh Intl. Waste management and landfill Symposium, Cagliari, Italy, Oct.2007.

- Briens, Cedric et al: Biomass Valorization for Fuel and Chemicals Production – A Review, IJCRE, vol. 6, R2, Available at http://www.bepress.com/ijcre/vol6/R2/

- Canaki, M (2007): Bioresour. Technol, 98, 183-190

- Canaki,M and Gerpen, J.V (2006): American Soc. Agri. Engineers,44, 1429-1436

- Cardiff University (2005) Anaerobic Digestion Page, www.wasteresearch.co.uk,

- Chandrasekhar (2002): Policy and Prospects on MSW, Workshop on MSW in India, IIT, Delhi.

- Chakrabarty, P, et al. (1995): Solid waste disposal and the environment – a review, Indian Journal of Environmental Protection 15 (1), 39–43.

- CIWM, Chartered Institution of Waste Management (2003): 'Energy from Waste: A good Practice guide', November 2003.

- Chennai best.com www.chennaibest.com/discoverchennai/citylifestyle/feature10.asp

- Civic Exnora web: http://www.exnorainternational.org/about_exnora.shtml Corporation of Chennai web: http://www.chennaicorporation.com/swm/swm.htm

- Cointreau, Sandra J. (1982): "Environmental management of urban solid wastes in developing countries: a project guide." Urban Development Dept, World Bank. http://www.worldbank.org/html/fpd/urban//solid-wm/techpaper5.pdf

- Colon, M., Fawcett, B, (2006): Community-based household waste management: lessons learnt from EXNOR's zero waste management scheme in two south Indian cities. Journal of Habitat International 30 (4), 916–9

- Columbia (2001), Columbia Earth Institute IWM Project Report "Life after Fresh Kills: Moving Beyond New York City's Current Waste Management Plan", available from *earth@columbia.edu.*

- Community based Solid Waste Management: The Asian Experience, Waste Concern, Dhaka, Bangaladesh, (2000)

- CONAMA (National Environmental Commission) (2002): Residuos Solidos Domiciliaros Report, Santiago, Chile

- Connett, Paul (2006): Zero-waste – a global perspective (pdf) Recycling Council of Alberta Conference

- Cornell Waste Management Institute (1996): Dept of Crop and Soil Sciences, Cornell University, Q & A by Nancy Trautmann and Tom Richard

- CDP, Delhi, Ch-12: Sold waste Management: Prepared for Department of Urban Development, Government of Delhi, New Delhi. Consultant; IL & FS Ecosmart Ltd

- Central Pollution Control Board (CPCB) (2004): Management of Municipal Solid Waste. Ministry of Environment and Forests, New Delhi, India

- Coad, Adrian (Ed) (1997): Lessons from India in SWM, WEDC, Loughborough University, UK.

- CPCB (2000): Status of Solid Waste Generation, Collection, Treatment and Disposal in Metrocities, Series: CUPS/46/1999–2000.

- CPCB (2000): Status of Municipal Solid waste Generation, Collection, Treatment and Disposal in Class I Cities, Series: ADSORBS/31/1999–2000.

- CPHEEO (2000): Manual on MSWM, Ministry of Urban Development, Government of India.

- Craig Freudemrich: How landfill works http://howstuffworks.com/landfill.htm/printable. (retrieved on 21 Nov 09)]

- CREED : Center for Research in Environmental economics and Political Decision Making, Working Paper, series No.24; http://www.iied.org/creed

- Cruazon, B. (2007) History of anaerobic digestion, web.pdx.edu,

- Damodharan N., Robinson A., David E. and Kalas-Adams N (2003): Urban Solid Waste Generation and Management in India. Procs. Sardinia (2003), IX Int. Landfill Symposium, CISA, Cagliari.

- Department of Environment (DOE) (2004): Country paper-Bangladesh, SAARC workshop on SWM, Oct.2004, Dhaka

- Daskalopoulos, E. et al., (1998): "An integrated approach to municipal solid waste management." Resources, Conservation and Recycling. 24(1): 33-50.

- Das, D., et al., (1998): Solid state acidification of vegetable waste. Indian Journal of Environmental Health 40 (4), 333–342.

- Dayal, G. (1994): Solid wastes: sources, implications and management. Indian Journal of Environmental Protection 14 (9), 669–677.

- de Klundert, A, et al (2001): 'ISWM- A set of five tools for decision makers – Experiences from the Urban Waste Expertise Programme (1995-2001): ISWM – the Concept, Micro- and Small enterprises, Finanancial and Economic Issues, Community Partnerships, The organic Waste flow', Gouda, Netherlands: WASTE 2001

- Decentralized composting through Public- Private - Community partnerships: Experience of Waste Concern, Waste Concern, Dhaka

- Department of Economic and Social Affairs (1999): "United Nations World Urbanization Prospectus." http://www.un.org/esa/population/publications/wup2001/WUP2001re port.htm.

- Dhussa A.K, and Tiwari R.C, (2000): Waste to energy in India, BioEnergy News,4,1

- Diaz, Luis F, et al (1997): Managing SW in Developing Countries, J. Waste management, p43-45.

- Digester gas and landfill gas: California energy commission, www.consumerenergycenter.org/renewable/basics/biomass/landfill.ht ml), and www.energy.ca.gov/development/biomass/anaerobic.html.

- Discovering Anaerobic Digestion and Biogas, www.face-online.org.uk,

- Doan, Peter L. (1998): "Institutionalizing household waste collection: the urban environmental management project in Cote d'Ivoire." Habitat International, 22(1): 27-39.

- Doelle, H. W. (2001)Biotechnology and Human Development in Developing Countries, www.ejbiotechnology.info

- Dolk, Helen. (2002) "Methodological Issues Related to Epidemiological Assessment of Health Risks of Waste

Management." Environmental and Health Impact of Solid Waste Management Activities, p195-210.

- Ecocities: Demonstration of appropriate technologies to maximize the recycling and reuse of selected streams of waste. Available at www.ecocities-india.org/e14227/e14276/index_eng.html

- EcoSolutions Manila (2009): 'Lagon and Vista group Solid and Liquid Waste management Technology demonstrations', at http://ecosolutionsmanila.blogspot.com/2009/11/lagos-and-vista-group-solid-liquid.html

- El-Fadel, Mutasem et al: (1997) "Environmental impacts of solid waste landfilling." Journal of Environmental Management. 50(1): 1-25.

- Emily Walling et al: Municipal SW Management in Developing Countries: Nigeria, a case study, (http://www.dnr.cornell.edu/saw44/NTRES331/Products/Spring%20 2004/Papers/SolidWasteManagement.pdf)

- Emission Standards Division. (1995a): "Air emissions from municipal solid waste landfills—Background information for final standards and guidelines." United States Environmental Protection Agency. http://www.epa.gov/ttn/atw/landfill/bidfl.pdf.

- Emission Standards Division. (1991). "Air emissions from municipal solid waste landfills—background for proposed standards and guidelines." United States Environmental Protection Agency. http://www.epa.gov/ttn/atw/landfill/laurv1.pdf.

- Energy from Waste State-of-the-Art Report, Statistics 5th Edition, August 2006, International Solid Waste Association

- Environmental Resource Management Report (1996): Municipal Solid Waste Management Study for the Madras Metropolitan Area, ERM report, London W1M 0ER.

- Environmental Protection Agency (2003) at http://www.epa.gov/epr/index.htm

- Environmental Protection Agency at http:///www.epa.gov/epawaste/nonhaz/index.htm, retrieved on 16Nov09

- EPA: Municipal and Industrial Waste Division. (1995b): Decision-Makers Guide to Solid Waste Management. vol II: chap.8. http://www.epa.gov/epaoswer/non-hw/muncpl/dmg2/chapter8.pdf.

- EPA (1998), U.S. Environmental Protection Agency "Inventory of Sources of Dioxin in the U.S. (April 1998), www.epa.gov/nceawww1/diox/htm.

- EPA Review (1998) EPA Contract No. 68-D5-0028 Work Assignment No. 98-05, August7, 1998 (www.epa.gov). Environmental Working Group, "Analysis of mercury pollution from coal-burning plants", www.ewg.org

- EnergyAnswers Corp., (2008): SEMASS Resource Recovery facility – Technical description & Performance History, 2008.

- Energy and Materials recovery: at www.columbia.edu/cu/wtert

- Environment Agency Waste Technology Data Centre An independent UK government review of advanced waste treatment technologies

- Enviros (2006): Mechanical biological treatment website, *www.mbt.landfill-site.com*, Waste Technology Home Page,

- Esakku S (2006): Assessment of Reclamation and Hazard Potential of Municipal Solid Waste Dumpsites. Doctoral Thesis submitted to Anna University, Chennai, India.

- Eassaku S, et al (2007): Municipal solid waste management in Chennai City, India, Sardinia 2007, and Eleventh International Waste management and landfill Symposium, Cagliari, Italy, 2007.

- Estevez, Paula (2003): Management of MSW in Santiago, Chile – Assessing WTE possibilities.

- Evans, G. "Liquid Transport Biofuels - Technology Status Report", National Non-Food Crops' Centre.

- EU Biomass Action Plan–

  http://ec.europa.eu/energy/res/biomass_action_plan/doc/2005_12_07_comm_biomass_action_plan_en.pdf; Presentation webpage:

  http://ec.europa.eu/energy/res/biomass_action_plan/nationa_bap_en.htm

- Europa press release, December (2005) Renewable energy: European Commission proposes ambitious biomass and biofuels action plan and calls on Member States to do more for green electricity http://europa.eu/rapid/pressReleasesAction do?reference = IP/ 05 / 1546& format=HTML&aged=0&language=EN&guiLanguage=en

- Extension (2009): Vermicomposting Animal manure, April 2009

- Felizardo, P et al (2006): Waste Manage., 26, 487-494

- Fellner, J. et al,(2007), A New Method to Determine the Ratio of Electricity Production from Fossil and Biogenic Sources in Waste-to-Energy Plants, Environmental Science & Technology, 41(7) 25

- Fergusen, T. & Mah, R. (2006) Methanogenic bacteria in Anaerobic digestion of biomass, p49

- Friends of the Earth (2004) Anaerobic digestion Briefing Paper, www.foe.co.uk,

- Friends of the Earth (2008): Mechanical Biological Treatment Briefing, www.foe.co.uk

- Fukuda, H et al (2001): J.Biosci. Bioengg.,92, 405-416

- Garg, Ankur et al (2007): Public Private Partnership for SWM in Delhi – A Case Study, Proc. of Intl. Conf. on Sustainable SWM, Chennai, Sept. 2007

- Garrod, G, and Willis, K,. (1998) "Estimating lost amenity due to landfill waste disposal." Resources, Conservation and Recycling. 22: 83-95.

- Gasification: http://en.wikipedia.org/wiki/gasification

- Gasification case studies by the Environment Agency of England

- Garg, S., Prasad, B., (2003) Plastic waste generation and recycling in Chandigarh. Indian Journal of Environmental Protection 23 (2), 121–125.

- Ghose, M.K, et al., (2006) A GIS based transportation model for solid waste disposal – A case study on asansol municipality. Journal of Waste Management 26 (11), 1287–1293.

- Ghosh, C., (2004) Integrated vermin – pcsciculture – an alternative option for recycling of municipal solid waste in rural India. Journal of Bioresource Technology 93 (1), 71–75.

- Gladding, Toni. (2002) "Health Risks of Materials Recycling Facilities." Environmental and Health Impact of Solid Waste Management Activities. 53-72

- Glawe, Ulrich et al,: SWM in Least Developed Asian Countries – A Comprehensive analysis, Intl. Conf. on ISWM in SE Asian cities, Siem Reap, Cambodia.

- Global warming methane could be far more potent than carbon dioxide www.newmediaexplorer.org

- Green dot System: http://colby.edu/personal/t/thtieten/swm-germ.html

- Gui, M.M et al (2008): Energy, 33, 1646-1653

- Gupta, S., et al, (1998). Solid waste management in India: options and opportunities. Resource, Conservation and Recycling 24, 137–154.

- Gupta, P.K, et al, (2007). Methane and Nitrous Oxide Emission from Bovine Manure Management Practices in India. Journal of Environmental Pollution 146 (1), 219–224.

- Guwahati, Coimbatore, Navi Mumbai and Kollam Projects: Indian Infrastructure Vol II, Issue 7, Feb.2009

- Haas, M.J et al (2005): In 'The Biodiesel Handbook',Knothe, G et al (Eds.), AOCS Press, urbana, Ill., 42-61

- Halim, S.F.A eta l (2009): Bioresour. Technol.,100, 710-716

- Hanrahan, D, et al (2006): Improving Management of MSW in India – Overview & Challenges, World Bank, South Asia region, New Delhi.

- Hayami, S. "Waste Management Strategy: Tokyo Metropolis" Presented at UP National Center for Transportation Studies (NCTS), Payatas Seminar (2007).

- Headley, J (1998): An overview of SWM in Barbados, In 'SWM: Critical issues for developing countries' Ed. By Elizabeth Thomas-Hope, Canoe press, Kingston

- Hester, R. E. and R.M. Harrison, eds. (2002): Environmental and Health Impact of Solid Waste Management Activities. Cambridge: The Royal Society of Chemistry

- Hoornweg, D, et al. (1998): What a waste: Solid waste management in Asia. Urban Development Sector Unit East Asia and Pacific Region.

- Hoornweg, D, et al (1999): Composting and its applicability in Developing countries, Urban waste management working paper series 8, Washington DC: World Bank.

- Humanik, F. *et al.* (2007): Anaerobic digestion of animal manure, www.epa.gov,

- Iftekar E.A. et al (2005): a Report on 'Urban SWM scenario of Bangladesh: problems and prospects', Waste Concern Publications, Dhaka, at http://www.wasteconcern.org/

- Inanace B., Idris A., Terazono A., and Sakai S. (2004): Development of a Database of Landfills and Dumpsite in Asian countries. J. Mat. Cycles Waste Mang, Vol. 6, pp. 97-103.

- Integrated Solid Waste Management in Hyderabad: News in Waste Management World, Sept. 2010. Available at waste-management-world.com/index/display/article-display/......

- INFORM Inc., and Recourse Systems, Inc. Business Recycling Manual. (1991)

- IPCC (1997): Revised 1996 IPCC guidelines for National GH Inventories, Houghton, J. T, et al (Eds) – IPCC, IPCC/OECD/IEA, Paris, France.

- IPCC (2006): IPCC guidelines for National GH Inventories, Vol. 5: Waste; Ch 2 at V5_2_Ch2_Waste_data-2006IPCCguidelines.pdf

- Jain A.K, (2004): Solid Waste Management in Mumbai, All India Inst. Of Self-Government, Mumbai.

- Jala B.K et al (1997): The emergency priority for disposal,use and recycling of MSW in India, J. Waste Management, p17-18

- Jalan, R.K., Srivastava, V.K., (1995): Incineration, land pollution control alternative – design considerations and its relevance for India. Indian Journal of Environmental Protection 15 (12), 909–913.

- Jayarama Reddy, P (2011): Pollution and Global Warming, BS Publications, Hyderabad, & AP Council of Science & Technology, Hyderabad.

- Jesper, Ahrenfeldt (2007): Characterization of biomass producer gas as fuel for stationary gas engines in combined heat and power production, Ph.D. Thesis, Technical University of Denmark, March 2007

- Jha, M.K.et al, (2003) Solid waste management – a case study. Indian Journal of Environmental Protection 23 (10)1153–1160.

- Joseph, Kurian (2002): Perspectives of SWM in India, Intl. symposium on the technology of the Treatment & Reuse of MSW, Shanghai, China.

- Joseph, K., (2006) Stakeholder participation for sustainable waste management. Journal of Habitat International 30 (4), 863–871.

- Johannes Lehmann. "Biochar: The new frontier" http://www.css.cornell.edu/faculty/lehmann/biochar/Biochar_home.htm

- Johannessen, L.M, (1999): Observations of solid waste landfills in developing countries: Africa, Asia and Latin America, Urban & Local government Working paper, Series no.3, World Bank, Washington DC.

- Johannessen, L.M. and Boyer, G,(1999): Observations of SW landfills in developing countries: Africa, Asia and Latin American Urban Development Division Waste Management Anchor Team, World Bank

- Juniper (2005) MBT: A Guide for Decision Makers – Processes, Policies & Markets, www.juniper.co.uk, (Project funding supplied by Sita Environmental Trust).

- Kalogirou, Efstratios (2010): Technical Report from China, 2010, at www.wtert.org and www.wtert.gr

- Kansal, A., (2002): Solid waste management strategies for India. Indian Journal of Environmental Protection, 22 (4), 444–448

- Kasseva, M, E. and Mbuligue (2000), "Ramifications of solid waste disposal site relocation in urban areas of developing countries: a case study inTanzania." Resources, Conservation and Recycling. 28(1-2): 147-161.

- Khan, R.R., (1994): Environmental management of municipal solid wastes. Indian Journal of Environmental Protection 14 (1), 26–30.

- Kim, H.J et al (2004): Catalysis Today, 93-95, 315-320

- Knight, B, Hackleman, D, Yokochi, A (2006): Technical Report, 'Production of Biodiesel from Waste Cooking Oil', ChE414.

- Kreith, Frank, ed. Handbook of Solid Waste Management. New York: McGraw-Hill (1994)

- Kumar, D, et al, (2001): Leachate generation from municipal landfills in New Delhi, India. 27th WEDC Conference on People and Systems for, Sanitation and Health, Lusaka, Zambia, 2001.

- Kumar Sudhir (2000): 'Technology options for MSW-to-energy projects', TERI Information Monitor on Environmental science, 5, No.1, June 2000.

- Kumar, S and Gaikwad SA (2004): 'MSWM in Indian urban centres: An approach to Betterment', Urban Development Debates in the New Millennium, Editor: KR Gupta, Atlantic publs.and Distributors, New Delhi.

- Kumar, S: MSW Management in India- Present practices and Future challenges, Aug.2005.

- Kumar, S, et al (2009): Assessment of the status of municipal solid waste management in metro cities, state capitals, class I cities, and class II towns in India: An insight, Waste Management, 29, p883-895.

- Lal, A.K., (1996): Environmental status of Delhi. Indian Journal of Environmental Protection 16 (1), 1–11.

- Landfill - Method Types, http://science.jrank.org/pages/3805 /Landfill-Method-types.html#ixzz0YC61cwQ

- Landfill – Decomposition, http://science.jrank.org/pages/3806/ Landfill Decomposition.html#ixzz0YC6MnYNL

- Landfill - Operating Principles, http://science.jrank.org/pages /3807/Landfill-Operating-principles.html#ixzz0YC7cSRVz

- Landfill - Alternatives To Landfills http://science.jrank.org/pages /3808/Landfill-Alternatives-landfills.html#ixzz0YC87A2R) http://www.consumerenergy

- Landfill – Recycling, http://science.jrank.org/pages/3809/Landfill- Recycling.html#ixzz0YC8RqLRO

- Landfill - Composting http://science.jrank.org/pages/3810/Landfill- Composting.html#ixzz0YC8seeTa

- Lardinois, 1, and A. van de Klundert. Organic Waste, Urban Solid

Waste Series, Vol. 1. Amsterdam and Gouda: Tool, Transfer of Technology for Development and WASTE Consultants (1994)

- Ma, F et al (1999): Bioresour. Technol.,70, 1-15

- Mahadevia, D, et al (2005): New Practices of Waste Management – Case of Mumbai, working paper 35, school of planning, CEPT University, Ahmedabad.

- Mahadevia, D and Pathak, M (2005): Local Government Led SWM in a Metropolis: Hyderabad-Secunderabad, Working Paper No. 34,CEPT University, Ahmedabad, December 2005

- Mansoor Ali, Editor (2004): Sustainable Composting – Case studies and guidelines for Developing countries, Water, Engineering and Development Centre, Loughborough University, UK

- Manual on SWM: Central Public Health and Environmental Engineering Organisation (CPHEEO), Ministry of Urban development, Government of India

- Marchetti, J.M et al (2007a): Renewable Sustainable Energy Rev.,11, 1300-1311

- Marchetti, J.M et al (2007b): Fuel, 86, 906-910

- Marcin et al (1994): Source reduction strategies and technological change affecting demand for pulp and paper in North America, Proc. Of Third Intl. symposium, Sept.1994, Seattle, USA

- Maudgal, S., (1995) Waste management in India. Journal of Indian Association for Environmental Management 22 (3), 203–208

- Mazmanian, D, and Kraft, M, eds. (2001), Toward Sustainable Communities: Transitions and Transformations in Environmental Policy. MIT Press: Cambridge, MA.

- Medina, M (2006): Scavenger Cooperatives in Asia and latin America, Resources, 31, p51-69

- Methanogens, microbewiki.kenyon.edu,

- Michaels, T (2007): The 2007 ISWA Directory of WTE plants, ISWA, June 2007; http://www.wte.org/directory.shtml

- Ministry of Environment and Forests (MoEF), (2000): The Gazette of India. Municipal Solid Waste (Management and Handling) Rules, New Delhi, India.

- Mishra, S G, et al,(1992): Environmental Pollution: Solid Waste, Venus Publishing House, New Delhi.

- Mishra, S G, and D. Mani (1993): Pollution through Solid Waste, Ashish Publishing House, New Delhi.

- MNR Environment, Thai Government: Waste Management – Realisation of WTE and Beyond.

- Modi J, et al (2002): Wealth of wastes, Waste Recyclers for SWM – A study of Mumbai, All India Institute of Self-government, Mumbai.

- Mohn, J. et al, (2008), Determination of biogenic and fossil CO_2 emitted by waste incineration based on $^{14}CO_2$ and mass balances. Bioresource Technology, 99, 6471-6479

- Mor S, et al, (2006) Municipal solid waste characterization and its assessment for potential methane generation: a case study. Journal of Science of the Total Environment 371 (1), 1–10.

- MOUD Report (2005): Management of Solid Waste in Indian Cities, Ministry of Urban Development, Govt of India, New Delhi.

- Mungai, G (1998): SWM and its environmental impact in Kenya, In 'SWM: Critical issues for developing countries, ed.Elizabeth Thomas-Hope.

- Ministry of Environment, (2008): Waste and Recycling. Waste disposal and Recycling Measures. http://www.env.go.jp /en/recycle/manage/waste.html

- Municipal and Industrial Waste Division (1995), Decision-Makers Guide to Solid Waste Management, vol II: ch8. US Environmental Protection Agency. http://www.epa.gov/epaoswer/non-hw/muncpl/dmg2/chapter8.pdf

- Municipal solid waste power plants' at center.org/renewable /basics/biomass/muni.html

- Municipal Solid Waste, US Environmental Protection Agency. Jan. 3, 2008. http://www.epa.gov/garbage/facts.htm

- Municipal Solid Waste Management in Asia (2002): Asian Institute of Technology, ISWA & UNEP

- MSW Management (Sept- Oct.2006): 'The Role of WTE in MSWM' at www.mswmanagement.com/september-october-2006

- Mwesigye, Patrick et al (2009): African Review Report on Waste Management, 'Integrated Assessment of Present Status of environmentally-sound management of wastes in Africa', Prepared for UNIDO.

- Nagpur and Lucknow, Solid Waste Management Part II, Background Paper, 12th Finance Commission, by IPE (P) Ltd.

- NEERI (1996): 'Strategy paper on SWM in India', National Environment Engineering Research institute, Nagpur.

- NEERI Report (2006): Assessment of Status of MSWM in Metro cities, State capitals, Class I cities, and Class II towns

- Neal, Homer A., and J.R. Schubel. (1987) Solid Waste Management and the Environment. New Jersey: Prentice-Hall, Inc.

- Nie, K et al (2006): J. Mol. Catalysis B: Enzymatic, 43, 142-147

- Niessen, W. R (2007): WTE using Combustion and Conversion technologies – Powerful alternatives in Waste management, Presented at Illinois recycling & SWM conference and Trade show.Peoria, Ill.june 2007.

- Nigerianewsnow.com. (2003) http://www.unesco.org/csi/act/lagos/lagnews-now.htm.

- Nielson, M, et al (2006): PM emissions from CHP Plants <25MW$_e$ (DOC) National Environmental Research Institute, Denmark

- Nigerianewsnow.com. 2003, http://www.unesco.org/csi/act/lagos/lagnews-now.htm

- NPC (2005): Up gradation Plan for Existing Dumpsites at Perungudi and Kodungaiyur (Chennai), National Productivity Council report for Corporation of Chennai

- O'Brien and Swana (2006): Applied Research Foundation – Comparison of the Air emissions from WTE facilities to Fossil fuel Power plants, Technical Report. http://www.swana.org

- Onibokun, A, G. (1999). Managing the Monsters: Urban Waste and Governance in Africa. International Development Research Centre. Ottowa.

- Onibokun, A.G., Kumuyi, A.J. (2003). "Ibadan, Nigeria." International Development Research Centre: Science for Humanity. Ch 3

- Oyedele, D.J, et al, (2008): "Changes in soil properties and plant uptake of heavy metals on selected MSW dump sites in Ile-Ife, Nigeria. African Journal of Environmental Science and Technology 3(5), 107-115

- Pahn, A.N and T. M. Pahn (2008): Fuel,87, 3490-3496

- Pappu, A., Saxena, M., Asokar, S.R., (2007). Solid Waste Generation in India and Their Recycling.

- Plasma Arc waste disposal: http://en.wikipedia.org/wiki/Plasma_arc_waste_disposal

- Plasma Ars gasification: Juniper Consulting 2008

- Pollutec 2007 innovations page, Renewable energies category (http://www.environnementonline.com/pollu_innos/details_sc.asp?l=e&id=42)

- Pollution Control department (PCD), Thailand: 'Overview of Waste management in Thailand', AIT060509, Sec.4, SWM,Waste & Hazardous substances management Bureau

- Position Paper on PPP in SWM sector (2009): Department of Economic Affairs, Ministry of Finance, Government of India.

- Predpall, Daniel (2004): 'MSW Conversion technologies: A viable alternative to Disposal', MSW Management, May/June 2004

- PROBIOGAS (2007): Promotion of Biogas for Electricity and Heat production in EU countries, Published Final report, Project Coordinator – Department of Bioenergy, University of Southern Denmark.

- Psomopoulos, et al (2009): Waste Management, 29, p1718.

- Pyrolysis: http://en.wikipedia.org/wiki/pyrolysis

- Pyrolysis:http://www.abc.net.au/catalyst/stories/s2012892.htm.

- Pyrolysis and Other Thermal Processing, US DOE. http://web.archive.org/web/20070814144750/http://www1.eere.energy.gov/biomass/pyrolysis.html

- Raje, D.V.et al., (2001): An approach to assess level of satisfaction of the residents in relation to SWM system. Journal of Waste Management and Research 19, 12–19

- Rachel, O Adewuyi et al. (2009): MSWM in Developed and Developing Countries – Japan and Nigeria as Case studies, JGS Paper 973.

- Ramachandra T.V, and Bachamanda S (2007): Environmental audit of MSWM, Intl.J.Environmental Technology and Management, 7, pp369-391

- Rao, K.J., Shantaram, M.V., (1993): Physical characteristics of urban solid wastes of Hyderabad. Indian Journal of Environmental Protection 13 (10), 425–721.

- Rathi, S., (2006): Alternative approaches for better municipal solid waste management in Mumbai, India. Journal of Waste Management 26(10), 1192–1200.

- Recycling: Today's Challenge, Tomorrow's Reward: http://matse1.mse.uiuc.edu/polymers/prin.html

- Reddy, S., Galab, S., (1998): An Integrated Economic and Environmental Assessment of Solid Waste Management in India – the Case of Hyderabad, India

- Renewable Energy Framework, www.esru.strath.ac.uk

- Richards, B. (1991). "High rate low solids methane fermentation of sorghum, corn and cellulose". *Biomass and Bioenergy* 1: 249–260. doi:10.1016/0961-9534(91)90036-C.edit

- Richards, B. (1994). "In situ methane enrichment in methanogenic energy crop digesters". *Biomass and Bioenergy* 6: 275–274. doi:

- Richard, Tom: Municipal Yard Waste composting: The Compost process, , Cornell cooperative Extension, Municipal yard waste Composting Operator's fact Sheet #1 of 10, http://www.cwmi.css.cornell.edu/factsheets.htm, and Free Wikipedia

- Saifuddin, N et al (2009): E-Journal of Chemistry, 6(S1),S485-S495

- Sanitary landfill: "http://science.jrank.org/pages/3804/Landfill-Sanitary- landfill.html">Landfill - Sanitary Landfill

- Sannigrahi, A.K., Chakrabortty, S., (2002). Beneficial management of organic waste by vermicomposting. Indian Journal of Environmental Protection 22 (4), 405–408.

- Sarkar P J (2003): SWM in Delhi – A Social Vulnerable study, in

Proc. Third Intl. Conference on Environmental Health, Chennai, Eds: Martin J Bunch et al.

- Sasse, Ludwig (1988): Biogas Plants, A publication of GAT and GTZ, Germany

- SC Committee Report (1999): Report of the Committee on SWM in Class I cities in India, Constituted by the Hon. Supreme Court of India.

- Schubeler, P (1996): Conceptual framework for MSWM in low-income countries, UNDP, UMP working paper series, no.9, St, Galen, SKT,Switzerland

- Senkoro, Hawa (2003): 'SWM in Africa: A WHO/AFRO Perspective',CWG workshop, March 2003, Dar Es Salaam

- SEPA MBT Planning Information Sheet Fact Sheet for Scottish Planning Considerations

- Seo, Seongwon et al., (2004): "Environmental impact of solid waste treament methods in Korea. Journal of Environmental Engineering. 130(1): 81-89

- Sharholy M, et al., (2008): Municipal solid waste management in Indian cities – A review, Waste Management, 28, pp459-464

- Shannigrahi, A.S., Chatterjee, N., Olaniya, M.S., (1997): Physico-chemical characteristics of municipal solid wastes in mega city. Indian Journal of Environmental Protection 17 (7), 527–529.

- Sharholy, M.et al., (2005) Analysis of municipal solid waste management systems in Delhi – a review. In: Book of Proceedings for the second International Congress of Chemistry and Environment, Indore, India, pp. 773–777

- Sha Ato, R, et al, (2007): Survey of solid waste generation and composition in a rapidly growing urban area in central Nigeria. Waste Management 27, 352-358.

- Shackelford C.D. "Environmental issues in geotechnical Engineering". Department of Civil Enginnering, Colorado State University, Fort Collins, Colorado, USA.

- Sharholy, M.et al, (2007): Municipal Solid Waste Characteristics and Management in Allahabad, India. Journal of Waste Management 27 (4), 490–496.

- Shekdar, A.V., (1999). Municipal solid waste management – the Indian perspective. Journal of Indian Association for Environmental Management 26 (2), 100–108.

- Shekdar, A.V. et al. (1992): Indian urban solid waste management systems – jaded systems in need of resource augmentation. Journal of Waste Management 12 (4), 379–387.

- Sherman, Rhonda: North Carolina Cooperative Extension Service: Department of Biological & Agricultural Engineering North Carolina State University

- Sherman, Rhonda: Division of Pollution Prevention and Environmental Assistance: North Carolina Department of Environment, Health, and Natural Resources

- Shibasaki-Kitakawa, N.H et al (2007): Bioresour. Technol.,98, 416-421

- Shi-ling Hsu, ed. (1999):Brown fields and property values,(pdf) Economic Analysis and land use Policy, US Environmental Protection Agency

- Singhal. S and Pandey. S (2001): 'SWM in India – Status and Future Directions', TERI Information Monitor on Environmental Science, 6, p1-4

- Sita (2004): Sita Mechanical Biological Treatment Position Paper, *www.sita.co.uk*,

- Solid Waste Audit Report, Federal capital Territory, Abuja, Nigeria. (2004).

- Solid waste Management Part I, Background Paper, 12th Finance Commission, by IPE (P) Ltd.

- Solid waste management – Principles and terminologies, Prakriti, Centre for management studies, Dibrugarh University, at http://cmsdu.org

- Solid Waste Management: Issues and Challenges in Asia (2007): Report of the APO Survey on SWM (2004-2005)., at Apo-tokyo.org/ 00e-books/IS-22_solidWasteMgt/IS-22_SolidWasteMgt.pdf

- Solid Waste Management: CPREnvironmental Education Centre, Chennai, at www.cpreec.org/pubbook-solid.htm

- Solid waste management Department, Municipal Corporation of Greater Mumbai, at http://www.mcgm.gov.in/Departments/ swmanage.html

- Srishti Report (1998): Recycling Plastics in Delhi

- Srinivasan K (2005): Public, Private and Voluntary Agencies in Solid Waste Management: A case study in Chennai city. Master degree thesis submitted to Tata Institute of social sciences, India.

- Srivastava, P.K, et al., (2005): Stakeholder-based SWOT analysis for successful municipal solid waste management in Lucknow, India. Journal of Waste Management 25 (5), 531–537.

- Sonesson, U. et al. 2000. "Environmental and economic analysis of management systems for biodegradable waste." Resources, Conservation and Recycling. 28(1-2): 29-53.

- Survey on Scope of Privatization of SWM in India, Feb.2007, FICCI, New Delhi.

- Svoboda, I (2003) Anaerobic digestion, storage, olygolysis, lime, heat and aerobic treatment of livestock manures, www.scotland.gov.uk,

- Swan, Jillian R. et al. (2002) "Microbial Emissions from Composting Sites." Environmental and Health Impact of Solid Waste Management Activities, p 73-101.

- Talukder, M.M.R et al (2010): Energy Fuels, 24, 2016-2019

- Tchobanoglous George, Hilary Theisen, and Samuel Vigil. Integrated Solid Waste Management, Engineering Principles and Management Issues. New York: McGraw-Hill (1993)

- The Economist (2002), Three years of democracy; Nigeria, 363(8267): 58. Vision 2010 Committee (2003), Ecology and the Environment: Protect our environment for our unborn children, Vision 2010 Report. http://www.vision2010.org/vision-2010/ecology.htm

- The Financial Express (2005): A wiser approach to e-waste, at http://www.financialexpress.com/fe_full_story.php?content_id=1085 65

- Themelis, Nicholas J and Gregory, A (2002): Mercury emissions from high temperature sources in the NY/NJ Hudson Raritan Basin, Proc of NAWTEC 10, American Society of Mechanical Engineers, p 205-215

- Themelis, Nicholas J (2003): An overview of W-t-E industry, Waste management World, 40-47

- Themelis, Nicholas J (2003): An overview of the global waste-to-energy industry, WTERT, July-august 2003.

- Themelis, Nicholas J (2006): The role of Waste-to-Energy in the USA, Third world Congress of the confederation of European WTE Plants (CEWEP), Vienna, May 2006.

- Themelis, Nickolas J. and Zhixiao Zhang (2010): 'The importance of WTE to China and the global climate', Proceedings WasteEng 2010, Beijing, May 2010.

- Themelis, Nicholas J (2002): Integrated Management of SW for New York City, American Society of Mechanical Engineers, Proc. Of NAWTEC, 10, p 69-86.

- The Swiss Federation: 'Waste Management' – National Reporting to CSD 18/19 by Switzerland, at waste-switzerland.pdf.

- Toxics Link web: http://www.toxicslink.org/art-view.php?id=1

- Toxics Link: publication, 'E-waste in Chennai, time is running out' at http://www.toxicslink.org/

- Transfer of low-cost plastic biodigester technology at household level in Bolivia

- 'Trial to reverse global warming', BBC News. http://news.bbc.co.uk/1/hi/england/7993034.stm.

- UNCH (2000): Integrated Solid Waste collection system in the city of Olangapo, Philippines, Best Practices data Base; http://www.bestpractices.org

- United Nations Development Programme (UNDP) 1997 Report, Energy After Rio: Prospects and Challenges

- UNDP/WB RWSG-SA (1991): Indian experience on Composting as means of resource recovery. UNDP/WB Water supply and Sanitation Program South Asia, Workshop on Waste Management Policies, Singapore, July 1991.

- UNEP: Municipal Solid Waste Management: (http://www.unep.or.jp/ietc/estdir/pub/msw/index.asp)

- UNEP: 'Regional Overviews and Information Services – Asia' at http//www.unep.or.jp/ietc/ESTdir/Pub/MSW/RO/contents_Asia.asp

- UNEP: Municipal Solid Waste Management: www.unep.or.jp/Ietc /publications/spc/solid_waste_management/index.asp. Aug.2006

- UNEP (2003): Afghanistan – Post conflict Environmental Assessment, UNEP, ISBN:92-1-158617-8

- UNEP (2001a): State of environment Bhutan, UNEP, ISBN:92-807-2012-5

- UNEP (2001b): State of Environment, South Asia, UNEP,2001

- UNEP (2002): Africa Environment Outlook: past, present and future perspectives, http://www.unep.org/neo/210.htm

- UNEP/Grid Arendal, (2002) Contribution from waste to climate http://www.vital graphics.net/waste/html_file/42-43_climate_c…

- UNEP: News letter and technical publications:<MSWM> Sound Practices, at www.unep.or.jp/ietc/estdir/pub/msw/sp5/sp5_1.asp

- UNEP: Chap.3 – Overview of SWM in Kenya: from Final draft – 'Selection, Design and Implementation of economic Instruments in the Kenyan SWM Sector, at www.unep.org/pdf/Kenya_ waste_mngnt_sector/chapter3.pdf

- UNEP (1996): International Source Book on Environmentally sound technologies for MSWM, UNEP Tech. Publication,6, Nov.1996

- UN (2000): State of Environment in Asia and the pacific, ESCAP & ADB, UN Publications, New York, ISBN:92-1-120019-9

- United States Environmental Protection Agency, Decision Makers Guide to Solid Waste Management. Washington: US Environmental Protection Agency (1989).

- United States, Government of, Office of Technology Assessment (OTA). Facing America's Trash: What Next for Municipal Solid Waste? Washington: OTA (1989)

- Van Steenis, Dick (2005): Incinerators – Weapons of mass destruction (DOC), RIBA Conference.

- Velzy, C.O, and Grillo, L.M (2007): Waste-to-Energy Combustion, Hand-book of Energy efficiency and renewable Energy, Taylor-Francis group, 2007.

- Viability of Advanced Thermal Treatment of MSW in the UK, 49 by Fichtner Consulting Engineers Ltd

- Vision 2010 Committee (2003), "Ecology and the Environment: Protect our environment for our unborn children." Vision 2010 Report. http://www.vision2010.org/vision2010/ecology.htm

- Visvanathan, C and Tenzin Norbu (2006): Reduce, Reuse and Recycle: 3 Rs in South Asia, Paper presented at 3R South Asia Expert Workshop, Sept. 2006, Kathmandu, Nepal.

- Visvanathan, C and Glawe, Ulrich (2006): Domestic SWM in South Asian Countries – A comparative analysis, presented at 3R South Asia Expert workshop, Aug/Sept 2006, Kathmandu, Nepal.

- Visvanathan, C (2007): Sustainable SW Landfill Management in Asia, Phase II – Review report, submitted to SIDA.

- Vijayawada City Development Plan, Ch4-Solid waste management

- Vogel, Michael (2002): Home Composting (SWM Series), Montana State University Extension Service, Revised Aug.2003.

- Wagner, Leonard (2007): Research report, 'Waste-to-Energy (WTE) Technology', MORA Associates, July 2007.

- Wan Omar, W.N.N et al (2009): J. Appl. Sciemces, 9(17), 3098-3103

- Wang, Y et al (2003): Bioresour. Technol., 90, 229-240

- Wang, Y et al (2010): energy Fuels, 24, 2104-2108

- Waste Concern: Waste management and recycling in Bangladesh.

- Waste-to-Energy, www.epa.gov/epawaste/nonhaz/index.htm

- Waste-to-energy, http://en.wikipedia.org/wiki/waste_to_energy

- Waste to Energy: www.medicities.org/docs/12

- WTERT Brochure at WTERT_Brochure_2006.pdf

- Waste Management in India (2008): Swedish Energy and Environment Technology in India Program, Swedish Trade Council, New Delhi, December, 2008

- Waste Guide: 2010 MoEF, an initiative of the Indo-German-Swiss partnership, at www.e-waste.in/weee_basics/weee_hazards/

- WasteTec MBT in Germany: A historical abstract of Mechanical Biological Treatment in Germany.

- WWF-Pakistan (2000): Pakistan country report, Waste Not Asia conf.,Taipei, Taiwan, at http://www.noburn.org/regional/pdf/country/Pakistan.pdf

- Woomer, P.L, et al, (1994): The importance and management of soil organic matter in the tropics. In: Woomer, P.L., Swift, M.J. (Eds.), The Biological Management of Tropical Soil Fertility. TSBF and Sayce Publ, UK.

- World Bank (1996), "Restoring Urban Infrastructure and Services in Nigeria." http://www.worldbank.org/afr/findings/english/find62.h

- World Bank (2006): Improving Management of MSW in India – Overview & Challenges, by Hanrahan, David, et al, Environment Unit, South Asia Region, New Delhi.

- World Bank (August 2008): SWM in Latin American Continent – Actual and Future CH4 emissions and reductions.

- World Bank Report: Private Sector Participation in SWM in India.

- World Bank (May 1999): What a Waste, SWM in Asia.

- Worm Digest, www.wormdigest.org

- Wheles, E and Pierece, E (2004): Siloxanes in landfill and digester gas, [http://www.scsengineers.com/Papers/Pierce_2004Siloxanes_Update_Paper.pdf; www.scsengineers.com

- Yelda, S. and Kansal, S., (2003). Economic insight into MSWM in Mumbai: acritical analysis. International Journal of Environmental Pollution 19(5), 516–527.

- Zarbock, O (2003): Urban SWM – Waste reduction in Developing nations, M Sc dissertation, Michigan Technical University, at www.cee.mtu.edu/peacecorps/documents_july03/

- Zhang, Y et al (2003): Bioresour. Technol.,90, 229-240

- Zurbrugg, C (2002): 'Urban SWM in low-income countries of Asia, How to cope with the garbage crisis', Presented for 'Scientific committee on Problems of Environment (SCOPE)', Urban SWM review session, Durban, S.A., November 2002. Useful Websites:

- http://science.jrank.org/pages/3804/Landfill-Sanitary-landfill.html

- http://www.iec.tu-freiberg.de/conference/conference_05/pdf/21_Picard.pdf .

- www.alfagy.com,
- http://www.claverton-energy.com/38-hhv-caterpillar-bio-gas-engine-fitted-to-long-reach-sewage-works.html
- www.dmu.ac.uk/in/itc/ad.htm
- www.unu.edu
- www.adelaide.edu.au,
- http://www.teri.res.in/teriin/camps/delhi.htm 2007
- http://www.environment.about.com
- http://www.edugreen.teri.res.in
- http://www.indiaone.com
- http://www.unesco.org
- http://www.unep.org
- http://www.unfpa.org
- Reports from World Bank, OECD, ADB, JICA, US EPA, IIASA, EEA and other International Agencies.